STP 1070

Geotechnics of Waste Fills—Theory and Practice

Arvid Landva and G. David Knowles, editors

ASTM
1916 Race Street
Philadelphia, PA 19103

Library of Congress Cataloging-in-Publication Data

Geotechnics of waste fills-theory and practice / Arvid Landva and G. David Knowles, editors.
(ASTM STP ; 1070)
"ASTM publication code number (PCN) 04-010700-38."--T.p. verso
Includes bibliographical references and index.
ISBN 0-8031-1285-8
1. Sanitary landfills--Congresses. 2. Engineering geology--Congresses. 3. Groundwater flow--Congresses. I. Landva, Arvid, 1930- . II. Knowles, G. David, 1944- . III. Series: ASTM special technical publication ; 1070.
TD795.7.G46 1990
628.3'64--dc20 90-44937
CIP

NOTE

The Society is not responsible, as a body, for the statements and opinions advanced in this publication.

Peer Review Policy

Each paper published in this volume was evaluated by three peer reviewers. The authors addressed all of the reviewers' comments to the satisfaction of both the technical editor(s) and the ASTM Committee on Publications.

The quality of the papers in this publication reflects not only the obvious efforts of the authors and the technical editor(s), but also the work of these peer reviewers. The ASTM Committee on Publications acknowledges with appreciation their dedication and contribution of time and effort on behalf of ASTM.

Printed in Ann Arbor, MI
May 1992
Second Printing, Ann Arbor, MI
January 1994

Foreword

This publication, *Geotechnics of Waste Fills—Theory and Practice,* contains papers presented at the symposium of the same name held in Pittsburgh, PA on 10–13 Sept. 1989. The symposium was sponsored by ASTM Committee D-18 on Soil and Rock. Dr. Arvid Landva, Professor of Civil Engineering, The University of New Brunswick at New Brunswick, presided as symposium chairman. He was also editor of this publication, along with G. David Knowles, Malcolm Pirnie Inc., Albany, NY.

Contents

INTRODUCTION

The purpose of this symposium was to explore the geotechnical properties and behavior of waste fill materials and to compile them into one volume that could serve as a reference text on a subject that is not widely addressed in accessible literature.

The symposium was sponsored by ASTM Committee D18, Subcommittee D18.14, Geotechnics of Waste Management. STP 1070 will serve as a guide to Committee D18 members in their future efforts to address the problems of landfill geotechnics, such as stability of slopes, settlement of fills and groundwater (leachate) flow.

The term "waste fill" covers a wide range of materials, from mineral fills contaminated with relatively small amounts of organic or vegetal debris through woodwastes and various types of tailings and slimes, to domestic and industrial refuse. Two categories of fill have purposely been excluded from this symposium: tailings and hazardous wastes. Also, we excluded liners, artificial or natural, from the list of topics. All these three topics have been widely covered in previous conferences, symposia and numerous papers.

Our involvement in the geotechnics of waste management has shown that more geotechnical attention should be paid to such aspects as placing methods, field and laboratory sampling and testing, classification and in-situ improvement methods. These are the topics that we sought to explore at this symposium, and the 23 papers presented here provide a valuable data base for the solution of problems pertaining to those topics.

The symposium was divided into four sections:

Section I - Landfill investigations, design, construction and closure (seven papers)

Section II - Stabilization, compaction and consolidation (six papers)

Section III - Stability and settlement analysis (six papers)

Section IV - Case histories (four papers)

LANDFILL INVESTIGATIONS, DESIGN, CONSTRUCTION

Morris and Woods emphasize the significant changes caused by large settlements after closure of landfills. These changes may negate contouring and drainage plans. Settlements can be predicted, but local regulations may not allow steeper slopes, even if temporary. Case records indicate necessity of perimeter ditches, proper compaction, daily covers, retaining structures for ash fills, and limited size of working areas. A computer program for primary and secondary settlements is given.

Orr and Finch report on case studies of the effects of earthquakes on landfills. Their studies pertain to the October 17, 1989, earthquake in the South Bay area east of Santa Cruz, California. They find that the two most important factors are acceleration and duration rather than the more commonly used magnitude. The properties of refuse may dampen or attenuate the effects of earthquakes.

Lawrence and Boutwell claim that electro-magnetic (EM) surveys have much to offer. They describe a statistical technique they have developed to interpret EM data: a multivariate regression prediction (MVRP). Three cases are described, and it is concluded that the correlation is satisfactory. The MVRP-EM method is most practical when there are time or budget restraints. It is extremely cost-effective for reconnaissance work.

Gifford et al. report on a geotechnical investigation of an Albany, New York, landfill to be used as a building site by the City. The investigation is laid out with due regard to architectural and structural requirements. The foundation layout is designed to minimize settlements or to allow for them. Settlements are predicted on the basis of the nature of the landfill materials and a comparison plot of case records of long-term settlement in landfill.

Sharma et al. discuss various methods of dynamic laboratory and field tests, including applicability. They describe the down-hole geophysical method as used at a landfill site in Richmond, California. The site is underlain by the San Pablo Bay Mud. The down-hole method was chosen because only one boring is required at each location, which makes this method cost-effective. Dynamic shear moduli and Poisson's ratios are reported for refuse and for the Bay Mud.

Huang and Lovell present a very thorough geochemical and geotechnical analysis of several sources of bottom ash (incinerated refuse). This paper constitutes an excellent data base for researchers and users.

Landva and Clark describe a comprehensive field and laboratory investigation of various waste fills in Canada. A classification system is proposed, and index tests and properties are discussed and presented. Also described is equipment developed for the testing of waste fill materials, and geotechnical properties are reported and discussed.

STABILIZATION, COMPACTION AND CONSOLIDATION

Briaud et al. describe a new test (the WAK test) they have introduced to check soil stiffness improvements after dynamic compaction. The WAK test appears to be at a preliminary stage, but it also appears to be a promising test that can be used as a very fast quality control test on dynamic compaction jobs. The authors also present their proposed curve fitting technique and stiffness determination.

Acar et al. present a comprehensive study of boiler slag. They discuss the results of laboratory studies and field compaction tests conducted to evaluate its engineering and field compaction characteristics. This paper represents another valuable data base for the geotechnical behavior of incinerated refuse. Recommendations are given for the optimum design and construction procedures for slag fill.

Davies discusses the reject resulting from the reworking of colliery waste tips and their use in landfill. The mixing of the coarser reject with tailings presents problems for compaction, but these may be alleviated by the addition of cement. The author discusses the properties of the cement-stabilized waste and conclude that the stabilizing effects diminish with increasing effective stress and water content.

Koutsoftas and Kiefer report on a dynamic compaction study of a mine waste spoil. They find significant improvements in geotechnical properties to depths of 9 to 12m after compaction with a 16 tonne weight from 20m height. Most of the improvement occurred during the later phase of treatment. The authors point out that the depth of improvement is limited and that another cost-effective and rapid method of improving waste fills is preloading.

Soliman presents the results of extensive tests on lime fixed flyash and FGD sludge. His conclusions are of considerable interest: the strength of the fixed material increases with time, with density, and with the salinity of the water. Hence the material could be compacted into blocks and dumped in the ocean to create a reef.

Martin et al. report on a study to stabilize acidic hydrocarbon sludge lagoons by microencapsulating it in a matrix of clay, which is neutralized and cemented with a lime-flyash pozzolanic mixture.

STABILITY AND SETTLEMENT ANALYSES

Mitchell et al. draw attention to the potential failure surfaces along lining system interfaces and their possible control of the overall stability of hazardous waste fills. Residual friction angles as low as 5° are reported. They carry out a 3-D stability analysis of a slope failure in a hazardous waste repository and conclude that, even though it is possible to plan filling operations on the basis of adequate factors of safety, this may presently be difficult because of a lack of a suitable 3-D analysis method and because of uncertainties about seismic effects.

Edil et al. outline an analysis approach for the settlements of refuse along the lines of previous analysis methods used for peats and organic soils. They compare their analytical results with actual field measurements and conclude that refuse settlement can be modeled satisfactorily. Another interesting conclusion is that primary compression is largely completed during the filling operation; secondary compression is more evident once filling has stopped.

Singh and Murphy evaluate studies of shear strength properties and settlement characteristics of refuse and discuss the inadequacy of the Mohr-Coulomb theory to account for the large yet non-catastrophic deformations in refuse. They conclude that a slope failure may not be the most critical aspect, but rather settlement of the refuse and bearing capacity of the foundation soil. They draw attention to the lack of knowledge of the dynamic strength characteristics of refuse.

Siegel et al. report on a comprehensive geotechnical investigation and slope stability study of an instrumented landfill in Monterey Park, California. They conclude, among other things, that CPT's are not useful in refuse, other than to identify weak zones, and that direct shear test results should be used with caution, depending on the size of the apparatus. Their tentative calculated factor of safety of 1.2 is subject to further studies in view of the uncertainties in determining refuse strength and the potential for refuse strength to change with time. One important conclusion from an interpretation of the 1987 Whittier Narrows and the 1988 Pasadena earthquakes is that landfill can withstand moderate earthquakes with only minor repairs.

Tieman et al. draw attention to the future needs for piggyback additions to landfills and illustrate some of the benefits of vertical piggybacking. But they also point out that such expansions can be complicated to design and construct. A case record is described where subgrade reinforcement and slope stability improvement were required. Each piggyback expansion will be unique with its own set of design and construction considerations.

Duplancic presents a geotechnical evaluation of deformation monitoring data on a hazardous waste landfill. The data indicate that the landfill is deforming similarly to earthfill dams. Deflections are larger in the fill zone, but almost negligible in rock and native clay zones. The analyses presented show that standard geotechnical techniques can be used to monitor the performance, and standard geotechnical computational methods can be used - with care - for landfill stability analyses and deformation assessment.

CASE HISTORIES

Belfiore et al. present a conventional soil mechanics approach to sludge fill investigations, emphasizing the necessity of adapting and integrating conventional geotechnical tools with the aid of a comprehensive performance monitoring program. The key objective was to study the effects of compaction methods on an improvement of the landfilling operations. On the basis of the results of the two case history studies, they conclude that the high drained strengths measured in the laboratory are confirmed by the long-term behaviour of sludges landfilled with slopes up to 35° without any stability problems. Also their tests and measurements show the beneficial effects of waste compaction, such as significant volume reduction and improvement in strength and deformation properties.

Hinkle describes the use of a 30m deep closed landfill as a marine container storage. He demonstrates that landfill property can be reclaimed and put to profitable use. One important aspect is a proper seal, and the design and construction of this is described in detail.

Oakley studies the use of the cone penetrometer (CPT) in a chemically stabilized waste fill. On the basis of field observations of settlements in two fills, he finds that settlements calculated from CPT data are reasonably close to those measured. Calculated rates of settlement are generally within about ± 50% of those measured.

Coduto and Huitric monitored settlement and horizontal movements at different depths within a sanitary landfill. They found, following two years of monitoring, that vertical strain rates are independent of depth while horizontal movements on slopes are greatest near the surface and diminish with depth. No permanent displacement occurred during a Richter magnitude 6.1 earthquake.

CLOSURE

A broad spectrum of topics have been addressed by the contributors to this volume. Settlement is analysed in five papers, stability of slopes in two, field and laboratory investigations in seven (demolition landfill, bottom ash, refuse, boiler slag, and lime-fixed flyash and FGD sludge). The effects of earthquakes are outlined in three papers, and field pilot tests (MVRP-EM survey, down-hole geophysical, CPT) in four papers. Stabilization by different methods (cement, dynamic compaction, lime-fixed flyash, clay and lime-flyash pozzolanic mixture, compaction) are described in five papers. Other topics addressed are the inapplicability of the Mohr-Coulomb criterion, the possible non-criticality of slope stability in regular landfills (compressibility of refuse and bearing capacity of the foundation soil are perhaps more important), the uncertainty of the strength characteristics of refuse, precautions required when designing and constructing piggyback additions to landfills, and the importance of designing and constructing a proper seal on a landfill to be used as a building site.

With all these topics addressed by experts in their respective fields, this volume should be a useful handbook for design and construction on and in the very large number of closed landfills in North America and elsewhere.

ACKNOWLEDGEMENTS

We are pleased to acknowledge ASTM's efficient organization of this symposium and want to thank Kathy Green, Rita Harhut and Monica Armata for their help and patience. Dorothy Savini and Bob Morgan expertly organized the symposium.

We were guided throughout the extensive review process by Dr. Larry Jackson of GTEL Environmental Laboratories, Inc., who offered much help and advice.

We are grateful to the former chairman of D18, Woody Schockley, and to the present chairman, Dick Ladd, for permission to arrange this symposium.

Our secretarial staffs at the University of New Brunswick and at Malcolm Pirnie's assisted ably in the review process.

Arvid Landva
Department of Civil Engineering
University of New Brunswick
Fredericton, New Brunswick
Canada E3B 5A3

G. David Knowles
Malcolm Pirnie Inc
4 Corporate Plaza
Washington Avenue Extension
Albany, NY
USA 12203

Landfill Investigations, Design, Construction

Derek V. Morris and Calvin E. Woods

SETTLEMENT AND ENGINEERING CONSIDERATIONS IN LANDFILL AND FINAL COVER DESIGN

REFERENCE: Morris, D.V. and Woods, C.E., "Settlement and Engineering Considerations in Landfill and Final Cover Design," Geotechnics of Waste Fills - Theory and Practice, ASTM STP 1070, Arvid Landva and G. David Knowles, Editors, American Society for Testing and Materials, Philadelphia, 1990.

ABSTRACT: Design of municipal landfills for closure is complicated by the large settlement that normally takes place for long periods of time after abandonment. This means that landfill slopes can change significantly with time, negating careful contouring and drainage provisions. It is possible to try and forecast this, but regulatory considerations (for landfills of specific depths especially) may hinder adequate design for full expected post-closure settlement. Preferred management techniques are outlined, and specific recommendations made for maintenance and settlement.

KEYWORDS: landfills, municipal waste, fill closure, waste fills, fill settlement.

Municipal landfills are designed with many constraints, both technical and legal, which cover not only operation, but also impact significantly on final closure. Poorly designed or operated landfills are often more likely to show signs of distress after abandonment, when little emphasis is placed on control and monitoring, than during operation, when significant attention is paid to safe compliance.

One of the most awkward technical post-closure considerations is the large amount of fill settlement that can take place for many years after abandonment. Predicting so much settlement is difficult analytically, as municipal fill undergoes large amounts of secondary consolidation, not easily incorporated into traditional settlement calculations. Moreover the regulatory situation may make it difficult legally to develop a closure plan that will continue to perform satisfactorily for an indefinite period.

Drs. Morris and Woods are Associate Professor and Professor in Civil Engineering at Texas A&M University, College Station, TX 77843.

Regulatory Considerations

Frequently the most intractable problem is administrative, in so far as final slopes are usually severely prescribed by regulation, and are in many cases not permitted to be formed at slopes that might be indicated by strict geotechnical design considerations. Regulatory practice in Texas (and many other states) is to classify municipal landfills according to whether storage is above or below ground. The operating plan of landfills licensed for below ground disposal will generally specify below-ground disposal only. As a result potential problems arise in the geotechnical design of the final cover, since this must be sloped to increase the runoff coefficient and minimize infiltration. However the slope of the final cover of below-ground landfills is normally limited by permit, in Texas to between 2% and 6% at the time of closure, irrespective of the recommendations of engineering analysis. In some states the maximum slope is 5%.

The situation is complicated further by the fact that many landfill operators choose to ignore the compression (or consolidation) of the municipal waste during operation, because they stand to benefit from doing so. Some state taxes are levied in theory on the basis of a unit rate per mass of deposited fill (e.g. 5¢ per ton). However rigorous weighing of every full and empty truckload is uneconomic, so in reality long-term measurements are made (usually by surveying) of deposited volume. These are then converted into an equivalent mass of fill, using an assumed density. In Texas this is currently in the range of 500 to 600 kg/m^3, which is generally a significant underestimate of actual municipal waste densities, particularly after some compression has taken place or under significant depths of overburden. This means that the deposited tonnage computed in this fashion is normally less than actually stored. More realistic values of density should undoubtedly be used by landfill operators, but since it would almost certainly result in a higher tax assessment, this represents a positive disincentive to support more accurate analysis. It would be preferable if taxes were levied on a strictly volumetric basis, thereby encouraging compaction.

Three case histories are discussed that demonstrate the consequences of improper design and operation, as follows:-

Case A. A municipally owned landfill that served approximately 70,000 people was closed approximately fifteen years ago. The location had an average rainfall of 1.25 m per year and a pan evaporation rate of 1.7 m per year. The landfilled area was about 150,000 m^2 and was a "below ground" landfill that utilized an area method of construction. The initial slopes were approximately one percent. Three years after closure 80,000 m^2 of the landfill cover would briefly hold ponded water after a rain. The water did not stay long because there were cracks or fissures open enough to expose the solid waste below the final cover. A portion of the final cover had settled below the original ground level. Leachate springs formed at the interface of the natural ground and the final cover and springs were almost continuous around the perimeter of the landfill. Leachate flow continued during dry periods; however, flow rates increased during wet weather.

The leachate flow rate was higher than could be accounted for assuming a zero runoff coefficient over the landfilled area. This caused a concern that there might be a groundwater spring under the landfill. Closer examination showed that the source of excess leachate was from an additional area of 150,000 m^2 that discharged runoff onto the landfilled area and into the solid waste through the portion of the final cover that had settled below the natural ground level. When this source of water was eliminated the leachate flow became intermittent. Fortunately, the city owned enough land adjacent to the landfill, that they were able to rebuild the final cover without too much difficulty to a two percent grade, at which point leachate flow ceased. Within another three years, however, the repaired cover had again settled enough to cause further ponding, open fissures and new leachate flow. The final cover remained a high maintenance item for nearly ten years.

The mistakes made during the operation of this landfill that contributed to the difficult closure included:

* no perimeter ditch to prevent runoff water from adjacent areas from reaching the landfilled area.

* below ground disposal only with no berm around the landfill area. The completed final cover had a slope of only about one percent which caused the final cover to be roughly parallel to the original ground. The general settlement caused the cover to sink below the natural grade in some places.

* the solid waste in the landfill was not compacted properly causing a great deal of differential settlement. The differential settlement caused the final cover to rupture in many locations.

* the final closure was not given appropriate priority during the operation of the landfill.

Case B. A privately owned municipal landfill that served approximately 30,000 people was closed approximately two years ago. It was in an area with 0.81 m of precipitation and 2.0 m of pan evaporation annually. The landfilled area was approximately 100,000 m^2, and had both "below ground" and "above ground" disposal.

Approximately three years before closure none of the landfill had received final cover, the slopes were less than two percent, and ponding water was extensive when it rained. Fortunately, no leachate was ever observed at this landfill. The solid waste was not being properly compacted. At this point a closure plan was developed for the landfill that included the purchase of compaction equipment, increasing the slope on the final grades to six percent, installing monitoring wells, and placing the final cover over completed areas as soon as they were finished.

In some areas that were being filled there was as much as 6 m of poorly compacted solid waste. Some of these areas settled as much as 1.5 m when solid waste and final cover were added. Several areas involving a total area of about 20,000 m^2 had to be refilled three times to hold a six percent slope until the landfill was closed.

Within two years after closure the six percent slopes have been reduced to about four percent due to settlement. No ponding has occurred. No leachate has been observed.

Case C. An industrial landfill is located in an area with 1.0 m of precipitation and 1.8 m of pan evaporation per year. It is an "above ground" landfill with a high plasticity clay liner and no berm, and is a "monofill" containing only ash and a final cover of a low plasticity clay.

The clay liner was constructed over the natural grade. The slopes of the natural ground ranged from two to four percent with several dry drainage channels. Only minor modification of the existing topography, including the channels, was undertaken when the liner was constructed. Perimeter ditches prevented surface water from entering the landfill area.

During construction immediately after a rain approximately 40,000 m^3 of ash washed out of the landfill. Fortunately it was captured in the surrounding sedimentation ponds.

The mistakes made in this operation included:

* No retaining structure was provided for this landfill.

* Did not use daily cover

* The working area with no final cover was too large. This resulted in low runoff and high infiltration. The rain water entered the ash, flowed downward to the liner, thence to the channel. As this water flowed out from under the fill into the perimeter ditches the ash slope became unstable and significant volumes of ash escaped from the site.

SIDE SLOPE DESIGN

One of the biggest problem areas in practice occurs at the edge of landfill sites, where local ponding and erosion often takes place, as shown in Fig. 1. Side slopes that may initially have been uniform, will tend to become non-uniform after sufficient post-closure settlement. This is true even if the fill is idealized as undergoing an equal amount of compressive strain across the site, as the change in slope at the base of the excavation will inevitably be reflected in a concave change in slope of the final settled surface.

The effect is magnified by two further complications. Firstly, compaction of fill is normally more difficult at the edge of a site, and subsequent settlement therefore greater. Secondly, the average effective depth of overburden there will be less than in the center of the site, so that the fill there will be less dense and more compressible, and compressive strains of the edge will be more than in the middle.

Solutions to this local problem are not easy, but possible techniques include thickening the soil cover there to compensate. It is in principle also a good idea to steepen slopes close to the edge, although care must be taken to prevent erosion. Slopes of overall slope greater than 6% can perform safely, but should be designed with proper provision for safe drainage - ideally drainage channels or ditches, and a slope should preferably also be terraced as a series of small slopes, to minimize potential erosion.

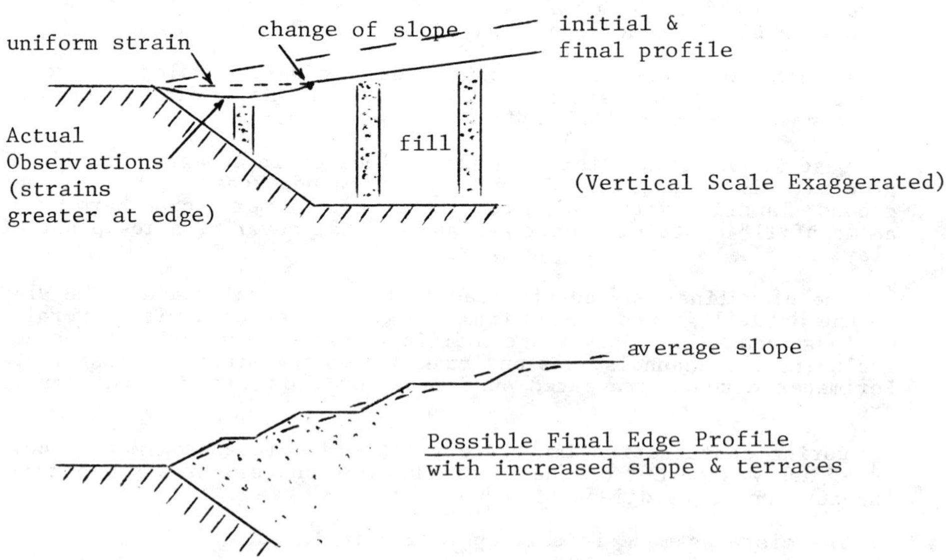

FIG. 1 - Settlement at Side of Landfill

ANALYSIS OF EFFECT OF DAILY COVER

An additional practical complication, that the following computer program addresses, but which is normally ignored, is the settlement behavior of daily cover. Placement of a daily cover of inorganic soil over waste fill is standard practice at most landfill sites. It is done in order to keep waste from blowing away, restrict access to rodents, birds and insects and provide additional overburden pressure. Typical placement procedures consist of 0.6 m of compacted waste and 0.15 m of soil cover. A simple settlement analysis would probably assume that this intermediate zone of material would settle as independent layers between the much thicker layers of waste, remaining largely intact and undergoing some consolidation settlement of its own.

In fact this conceptual model of cover soil behavior does not correspond to reality. Although this inert soil component initially occupies in this case 20% of the total fill volume, observations at operating landfills indicate that the proportion becomes significantly reduced as settlement proceeds. This is due to a small extent to the compression of the soil under self-weight, plus (most importantly) migration of this material into the voids in the adjacent solid waste as shown in Fig. 2.. The net result is that this soil layer can reduce to less than one-quarter of the original volume. The resultant density of the fill is therefore considerably more than would be expected if this effect was ignored.

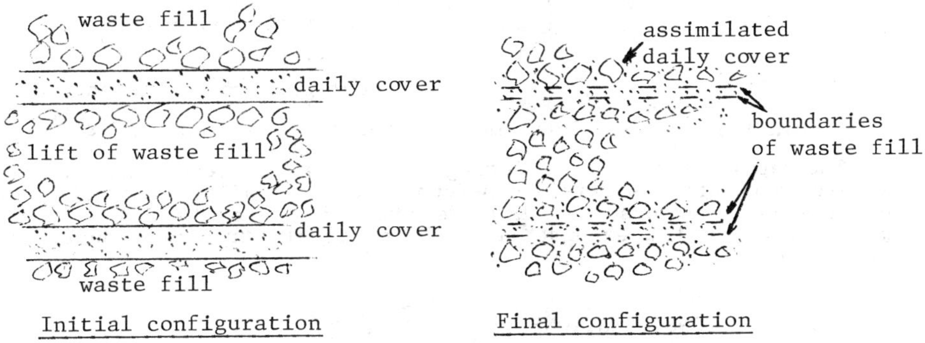

FIG. 2 - Absorbtion of daily cover into waste fill

<u>Further design considerations:</u> This effect has implications also for daily operation and design. It is often suggested that one way of increasing the storage efficiency of a landfill is to increase waste lift thicknesses relative to daily cover thickness, or even to eliminate daily cover completely. The suggestion is superficially appealing as it would seem to minimize dilution of valuable storage volume with natural soil. In addition the decreased overall density (since the soil cover is typically much denser than solid waste by a factor of 2 to 4) should reduce settlement values of the final landfill, and make design of final closure and abandonment easier. These arguments are less convincing in reality, however.

<u>Effect on storage volumes:</u> In fact the net loss of storage due to the presence of the extra daily cover of soil, is not large in practice, for the above reasons. A soil layer that may initially have occupied 20% of the overall disposal volume, will, after assimilation into adjacent fill, occupy only about 5% of the overall fill volume at depth. This no longer represents a serious economic loss for the operator. Indeed the corresponding increase in overall fill density will assist in self-weight compression of the landfill as a whole, although this will be to some extent offset by the reduced compressibility of the

adjacent fill. The net result is to largely compensate for even this small loss of storage volume, which is in fact relatively insensitive to the thickness of daily cover used.

<u>Effect on settlement:</u> The second argument, that reduction of daily cover will decrease post-construction settlement by reducing self-weight densities, is true, but its applicability is debatable. Overall fill settlement may be less as a result of lower self-weight stresses, but not by as much as might be supposed, since the absence of the soil "infilling" of the voids in the solid waste will cause the overall compressibility to be increased. Furthermore the time taken for settlement will generally be increased, such that a larger proportion of the settlement is liable to take place after closure. If this is the case, then it is possible that post-closure settlement of the landfill may actually be more than before, although pre-closure settlement may have been reduced somewhat, as illustrated in the accompanying Fig. 3.

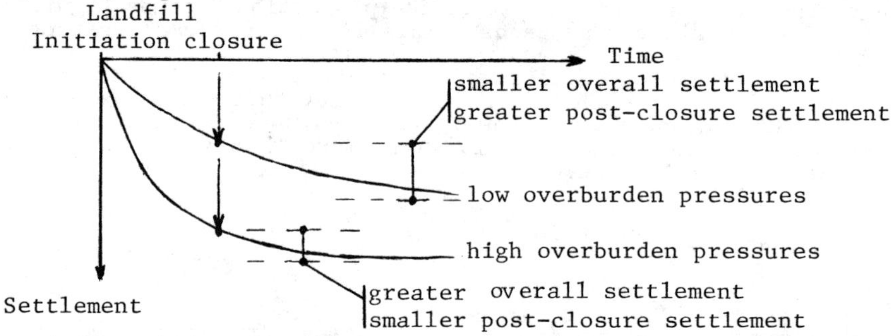

FIG. 3 - Possible Settlement Curves for Dense and Light Fills

The result of these conflicting desires is that recommended practice should be to utilize fairly generous layers of daily cover soil. This considerably reduces fire hazards etc., and greatly eases landfill management, at the expense only of relatively trivial reductions in storage capacity.

GEOTECHNICAL CONSIDERATIONS

Among the practical problems associated with the utilization of the storage volume created by the fills, settlement may be the most significant engineering problem because of its large magnitude and long-term nature. Representative field data is very important in this context.

In order to enable this calculation to be performed iteratively with sufficient accuracy, and incorporation of both primary and secondary consolidation, it is convenient to use a computer program. A relatively comprehensive and yet easy to use version is outlined in this paper.

In this program, initial primary compression of solid wastes under selfweight as well as long-term secondary settlement taking place for a specified period of time after construction are computed. Also included in the program is a routine for calculation of unit weights at different depths of fill, useful if organic content is known.

PROGRAM FOR FILL SETTLEMENT COMPUTATION

The method used in calculation of fill settlement and density with time, is similar to that outlined by Sowers [1], which has been widely discussed and used for comparison purposes by others ([2], [3]).

The method is based on the following assumptions, principles and procedures, which represent the major limitations on the use of this model, as stated herein :-

Initial Primary Compressions

(1) The initial primary mechanical compression due to changes in overburden pressure occurs rapidly with little or no pore pressure build-up. The initial and primary phases are complete in less than a month.

(2) The primary settlement occurring in an arbitrary waste layer due to construction of an additional layer can be expressed by the well-known equation:

$$S = H\frac{C_c}{1+e_o}\log\frac{P_o+dP}{P_o}$$

where,

S = primary compression occurring in the layer under consideration

H = initial (before-compression) thickness of the waste layer under consideration

C_c = primary compression index. C_c is assumed to be proportional to the initial void ratio of the layer (C_c = COEFF1*e_o, see Sample Results for typical COEFF1 values).

e_o = initial (before-compression) void ratio of the layer.

P_o = existing overburden pressure acting at the mid level of the layer.

dP = increment of overburden pressure at the mid level of the layer under consideration from the construction of an additional layer (100% of pressure increase at the top new layer is assumed to be transferred to the layer under consideration).

(3) The daily and final cover soils are not assumed to undergo compression due to overburden pressures. However, the thickness of a daily soil cover is assumed to reduce to one-fourth its original thickness after construction, due to the migration of soils into void spaces in waste layers. For very clayey soil cover, this value may be greater. This is an empirical assumption based on limited field observations, discussed previously in the section entitled "Analysis of Effect of Daily Cover".

(4) The above computation procedures are repeated for each waste layer and for each construction stage, and the contribution of that layer to the overall compression in a particular construction stage is determined. The sum of the contributions of each construction stage then constitutes the total compression achieved up to that construction stage under consideration.

Long-term Secondary Settlements

(1) Settlement of waste fills continues at substantial rates after construction. The settlement occurring in an arbitrary layer for a certain period of time after completion of the landfill can be expressed by the following equation:

$$SS = H\frac{C_s}{1+e_o}\log\frac{t_2}{t_1}$$

where,

SS = long-term secondary settlement occurring in the layer under consideration, between time periods t_1 and t_2.

H = initial (before settlement) thickness of the waste layer under consideration.

C_s = secondary compression index. C_s is assumed to be proportional to the initial void ratio of the layer (C_s = COEFF2*eo, see Sample Results for typical COEFF2 values).

e_o = initial (before settlement) void ratio of the layer.

t_1 = starting time of the time period for which long-term settlement of the layer is desired (t_1 = 1 month preset in the program).

t_2 = ending time of the time period for which long-term settlement of the layer is desired.

(2) The above computation procedure is repeated for each waste layer for the given time period. The sum of the contributions of each layer then constitutes the total long-term settlement achieved during that time period.

SAMPLE RESULTS

An example of such a computation is shown as follows. The following input data is required:

NPROB - number of problems to be analyzed, 1 in this case.

TITLE - self-explanatory, up to 60 characters for each problem.

NLAYER - number of waste layers (or lifts) placed, up to 40.

IFLAG - 0 if a compression index is to be computed from a COEFF1.

- 1 if a unit weight to be entered as GAMMAX.

HWASTE - original thickness of a waste layer (or lift), before compression.

HSOIL - original thickness of daily soil cover (assumed to be reduced to one-quarter of its original thickness after construction).

HCOVER - thickness of final soil cover (assumed to remain constant).

GAMMAW - initial compacted unit weight of waste, typically 3 to 6 kN/m^3 depending on composition and compaction techniques [2].

GAMMAS - initial unit weight of soil cover, typically 20 kN/m^3.

WC - initial water content of waste, as a decimal.

SG - specific gravity of waste, typically 1.5.

DUMMY - equal to COEFF1 if IFLAG = 0, typically 0.15 to 0.9 and increases with organic content in the waste [1].

- equal to GAMMAX if IFLAG = 1, expected unit weight of waste under full overburden at the site, if this is preferred to entering a value of COEFF1. See Fig. 4 for recommended values.

NTIME - number of time intervals for which a settlement calculation is desired (maximum of 20).

COEFF2 - secondary compression index divided by voids ratio, typically 0.03 to 0.09 and increases with conditions favorable to decomposition [1].

TIME(I) - time intervals after landfill completion for which a settlement calculation is desired (in months).

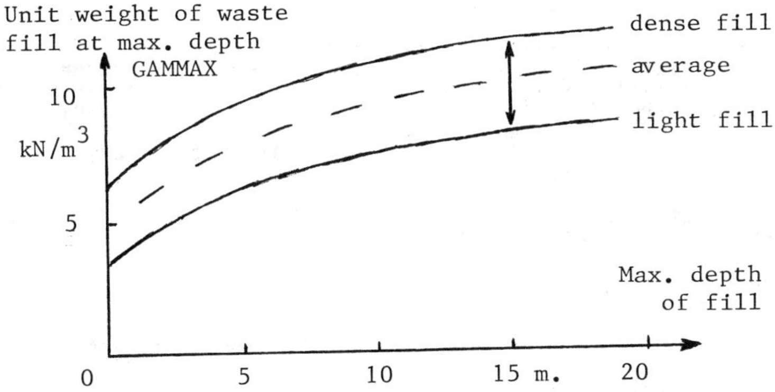

FIG. 4 - GAMMAX versus landfill depth

For a sample problem with 20 layers of waste fill placed initially 750 mm thick at a unit weight of 3.75 kN/m^3, water content of 0.15 and specific gravity 1.5, with 150 mm of soil cover of unit weight 19 kN/m^3 and a final 600 mm soil cover, the following results are obtained, if the primary compression of the waste can be described by COEFF1 = 0.8 and the secondary compression by COEFF2 = 0.06:

Initial as placed total thickness at site	=	18.6 m
Initial compression of site	=	7.3 m
Initial landfill thickness	=	11.3 m
Unit weight of waste at surface	=	3.8 kN/m^3
Unit weight of waste at bottom of landfill	=	8.7 kN/m^3
Long-term settlement, after 1 year	=	0.31 m
Long-term settlement, after 10 years	=	0.60 m
Long-term settlement, after 100 years	=	0.89 m

This program enables important settlement parameters to be calculated rapidly and to an accuracy impossible with hand-calculation, and the authors have found it to be a useful tool for analysis and design.

REFERENCES

[1] Sowers, G.F., "Settlement of Waste Disposal Fills," Proceedings 8th International Conference on Soil Mechanics and Foundation Engineering, Moscow, 1973, pp. 207-210.

[2] Yen, B.C. and Scanlon, B., "Sanitary Landfill Settlement Rates," Journal of the Geotechnical Division, ASCE, Vol. 101, No. GT5, 1975, pp. 475-487.

[3] Rao, S.K., Moulton, L.K., and Seals, R.K., "Settlement of Refuse Landfills," Proceedings, Conference on Geotechnical Practice for Disposal of Solid Waste Materials, University of Michigan, Ann Arbor, June 13-15, 1977, published by ASCE, pp. 574-598.

APPENDIX I - PROGRAM SOURCE CODE LISTING (FORTRAN 77)

```
C*****************************************************************
C     PROGRAM SETT87 -  ANALYSIS OF SANITARY LANDFILL SETTLEMENTS
C*****************************************************************
      DIMENSION S(40,40),SS(40,20),H(40,40),HH(40,40),E(40,40),
     *          CC(40,40),GAMMA(40,40),P(40,40),
     *          S1(40),SS1(20),SUM(40),GAVE(40),HTOT(40),HHTOT(40),
     *          TIME(20),TITLE(15)
C
      OPEN (UNIT=5, FILE='S.DAT', STATUS='OLD')
      OPEN (UNIT=6, FILE='S.OUT', STATUS='NEW')
C
      READ  (5,100) NPROB
      DO 1  IPROB=1,NPROB
C
C**** COMPUTATION OF INITIAL SETTLEMENTS *************************
C
      READ  (5,105) TITLE
      READ  (5,100) NLAYER,IFLAG
      READ  (5,110) HWASTE,HSOIL,HCOVER,GAMMAW,GAMMAS,WC,SG,DUMMY
      COEFF1=0.
      IF(IFLAG.EQ.0) COEFF1=DUMMY
      IF(IFLAG.EQ.1) GAMMAX=DUMMY
100   FORMAT(2I5)
105   FORMAT(15A4)
110   FORMAT(8F10.3)
C
      VOID = 62.4*27.0*SG*(1.0+WC)/GAMMAW - 1.0
C
      WRITE (6,220)
      WRITE (6,200) IPROB,TITLE
      WRITE (6,210) NLAYER,HWASTE,HSOIL,HCOVER,GAMMAW,GAMMAS,
     *              WC,SG,VOID
      IF (IFLAG.EQ.0) WRITE (6,211) COEFF1
      IF (IFLAG.EQ.1) WRITE (6,212) GAMMAX
200   FORMAT(//' (PROBLEM #',I2,')' ,4X,15A4   //' ***  INITIAL',
     *    ' COMPRESSION OF SOLID WASTE FILLS  ******************')
210   FORMAT(//' ===  INPUT DATA  ==='/
     * /' TOTAL NO. OF LAYERS (NLAYER)                          =',I10,
     * /' INITIAL THICKNESS OF A SOLID WASTE LAYER (HWASTE)  =',F10.2,
     *  ' FT',
     * /' INITIAL THICKNESS OF A DAILY SOIL COVER (HSOIL)     =',F10.2,
     *  ' FT',
     * /' INITIAL THICKNESS OF FINAL SOIL COVER (HCOVER)      =',F10.2,
     *  ' FT',
     * /' INITIAL UNIT WEIGHT OF SOLID WASTE (GAMMAW)         =',F10.2,
     *  ' LB/CY',
     * /' INITIAL UNIT WEIGHT OF SOIL (GAMMAS)                =',F10.2,
     *  ' LB/CF'
     * /' INITIAL MOISTURE CONTENT OF SOLID WASTE (WC)        =',F10.2,
     * /' INITIAL SPECIFIC GRAVITY OF SOLID WASTE (SG)        =',F10.2,
     * /' COMPUTED INITIAL VOID RATIO  OF SOLID WASTE (VOID) =',F10.2)
211   FORMAT(' ESTIMATED PRIMARY COMPRESSION INDEX (COEFF1*Eo)',
     *         4X,'=',F7.3,' Eo')
212   FORMAT(' EST''D MAXIMUM UNIT WEIGHT OF SOLID WASTE (GAMMAX)  =',
     *   F10.2,' LB/CY')
220   FORMAT(1H1)
```

```
      PW = (GAMMAW/27.)*HWASTE
      PS =  GAMMAS*HSOIL
      IF (IFLAG.EQ.0) GO TO 13
      WRITE (6,220)
      WRITE (6,215)
215   FORMAT(//' ===  DETERMINE REQUIRED PRIMARY COMPRESSION INDEX (CC)',
     *  '  ==='//' ITERATION    CC     ACHIEVED-GAMMA  REQUIRED-GAMMA'/)
13    CONTINUE
C
      DO 10 K=1,100
      DP = PW+PS
      DO 20 J=1,NLAYER
      IF (J.EQ.NLAYER) DP=PW+GAMMAS*HCOVER
      DO 30 I=1,J
      IF (I.EQ.J) GO TO 25
      P0 = (0.5*PW+PS) + (J-I-1)*(PW+PS)
C
      S(I,J) = CC(I,J-1)/(1.+E(I,J-1)) * H(I,J-1) * ALOG10((P0+DP)/P0)
      H(I,J) = H(I,J-1) - S(I,J)
      HH(I,J)= H(I,J) + HSOIL/4.
      E(I,J) = E(I,J-1) - S(I,J)*(1.+E(I,J-1))/H(I,J-1)
      GAMMA(I,J) = GAMMA(I,J-1)*H(I,J-1)/H(I,J)
      GO TO 26
C
25    S(I,J)=0.
      H(I,J)=HWASTE
      HH(I,J)=HWASTE+HSOIL/4.
      IF(J.EQ.NLAYER) HH(I,J)=HWASTE+HCOVER
      E(I,J)=VOID
      GAMMA(I,J)=GAMMAW
C
26    CC(I,J) = COEFF1*E(I,J)
      P(I,J) = (J-I+1)*(PW+PS)
      IF(J.EQ.NLAYER) P(I,J) = P(I,J) + (HCOVER-HSOIL)*GAMMAS
30    CONTINUE
20    CONTINUE
C
      IF (IFLAG.EQ.0) GO TO 16
C
      GG = GAMMAX-GAMMA(1,NLAYER)
      WRITE (6,216) K,COEFF1,GAMMA(1,NLAYER),GAMMAX
216   FORMAT(I5,F10.2,'Eo',2F15.1)
      IF (GG-10.) 14,14,15
15    COEFF1 = COEFF1 + 0.02
10    CONTINUE
C
      WRITE (6,217) K-1
217   FORMAT(///' &&&&&&&&  CLOSURE NOT ACHIEVED IN',I5,
     *            ' ITERATIONS  &&&&&&&&&&&&&&&&&&&&')
      GO TO 1
C
14    WRITE (6,218) COEFF1
218   FORMAT(/' ADOPTED PRIMARY COMPRESSION INDEX (CC) =',F7.2,' Eo')
16    CONTINUE
C
      WRITE (6,220)
      WRITE (6,230)
      K=NLAYER-2
      DO 40 J=K,NLAYER
      WRITE (6,235) J
      WRITE (6,240) (I,J,S(I,J), I,J,H(I,J), I,J,E(I,J),
     *               I,J,GAMMA(I,J), I,J,CC(I,J), I=1,J)
40    CONTINUE
230   FORMAT(///' ===  VALUES OF VARIOUS VARIABLES IN SELECTED',
     *' CONSTRUCTION STAGES  ==='//
     *6X,'S(I,J) = SETTLEMENT OF LAYER-I FROM  CONSTRUCTION OF LAYER-J'/
     *6X,'H(I,J) = THICKNESS  OF LAYER-I AFTER CONSTRUCTION OF LAYER-J'/
     *6X,'E(I,J) = VOID RATIO OF LAYER-I AFTER CONSTRUCTION OF LAYER-J'/
     *6X,'G(I,J) = UNITWEIGHT OF LAYER-I AFTER CONSTRUCTION OF LAYER-J'/
     *6X,'CC(I,J)= COMP.INDEX OF LAYER-I AFTER CONSTRUCTION OF LAYER-J')
235   FORMAT(//' (WHEN LAYER',I3,' COMPLETED)'/)
240   FORMAT(' S(',I2,',',I2,')=',F5.3,
```

```
     *        ' H(',I2,',',I2,')=',F5.3,
     *        ' E(',I2,',',I2,')=',F5.3,
     *        ' G(',I2,',',I2,')=',F6.1,
     *        ' CC(',I2,',',I2,')=',F5.3)
C
      SUM(1)=0.
      DO 50 J=1,NLAYER
      S1(J)=0.
      DO 55 I=1,J
      S1(J)=S1(J)+S(I,J)
55    CONTINUE
      IF(J.EQ.1) GO TO 50
      SUM(J)=SUM(J-1)+S1(J)
50    CONTINUE
C
      TOT1=0.
      TOT2=0.
      DO 60 J=1,NLAYER
      JJ=NLAYER+1-J
      TOT1=TOT1+H(JJ,NLAYER)
      TOT2=TOT2+HH(JJ,NLAYER)
      HTOT(JJ)=TOT1
      HHTOT(JJ)=TOT2
60    CONTINUE
C
      GTOT=0.
      DO 65 J=1,NLAYER
      JJ=NLAYER+1-J
      GTOT=GTOT+GAMMA(JJ,NLAYER)*H(JJ,NLAYER)
      GAVE(JJ)=GTOT/HTOT(JJ)
65    CONTINUE
C
      WRITE (6,220)
      WRITE (6,245)
      DO 70 J=1,NLAYER
      THICK= J*HWASTE
      PCT1 = SUM(J)/THICK*100.
      WRITE (6,250) J,THICK,SUM(J),PCT1
70    CONTINUE
245   FORMAT(//' ===  INITIAL COMPRESSION VS. CONSTRUCTION',
     *    ' STAGES  ==='//
     *    ' WHEN           AS-COMPACTED   INITIAL         PERCENT'/
     *    ' () LAYERS      WASTE DEPTH    COMPRESSION     COMPRESSION'/
     *    ' COMPLETED      (FT)           (FT)            (%)'/)
250   FORMAT(I5,3F15.2)
C
      WRITE (6,220)
      WRITE (6,255)
      DO 75 J=1,NLAYER
      JJ=NLAYER+1-J
95    RATIO = GAVE(JJ)/GAMMA(JJ,NLAYER)
75    WRITE (6,260) JJ,HTOT(JJ),HHTOT(JJ),P(JJ,NLAYER),GAVE(JJ),
     *                GAMMA(JJ,NLAYER),RATIO
255   FORMAT(//' ===  VARIATION OF UNIT-WEIGHTS OF WASTE WITH DEPTH',
     * ' AND PRESSURE'/6X,'(AFTER INITIAL COMPRESSION COMPLETED)',
     * '  ================================'//
     * ' LAYER#   ACCUMULATED ACCUMULATED ACCUMULATED ACCUMULATED',
     * ' UNIT-WEIGHT  RATIO'/
     * ' (BOTTOM  DEPTH OF    DEPTH OF    OVERBURDEN  AVERAGE    ',
     * ' OF WASTE    (AVE.U-W'/
     * ' LAYER=1) COMPACTED   COMPACTED   PRESSURE OF UNIT-WEIGHT',
     * ' AT EACH       /U-W)'/
     * '          WASTES      WASTE&SOIL  WASTE&SOIL  OF WASTE   ',
     * ' LAYER'/
     * '          (FT)        (FT)        (PSF)       (PCY)      ',
     * ' (PCY)'/)
260   FORMAT(I5,2F12.2,3F12.1,F12.3)
C
C**** COMPUTATION OF LONG-TERM SETTLEMENTS ********************
C
      READ  (5,120) NTIME
      IF(NTIME.EQ.0) GO TO 1
      READ  (5,125) COEFF2
```

```
      READ  (5,130) (TIME(I),I=1,NTIME)
120   FORMAT(I5)
125   FORMAT(F10.3)
130   FORMAT(16F5.1)
C
      WRITE (6,220)
      WRITE (6,265)
      WRITE (6,270) NTIME,COEFF2
      DO 80 J=1,NLAYER
      WRITE (6,275) J,E(J,NLAYER),H(J,NLAYER)
80    CONTINUE
      WRITE (6,280) HTOT(1)
C
      TIMEINIT=TIME(1)
      DO 85 K=1,NTIME
      SS1(K)=0.
      DO 86 J=1,NLAYER
      VOIDO = E(J,NLAYER)
      CS = COEFF2*VOIDO
      SS(J,K) = CS/(1.+VOIDO) * H(J,NLAYER) * ALOG10(TIME(K)/TIMEINIT)
      SS1(K) = SS1(K)+SS(J,K)
86    CONTINUE
85    CONTINUE
C
      WRITE (6,220)
      WRITE (6,285)
      DO 90 K=1,NTIME
      PCT2 = SS1(K)/HTOT(1)*100.
90    WRITE (6,290) TIME(K),SS1(K),PCT2
265   FORMAT(//' ***  LONG-TERM SETTLEMENTS OF SOLID WASTE FILLS',
     *       '  ***************'//)
270   FORMAT(' ===  INPUT DATA  ==='//
     *      ' NO. OF TIME SEGMENTS (NTIME)                 =',I10/
     *      ' SECONDARY COMPRESSION INDEX (COEFF2*Eo) =',F7.3,' Eo'//
     *      ' LAYER #        VOID RATIO            WASTE THICKNESS'/
     *      '                AFTER INITIAL         AFTER INITIAL   '/
     *      '                COMPRESSION           COMPRESSION     '/
     *      '                COMPLETED(Eo)         COMPLETED(FT)   '/)
275   FORMAT(I5,2F20.2)
280   FORMAT(/' TOTAL',19X,F20.2)
285   FORMAT(//' ===  RESULTED LONG-TERM SETTLEMENTS  ==='//
     *         ' TIME ELLAPSED (MONTH)    SETTLEMENT (FT)',
     *         '    PERCENT SETTLEMENT (%)'/)
290   FORMAT(F15.1,F20.2,F20.2)
C
C     PRINTOUT SUMMARIZED RESULTS
C
      H0=HWASTE*NLAYER
      SA=SUM(NLAYER)
      SB=SS1(NTIME)
      SC=SA+SB
      RA=SA/H0*100.
      RB=SB/HTOT(1)*100.
      RC=SC/H0*100.
      WRITE(6,220)
      WRITE(6,295) GAMMAW,GAMMAX
      WRITE(6,296) H0,SA,SB,SC,RA,RB,RC
295   FORMAT(' ***  SUMMARY  *******'///
     *       ' INITIAL UNIT WEIGHT =',F10.2,' PCY'/
     *       ' MAXIMUM UNIT WEIGHT =',F10.2,' PCY'//)
296   FORMAT(//' AS-PLACED WASTE DEPTH(H0)      =',F10.2,' FT'//
     *         ' PRIMARY   SETTLEMENT (SA)      =',F10.2,' FT'/
     *         ' SECONDARY SETTLEMENT (SB)      =',F10.2,' FT'/
     *         ' TOTAL     SETTLEMENT (SA+SB)   =',F10.2,' FT'//
     *         '  SA    / H0                    =',F10.2,' %'/
     *         '  SB    / H0-SA                 =',F10.2,' %'/
     *         ' SA+SB / H0                     =',F10.2,' %')
C
1     CONTINUE
      CLOSE (5)
      CLOSE (6)
      STOP
      END
```

William R. Orr[1] and Michael O. Finch[2]

SOLID WASTE LANDFILL PERFORMANCE DURING THE LOMA PRIETA EARTHQUAKE

REFERENCE: Orr, W. R., Finch, M. O., "Solid Waste Landfill Performance During the Loma Prieta Earthquake," Geotechnics of Waste Fills--Theory and Practice, ASTM STP 1070, Arvid Landva, G. David Knowles, editors, American Society for Testing and Materials, Philadelphia, 1990.

ABSTRACT: The Loma Prieta earthquake (M_L=7.1) produced only minor damage to solid waste landfills located near the epicenter. Site personnel were interviewed using a questionnaire to determine Modified Mercalli Intensity customized for landfills. Questionnaire results showed maximum intensities of VII near the epicenter and VI on bay muds. Estimated Peak Horizontal Accelerations for these sites ranged from 0.10 to 0.45g. Earthquake acceleration and duration appear to be the most important factors for predicting the seismic behavior of solid waste landfills.

KEYWORDS: earthquake, solid waste landfill, seismic stability

INTRODUCTION

On October 17, 1989, a strong earthquake shook the South Bay Area centered in the Santa Cruz Mountains north of Watsonville and east of the city of Santa Cruz. The tremor measured 7.1 in Richter Magnitude and was felt as far away as Los Angeles nearly 350 miles (563 km) from the epicenter. The earthquake heavily damaged many buildings and roads in the San Francisco Bay and Monterey Bay regions. Utility services

---

[1] Chief, Standards and Regulations Division, California Integrated Waste Management Board, 1020 Ninth Street, Sacramento, CA 95814.

[2] Engineering Geologist, Standard and Regulations Division, California Integrated Waste Management Board, 1020 Ninth Street, Sacramento, CA 95814.

including electricity, gas, and telephone were disrupted over a wide area for an extended period of time.

In California, solid waste landfills are required to be designed to withstand a Maximum Probable Earthquake (MPE) [1]. The MPE is the maximum earthquake a site is likely to experience during a 100-year interval and no lower than the maximum that has occurred during historic time [2]. California requires final site face slopes steeper than 3 : 1 (horizontal to vertical) to have a factor of safety of at least 1.5 under dynamic conditions, or demonstrate that the design can accommodate the seismic displacement. Proposed federal regulations [3] expected to be finalized by mid-1990 will require resisting the horizontal acceleration in lithified material with a 90 percent or greater probability that the acceleration will not be exceeded in 250 years. The term "lithified material" is not defined in the proposed rule. Potentially, earthquakes can disrupt cover material, environmental control systems (including leachate and landfill gas control systems), monitoring systems (including ground water and landfill gas monitoring systems), and induce slope failures that expose wastes or bury structures. Because of the potential for damage to solid waste facilities, the California Integrated Waste Management Board (Board) staff conducted damage surveys.

BOARD STUDY

On October 18, 1989, Board staff contacted all local enforcement agencies (LEAs) within the 12 counties most affected by the earthquake. LEAs are entrusted with solid waste facility compliance within their respective jurisdictions as designated under State law. These counties included: Alameda, Contra Costa, Marin, Monterey, Napa, San Benito, San Francisco, San Mateo, Santa Clara, Santa Cruz, Solano, and Sonoma.

Board staff conducted field examinations on October 19 and 20, 1989, of 10 solid waste landfills in the area impacted by the earthquake. Staff selected locations before the extent of damage to solid waste landfills was fully known. In anticipation of where damage would likely occur, locations both near the epicenter and on "bay mud" soils in the South San Francisco Bay Area were examined. "Bay mud" locations were chosen to ascertain the affect of weak foundations on landfills. "Bay mud" ranges in thickness from 0 to 80 feet (24 m) with a water content of typically 100 to 300%, a unit weight of approximately 60 to 90 pounds/cubic foot (0.96 to 1.44 g/cm^3), and a cohesion of 400 to 1000 pounds/square foot (19 to 48 kN/m^2). Two teams of four staff each, consisting of engineering geologists and facility inspectors, investigated these sites for damage and interviewed waste facility personnel. A questionnaire was developed for solid waste landfills similar to those used by the United States Geological Survey and the California Division of Mines and Geology to determine Modified Mercalli Intensities during an earthquake [4]. The survey results were compared with the Mercalli index and values were assigned. Results of these questionnaires are presented in Table 1. The locations of the sites are given in Figure 1.

The "bay mud" sites examined were: Durham Road Sanitary Landfill, Newby Island Sanitary Landfill, Shoreline Sanitary Landfill, and Zanker Road Sanitary Landfill. The epicentral sites were: Ben Lomond Solid Waste Disposal Site, Buena Vista Disposal Site, Crazy Horse Sanitary Landfill, Ox Mountain Sanitary Landfill, Santa Cruz City Sanitary

TABLE 1 **QUESTIONNAIRE RESULTS**

SOLID WASTE FACILITY	NUMBER OF RESPONSES	REPORTED SHAKING Moderate	Strong	Violent	MODIFIED MERCALLI INTENSITY	ESTIMATED* PEAK ON-SITE HORIZONTAL ACCELERATION
Buena Vista	3			3	VII	0.45g
Ben Lomond	2			2	VII	0.35g
Watsonville	1			1	VII	0.35g
Santa Cruz	2			2	VII	0.30g
Crazy Horse	3	1	1	1	VI	0.25g
Zanker Road	1		1		VI	0.20g
Newby Island	2		2		VI	0.15g
Shoreline	1		1		VI	0.15g
Durham Road	12	6	6		VI-V	0.10g
Ox Mountain	1	1			V	0.10g

* Based on Figure 3.

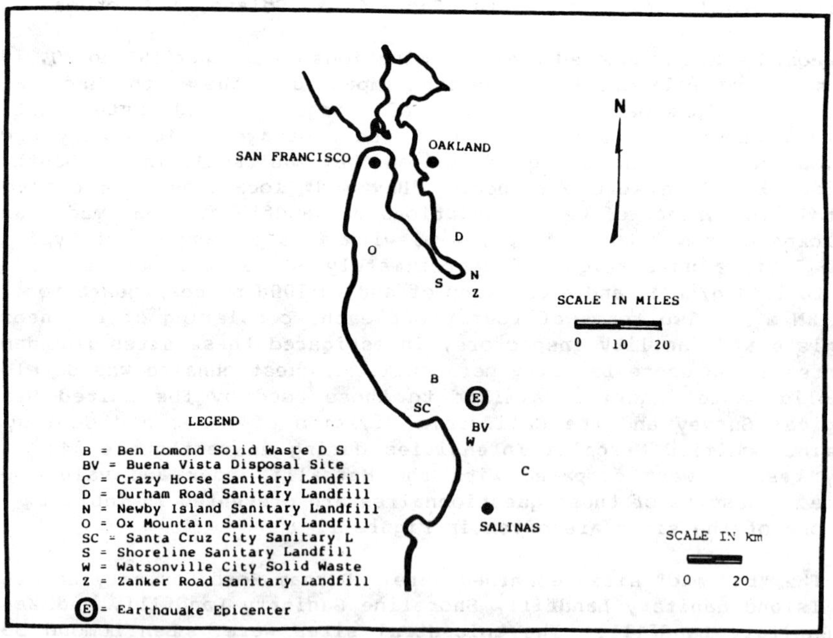

FIGURE 1 Location of Surveyed Solid Waste Landfills.

Landfill, and Watsonville City Solid Waste Disposal Site. None of these facilities are equipped with liners.

RESULTS

Results of the LEA surveys and field visits concluded that the State's solid waste landfills experienced only minor damage from the earthquake. The most common type of damage included minor cracking of landfill slope surfaces. Staff found little to distinguish earthquake-induced cracks from normal settlement cracks caused by the consolidation and decomposition of wastes. Obvious signs of liquefaction, such as sand boils, were not noticed. No failures were observed through landfill slopes.

As explained below, many landfill gas recovery systems were temporarily affected. First, power losses in the distressed areas caused pumps used in gas recovery and leachate control systems to shut down. Second, several landfill gas powered generators turned off automatically due to motion detectors. Furthermore, several above-ground pipes comprising gas recovery systems broke under the earthquake stress. However, all gas recovery and leachate control systems were repaired and back in operation within 24 hours of the earthquake. Underground structures were not examined in this survey. No changes in the quantities of leachate and landfill gas recovery were reported by facility operators after the earthquake.

A total of 13 solid waste landfills experienced minor damage from the earthquake. The following summary identifies the solid waste landfills visited by the Board staff and documents all damages reported to solid waste landfills by LEAs in each county surveyed:

Alameda

The Board's field team visited the Durham Road Landfill and did not observe any damage. No damage was reported at any Alameda County site.

Contra Costa

No damage was reported to solid waste landfills in Contra Costa County.

Marin

The Redwood Sanitary Landfill experienced damage to its gas recovery system in the form of minor leaks at surface pipe joints. No other damage was reported in Marin County.

Monterey

The Board's field team visited the Crazy Horse Landfill and did not observe any damage. However, the facility operator reported that the gas recovery system was down for 5 hours due to power failure. No other problems were reported in Monterey County.

Napa

No damage was reported in Napa County.

San Benito

No damage was reported in San Benito County.

San Francisco

No damage was reported to solid waste landfills in San Francisco County.

San Mateo

The Board's field team observed minor settlement cracks at the Ox Mountain Landfill. In addition, the site owner reported damage to surface pipe joints with the gas recovery system at the closed Junipero Serra Disposal Site. No other damage was reported in San Mateo County.

Santa Clara

The LEA reported minor slope cracking at the Pacheco Pass Landfill, the City of Palo Alto Disposal Site, and the Guadalupe Disposal Site. In addition, on-site trailers were displaced off their foundations at the Guadalupe Disposal Site and the Zanker Road Landfill.

The Board's field team observed damage to a 2-foot (0.61 m) wide gunite-lined drainage ditch at the Newby Island Disposal Site. They also reported a small grass fire at the Shoreline Regional Park Landfill. Sparks from a damaged high voltage tower adjacent to the site apparently started the fire, which was extinguished within ten minutes of the earthquake.

Santa Cruz

The Board's field team visited the City of Santa Cruz Landfill, the Ben Lomond Disposal Site, the Buena Vista Landfill, and the City of Watsonville Landfill.

The City of Santa Cruz Landfill appeared to be the most severely damaged site. Staff members observed minor cracking at several points along the edge of the landfill, some cracks appeared to be leaking landfill gas. Moderate cracks were also observed in the dikes surrounding adjacent septage and leachate collection ponds. Because of a power loss, the gas recovery system was down for 24 hours.

Staff observed minor cracking at the Ben Lomond Disposal Site. The cracks were located at points along the edge of the site and along the slope benches.

The Buena Vista Landfill also experienced minor slope cracking. There was minor cracking to a buried landfill gas header line that was being excavated to re-locate above ground. The cracks were temporarily patched pending replacement as part of the re-location. In addition, the gas recovery system was down for 24 hours due to power failure.

A small landslide on virgin ground crushed and buried drainage structures at the toe of the City of Watsonville Landfill.

No other damage was reported in Santa Cruz County.

Solano

No damage was reported in Solano County.

Sonoma

No damage was reported in Sonoma County.

As shown in Table 1, the Modified Mercalli Intensities at the 10 inspected facilities ranged from V to VII based on the responses to questionnaires given to site personnel.

Earthquakes of intensity V are felt by nearly everyone sometimes breaking dishes, unstable objects, and windows. A few instances of cracked plaster may be observed. Disturbances of trees, poles, and other large objects sometimes are noticed.

Intensity VI earthquakes are felt by all and many people run outdoors. Heavy furniture is moved. A few instances of fallen plaster or damaged chimneys are reported.

Intensity VII earthquakes cause all people to run outdoors. Poorly built or badly designed structures sustain considerable damage. The earthquake is apparent to motor car drivers.

It had been our intention to compare the results of surveys taken for those on landfilled ground versus natural ground at the time of the earthquake. However, because of the rural location of the epicenter and time of the earthquake most of the landfills were closed for the day. The remaining personnel on-site were typically in maintenance shops and other buildings. As expected the higher intensities were recorded near the earthquake epicenter with lower values at greater distances. Figure 2 depicts a contour map of these intensities recorded at the solid waste landfills. Using the approximate relationship of closest horizontal distance from the zone of energy release to peak horizontal acceleration for a Magnitude 7.1 earthquakes, as shown in Figure 3 [5], the peak horizontal acceleration were estimated for the 10 inspected landfills. These accelerations ranged from 0.10 to 0.45g and are presented in Table 1.

In contrast strong motion detectors not located on landfills

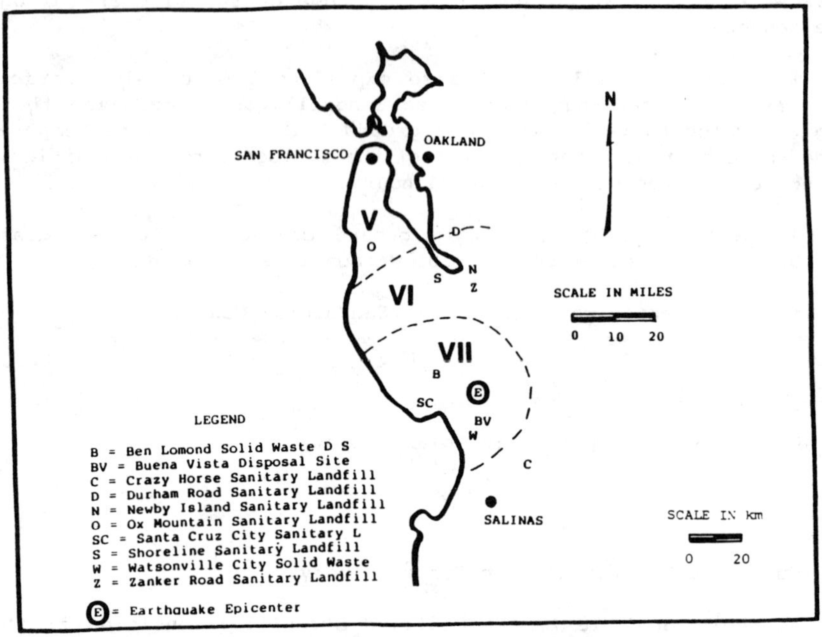

FIGURE 2 Extent of Earthquake Intensities at Landfills.

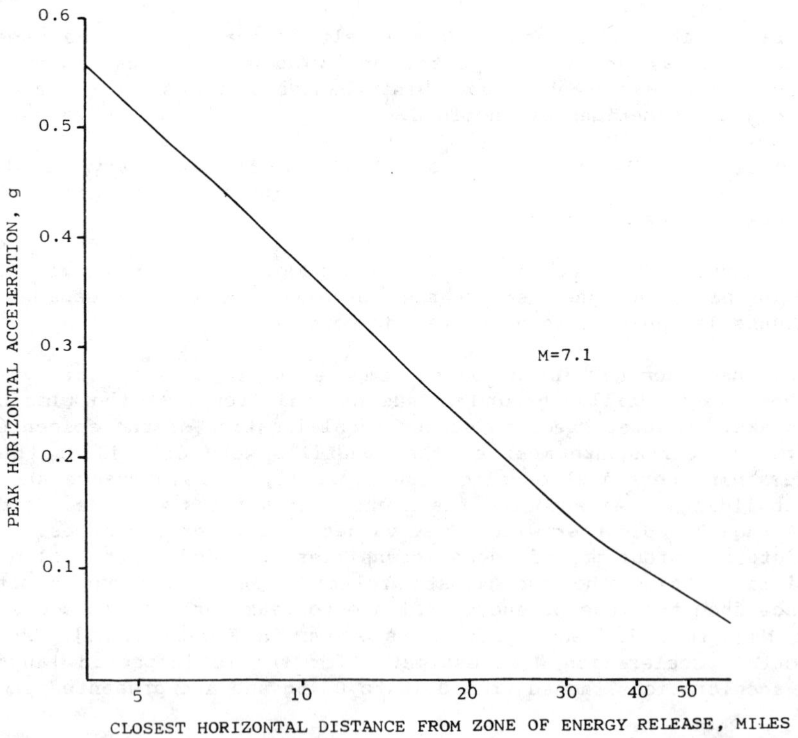

FIGURE 3 Approximate Relationship Between Distance and Acceleration for Magnitude 7.1 Earthquake [5]. One mile equals 1.609 km.

recorded peak horizontal accelerations within 20 miles (32 km) of the epicenter ranging from 0.40 to 0.54g for this earthquake [6]. Detectors located near San Francisco Bay recorded accelerations of 0.06 to 0.33g [6].

CONCLUSIONS

Although earthquakes are measured in Magnitude, the two most important factors for landfill design appear to be acceleration and duration. Studies of the 1987 Superstition Hills Earthquakes showed that liquefaction occurred only after an extensive duration (about 45 seconds) of seismic shaking [7]. The short duration of the Santa Cruz Mountains Earthquake (about 15 seconds [6]) may have prevented more extensive liquefaction and foundation failure at solid waste landfills.

Secondarily, the limited surface damage, given the peak horizontal accelerations estimated for the 10 solid waste landfills suggests that the properties of solid waste may tend to dampen or attenuate the effects of earthquakes. The authors hope to place strong motions instruments on landfilled and natural ground for confirmation of this hypothesis.

Based on this study several recommendations can be made to limit the impact of earthquakes on solid waste landfills located in seismic zones: 1) locate gas and leachate recovery lines above ground for ready inspection and access if a leak or rupture occurs, flexible designs will better accommodate seismic events; 2) use backup generators to keep environmental control systems functioning; and 3) do not over steepen cut slopes in proximity to landfill structures to prevent damage from landslides.

Although the Loma Prieta Earthquake was a major seismic occurrence it did not approach the destructive potential of a great earthquake. Little data exists for solid waste landfill performance during a great earthquake such as the 1906 San Francisco event which would represent the MPE for many of these solid waste landfills.

Solid waste landfills may contain toxic and/or explosive gases and fluids. The escape of these waste components may pose an immediate threat to public health and the environment. For these reasons solid waste landfills should continue to be prudently protected against failure during earthquakes.

REFERENCES

[1] California Integrated Waste Management Board, 1989, California Code of Regulations, Title 14, Division 7, Chapter 5, Article 7.8.

[2] California Division of Mines and Geology, 1975, Recommended guidelines for determining the maximum credible and the maximum probable earthquakes, Sacramento, 2p.

[3] Environmental Protection Agency, 1988, Code of Federal Regulations, Part 258 (proposed).

[4] Wood, H. O., and Neumann, F., 1931, "Modified Mercalli Intensity Scale," Bulletin of the Seismological Society of America, vol. 12, No. 4, December 1931, pp. 115-137.

[5] Seed, H. B., and Idriss, I. M., 1982, Ground Motions and Soil

Liquefaction During Earthquakes, Monograph, Earthquake Engineering Research Institute, University of California, Berkeley, CA, 87p.

[6] Shakal, A., Huang, M., Reichle, M., Ventura, C., Cao, T., Sherburne, R., Savage, M., Darragh, R., and Petersen, C., 1989, CSMIP Strong-Motion Records from the Santa Cruz Mountains (Loma Prieta), California Earthquake of 17 October 1989, Report No. OSMS 89-06, California Department of Conservation, Division of Mines and Geology, Office of Strong Motion Studies, Sacramento, CA, 196p.

[7] Holzer, T. L., 1988, Personal Communication, U. S. Geological Survey, Menlo Park, CA.

Thomas A. Lawrence and Gordon P. Boutwell

PREDICTING STRATIGRAPHY AT LANDFILL SITES USING ELECTROMAGNETICS

REFERENCE: Lawrence, T. A., and Boutwell, G. P., "Predicting Stratigraphy at Landfill Sites Using Electromagnetics," Geotechnics of Waste Fills - Theory and Practice, ASTM STP 1070, A. Landva and G. D. Knowles, Editors, American Society for Testing and Materials, Philadelphia, 1990.

ABSTRACT: Stratigraphic mapping has been done at several landfill sites using a statistical method of electromagnetic (EM) data interpretation which correlates the stratum depth with EM data at boreholes. The mathematical relationship developed by multivariate regression analysis is then used to predict stratigraphy at EM survey points. This method of Multivariate Regression-Prediction (MVRP) has been used successfully at a number of landfill sites for subsurface sand channel delineation, paleokarstic feature locations, and stratigraphic mapping between boreholes. MVRP results are in terms of depth or thickness, and do not require further data reduction.

KEYWORDS: surface geophysics, electromagnetics, conductivity, Multivariate Least-Squares Regression

An electromagnetic (EM) survey has much to offer when facing time or budget constraints. For example, for a particular landfill geotechnical investigation, a two-day, $2000.00 EM survey with 90 survey points was chosen over drilling 12 boreholes at $1500.00 each. However, most methods of EM data interpretation cannot produce hard data such as stratum thickness or contaminant concentration directly from survey results. A statistical technique for interpretation of EM data has been developed which allows correlation of actual field data with EM survey data [1]. A multivariate regression analysis is used to relate field data and EM readings at known points. The resulting mathematical relationship is then used to predict values at other survey locations. This interpretation procedure has been used successfully in a number of stratigraphic investigations at solid and hazardous waste landfills.

Thomas Lawrence is Chief Hydrogeologist and Dr. Boutwell is President at Soil Testing Engineers, Inc., 316 Highlandia Drive, Baton Rouge, Louisiana 70810.

INTERPRETATION TECHNIQUES

An EM field survey produces readings of apparent conductivity at various points. These readings are a measurement of terrain conductivity, which is a function of the conductivity of the soil or rock material, porosity and saturation state of the material, and the conductivity of the pore fluid for each stratum [2]. These conductivity readings alone are not diagnostic of layer geometries and in most cases are simply used to observe lateral and vertical variations.

It is possible to calculate strata thicknesses by relating layer conductivities and the cumulative response of the layers to the apparent conductivity [3]. This method is limited in range of conductivity and requires trial and error to arrive at a reasonable combination of layer depth and conductivity.

A commercially available forward-inverse method constructs a theoretical layered-conductivity curve to avoid equivalency problems, and predicts layer conductivities from instrument readings. The forward method uses Hankel transforms [4] to produce the sounding curves, which are then inverted to produce layer conductivities. This method requires a field data set, consisting of layer thicknesses, conductivities, and instrument readings for a particular location to produce the curves. It assumes stratigraphy is relatively uniform in order to relate the theoretical curve to other points. The method produces layer conductivities from instrument readings at the other survey points.

What is needed is a method of EM data interpretation which produces quantifiable results in terms of stratigraphy from actual field data, with a definable precision. This has been accomplished by using the commonly accepted Multivariate Least Squares Regression analysis to establish a relationship between data taken at known points and the EM readings at those locations, which is then used to predict the dependent variable at other EM survey points. The results of this interpretive technique are in terms of thickness or depth, with no further data reduction necessary. This method is advantageous in the fact that the EM survey serves as an extension of hard data, rather than a method which must be correlated with additional hard data or another geophysical method.

MULTIVARIATE REGRESSION-PREDICTION

The overall objective of most surface geophysical surveys is to define some parameter such as contaminant concentration or depth to stratum at the survey points. The simple contouring method alone cannot account for the effects of spatial variations in conductivities on the total response. A layered analysis can, but only for simple subsurface geometries. For both techniques, the final conductivities assigned to a stratum are then related in some way to contaminant concentration, depth, or thickness of the layer.

Multivariate Regression-Prediction (MVRP) assumes that there is some mathematical relationship between the dependent variable (e.g., depth of stratum) at a known point or borehole and the measured responses for various coil spacings and orientations. Regression analysis provides a "best-fit" relationship in the form of a prediction equation defining the dependent variable at these "hard data" points as a mathematical function of the various strengths of the independent variables

$$f(Z) = A + B \cdot g(X1) + C \cdot h(X2) + \ldots \tag{1}$$

where

Z = Dependent Variable
X1,X2 = Independent Variables
A,B,C = Regression Constants

Equation 1 is then used to predict the dependent variable at other EM survey points. This method is not limited to linear relationships, but can easily be extended to functions of the variables (e.g., Log, Square, Exponent, etc.). The mathematics are the same, since going to functions is merely a scale change.

The accuracy of these predicted values is evaluated through the regression parameters, Coefficient of Correlation (Cc) and Standard Estimate of Error (SEE). The former indicates the reliability of the prediction, the latter its precision. A qualitative guide to reliability is [5] given in Table 1. A true Level of Confidence in the Coefficient of Correlation can also be determined using the "Student's t" procedure [6].

TABLE 1 -- Correlation Reliability

Cc (abs. value)	Strength of Relationship
Less than 0.20	Slight, almost negligible
0.20 - 0.40	Low correlation, definite but small
0.40 - 0.70	Moderate correlation, substantial
0.70 - 0.90	High correlation, marked relationship
0.90 - 1.00	Very high correlation, very dependable

The Cc is a measure of the linear relationship of the variables along the regression line which is defined by the prediction equation. Obviously, predictions at other points will be more accurate if the linear relationships between independent and dependent variables is high (high Cc). Also, the greater the number of hard data points available, the higher the degrees of freedom. This allows a closer Level of Confidence to be made statistically.

The SEE is a measure of scatter of observed data around the predicted regression line. The closer the observed values are to the predicted values, the lower the SEE and greater the precision of the predictions. SEE can be used as a quantitative evaluation of the precision of the predictions using the Two-Sided Tolerance Test (TSTT). The Tolerance Test procedure [7] determines a level of confidence that a certain proportion of actual values fall within the range defined by the predicted values, SEE, and Tolerance Factor (K)

$$f(Z_p) - K*SEE \leq f(Z_p) \leq f(Z_p) + K*SEE \quad (2)$$

where

Z_p = Predicted value of Z

MVRP analysis has been used with Geonics, Ltd. instruments, but can be used with any instrument which can give multiple readings at each station, such as D.C. resistivity. It has been especially successful when used with the EM34-3 because of the instrument's varied near-surface and depth response with horizontal and vertical dipoles, in addition to different reading depths with increased coil spacings. These two instrument features are often useful in creating the prediction equation. The dipole orientation that has a stronger near-surface response can serve as a means of subtracting the effects of shallow soil or rock strata. The regression equation gives these readings a smaller, or even inverse, relationship to the dependent variable, while the readings with a strong depth response are given a more positive relationship.

When dealing with layers less than one meter in thickness, especially at depth, it may be more prudent to use D.C. resistivity, which allows a greater number of depth readings at various electrode spacings. The ultimate objective is to bracket the zone of interest with a number of readings, so a regression equation is created which realistically predicts the dependent variable.

Experience has shown that when performing the Multivariate Regression Analysis, a higher Cc and lower SEE can be attached if the logarithm of the independent variables (EM data), is used. In this case, the SEE is expressed as a logarithm.

CASE STUDIES

To demonstrate the practicality of the MVRP method, a series of examples are presented. The Geonics, Ltd., EM-31, EM34-3, and EM34-3XL were used in these case studies.

Case #1

Preliminary boreholes drilled at a landfill site in Northwest Florida indicated a possible sand channel along one side, which was thought to be a buried stream channel. Typical stratigraphy at the site was a 7 meter-thick silty/clayey sand underlain by a clay layer extending to the top of a limestone stratum near the 12 meter level. Three boreholes along the west side indicated 12 to 30 meters of sand beneath the surface veneer. An EM survey was used to delineate the areal extent and thickness of the sand stratum. Based on simple contouring of the surface apparent conductivity (Figure 1), the sand body was found to be not a linear channel, but a series of discontinuous anomalies (shaded zones). These anomalies were most likely paleosinks, old karstic solution features since filled in by sediments.

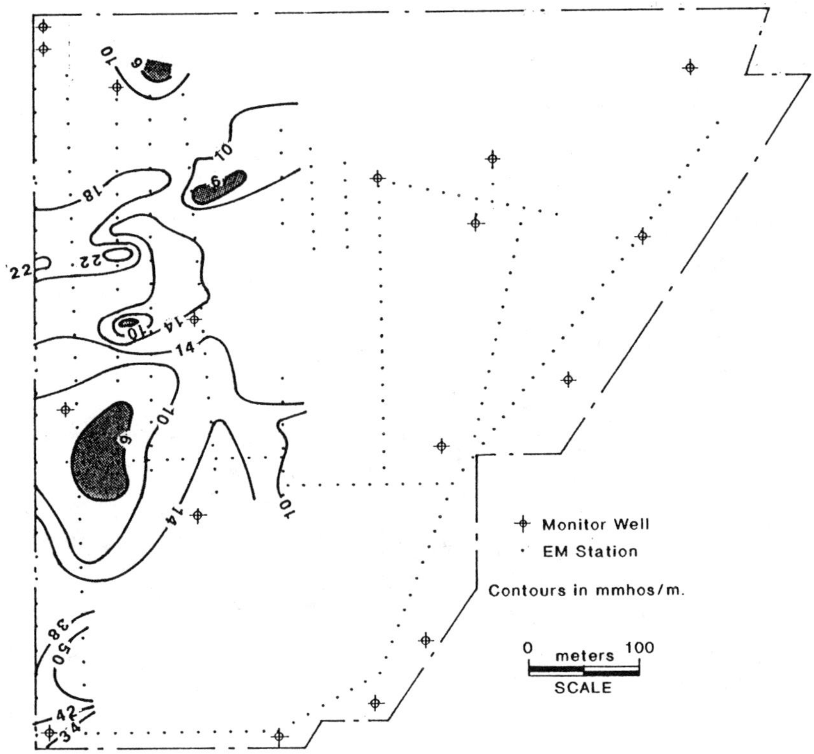

Figure 1---Contoured EM Field Data.

A MVRP analysis was performed using EM readings taken at 13 borehole locations (hard data points). Using varied dipole orientations and spacings, four conductivity readings (independent variables) were taken at each EM survey point. Two regressions then were used to predict the thickness of the sand bodies and the depth to the limestone stratum at the other EM survey points. The accuracy of the predictions are given in Table 2.

TABLE 2 -- Accuracy of Predictions (Case 1)

Predicted Value	Correlation Coefficient	Standard Estimate of Error
Sand Thickness	0.90	0.23 (log)
Limestone Depth	0.98	0.10 (log)

While the correlation for predicted sand thickness is very high, the precision as analyzed by TSTT was only fair. There was a 90% probability that the real thicknesses would not vary from those predicted by a factor of more than 2.4. The actual mean error was 2 meters on an average thickness of 7 meters. Figure 2 is a map of the clay stratum elevation (land surface elevation minus predicted sand thickness).

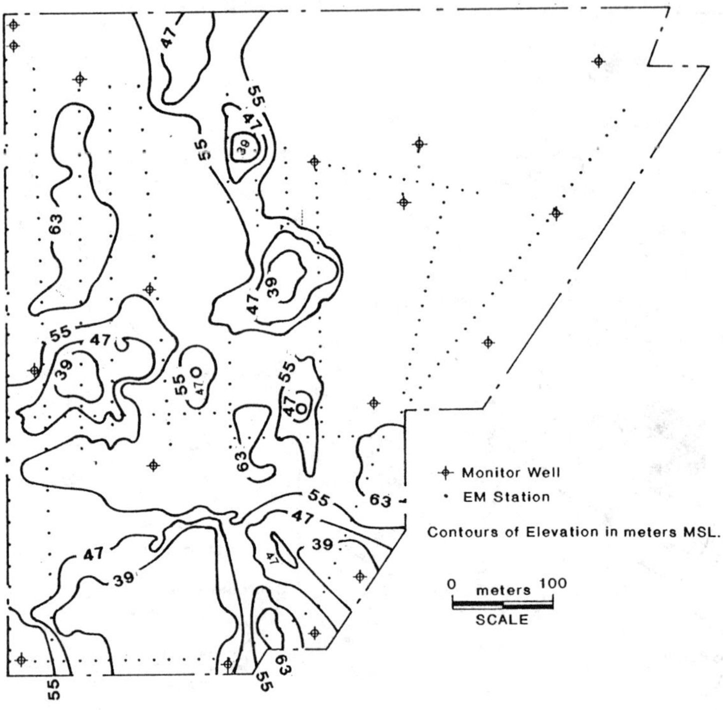

Figure 2---Elevation of Top of Clay.

Subsequent borings in the vicinity of several "predicted" sinkholes found that actual depth to limestone was within closer tolerances to the predicted depths. However, the filling material in the paleosinks ranged from clayey sand to clay. Therefore, the lack of high precision in predicted sand thickness was representing the variation in sediment size and conductivity.

<u>Case #2</u>

A RCRA landfill site in Louisiana is predominantly underlain by Pleistocene clays, with a thin, discontinuous silt and sand layer about 5 meters below land surface. Several preliminary borings in one section of the site encountered sand thicknesses in excess of 3 meters. From these borings, it was suspected that a narrow, near-surface sand channel existed in the immediate area. Ordinary contouring between boreholes less than 60 meters apart indicated the channel to be about 120 meters wide; the mean observed depth to base-of-sand was 8 meters.

An EM survey was used to delineate the actual extent and depth of the sand between boreholes. Eight borehole locations were used for the hard data base to produce the regression equation. From this MVRP analysis, it was predicted that the actual sand channel was about 60 meters wide (Figure 3). Mean error between depths as predicted and observed in the boreholes was 1 meter. In other words, the mean scatter of observed data points around the predicted regression line was 1 meter. The maximum error was a prediction of 10 meters when the actual depth was 12 meters. A comparison of strata interpolated between boreholes and predicted strata at EM stations 20 meters apart shows the greater detail available, and the high correlation between predicted and observed data at boreholes.

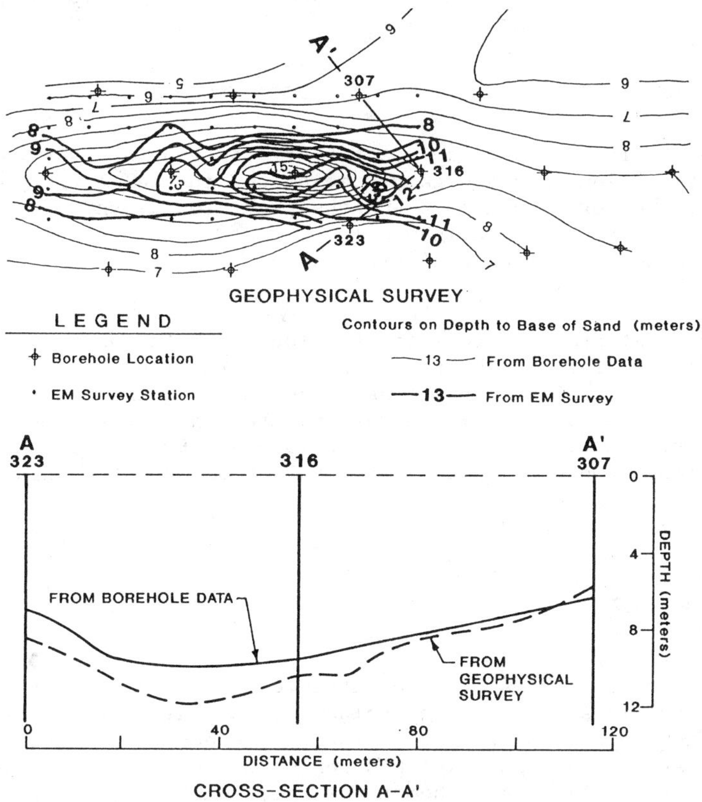

Figure 3---Extent of Sand Channel.

Case #3

During the preliminary site investigation for a proposed sanitary landfill in Central Louisiana, new State regulatory requirements for borehole spacings were proposed. In order to satisfy these requirements, rather than spend time and money adding supplemental borings, a geophysical survey was proposed for the section of the site which had already been drilled. This would determine if there were any structural inconsistencies not detected by the borings at the coarser spacing.

Typical stratigraphy at the site consisted of 10 to 15 meters of clay, 3 meters of silt, and a lower clay layer. MVRP analyses of the EM survey were performed using 28 hard data points (boreholes), and the resulting relationships were used to predict the depths of the upper three strata. Precision of the predictions are in Table 3.

TABLE 3 -- Accuracy of Predictions (Case 3)

Prediction	Coefficient of Correlation	Standard Estimate of Error
Base of First Clay	0.96	0.10 (log)
Base of First Silt	0.98	0.10 (log)
Base of Second Clay	0.99	0.08 (log)

The SEE indicates a low scatter of observed values around the linear regression, and a high degree of confidence in the predicted values at the survey points. For the predicted depth to upper clay base, there was a 95% confidence level that 95% of the predicted values did not vary from real thicknesses by a factor of more than 1.4.

The stratigraphic predictions, along with the hard data points, were contoured for the desired stratum information (Figure 4). What appeared to be a silt channel was detected along the eastern boundary of the survey area, incised in the more regular lower clay stratum. A cross-sectional comparison of boring logs and EM results shows a good correlation between contouring and MVRP predicted stratigraphy (Figure 5).

CONCLUSIONS

The common methods of data interpretation available for surface electromagnetic surveys are not always practical or reliable for use in stratigraphic mapping. A statistical method of EM data interpretation has been developed for use in stratigraphic and groundwater studies. A mathematical relationship is determined between a dependent variable, such as stratum depth, and EM readings at known points, using Multivariate Least-Squares Regression. Predictions of stratum depth can then be made at other EM survey points using this mathematical relationship. The advantages of using this Multivariate Regression-Prediction (MVRP) for data interpretation are the EM survey serves as an extension of hard data, rather than a method to be correlated with additional data or another method, and the results from MVRP are in terms of thickness with no further data reduction necessary.

MVRP-EM surveys are most practical when faced with time or budget constraints. They are also extremely cost-effective for preliminary

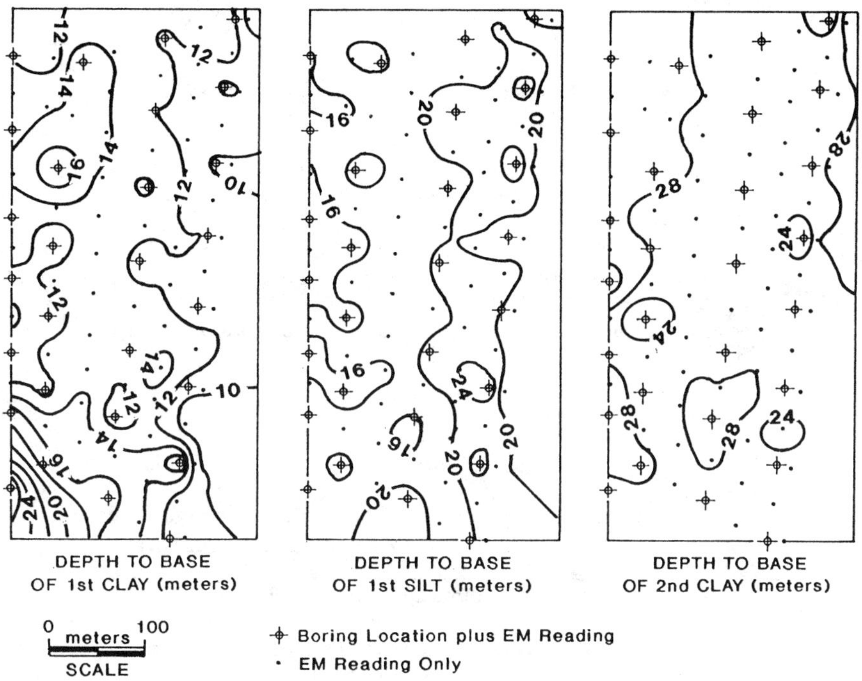

Figure 4---EM/MVRP Survey Results.

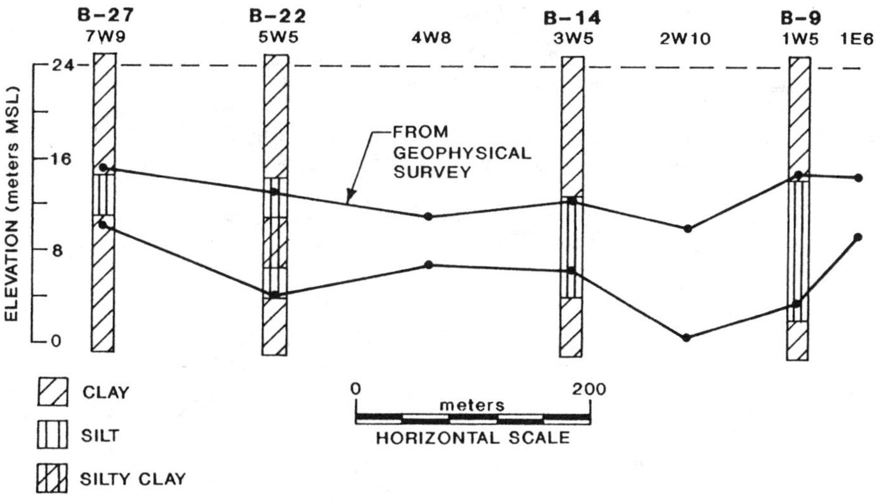

Figure 5---Predicted Stratigraphy versus Borehole Data.

reconnaissance work. For example, in the early stages of a landfill site selection, geologic maps indicated an abrupt change between clays from old backswamp deposits and channel sands at one particular site. For the price of an additional borehole, it was proposed that an EM survey be run between a minimum number of borings to find the exact location of the facies change. From this, an assessment of the suitability of the site for use as a landfill could be made. (For a general information survey, it is recommended that at least five hard data points be used.) Results of these and other studies [1] have shown that MVRP-EM analysis can be an efficient method for predicting subsurface conditions by extending the usefulness of a limited number of boreholes. Once the EM method is understood, MVRP can be applied to develop stratigraphic analyses at landfills and other sites.

REFERENCES

[1] Boutwell, G. P., and Lawrence, T. A., "Electromagnetic Data Interpretation Using Statistics," Proceedings of the FOCUS Conference on Eastern Regional Ground Water Issues, National Water Well Association, Dublin, Ohio, 1988.

[2] McNeill, J. D., Electrical Conductivity of Soils and Rocks, Geonics, Ltd., Technical Note TN-5, 1980.

[3] McNeill, J. D., Electromagnetic Terrain Conductivity Measurements at Low Induction Numbers, Geonics, Ltd., Technical Note TN-6, 1980.

[4] Anderson, W. L., Numerical Integration of Related Hankel Transforms of Orders 0 and 1 by Adaptive Digital Filtering, Geophysics, Vol.44, No.7, July, 1979, Pg. 1287-1305.

[5] Guilford, J. P., Fundamental Statistics in Psychology and Education, McGraw-Hill Book Company, Inc., New York, 1950.

[6] Li, J. C. R., Statistical Inference, Edwards Brothers, Inc., Ann Arbor, Michigan, 1964, pp.301-302.

[7] Miller, I., and Freund, J. E., Probability and Statistics for Engineers, Prentice-Hall, Englewood Cliffs, N.J., 1965.

Gregory P. Gifford, Arvid O. Landva, Vernon C. Hoffman

GEOTECHNICAL CONSIDERATIONS WHEN PLANNING CONSTRUCTION ON A LANDFILL

REFERENCE: Gifford, G. P., Landva, A. O., and Hoffman, V. C., "Geotechnical Considerations When Planning Construction on a Landfill," Geotechnics of Waste Fills - Theory and Practice, ASTM STP 1070, Arvid Landva and G. David Knowles, Eds., American Society for Testing and Materials, Philadelphia, 1990.

ABSTRACT: The closing of landfills has become commonplace. In many instances these closed facilities are desirable building sites because of close proximity to urban areas. Much of the subsurface investigation for foundation design parameters in landfills is more difficult and more expensive than for conventional sites underlain by soil or rock. A subsurface investigation of one particular landfill site is presented and discussed. The results obtained guide the decision making process for both further investigation and foundation type selection. The purpose of the paper is to inform the practicing engineer of one approach to the evaluation of geotechnical considerations of a landfill site.

KEYWORDS: landfill, foundations, settlement, dynamic compaction, deep foundations, adjustable floor slab.

INTRODUCTION

In recent years the closing of landfills has become common in New York State. Some of these sites are being used as sites for building by developers and government agencies. The purpose of this paper is to present and discuss an approach to the evaluation of geotechnical considerations of such a site.

G. Gifford is an assistant professor of civil engineering, Union College, Schenectady, NY, 12308. A. Landva is a professor of civil engineering, University of New Brunswick, Fredricton, NB, Canada, E3B5A3. V. Hoffman is principal of Vernon Hoffman, PE, 118 S. Ferry St., Schenectady, NY 12305.

The referenced site is the North Albany Demolition Landfill (NADL) in Albany, New York. The landfill was used primarily for disposal of demolition debris and excess soil. A secondary use was disposal of municipal solid waste.

SITE INVESTIGATION

Historical Overview

The NADL is situated on approximately 45 acres at the north end of the city of Albany. The site is bordered by Routes I-787 to the east, I-90 to the south, Erie Boulevard to the west, and the Albany County Wastewater Treatment Plant to the north. Access to the site is from Erie Boulevard. The site is part of the Hudson River flood plain, as the river flows to the south, immediately to the east of I-787.

During the early 1900's, a large river-water filtration plant operated on the site. The plant was abandoned during the 1930's. Several massive concrete foundations, tanks, and tunnel structures remain beneath the landfill. The tanks were reportedly demolished prior to beginning landfill operation.

The northern portion of the site was operated as a municipal solid waste landfill during the 1960's. The quantity and boundaries of the municipal refuse are unknown. However, certain records indicate that the southwest quadrant consists mostly of construction debris and limited areas of municipal refuse. The fill was placed with very little sorting or compaction. Soil cover was placed over the fill at the end of each day.

The thickness of the landfill was estimated from old site plans to vary between 4.6 and 9.2m. The water table was estimated to be near original grade. Occasional perched water tables exist within the landfill.

Preliminary Investigation

The primary purpose of the preliminary investigation phase was to evaluate the feasibility of building on the landfill. In the case of the NADL, the proposed construction involved a 4650m^2 Public Works garage/office facility with future plans to expand the complex. A steel structure with metal sheathing is planned. The maximum column load is estimated to be between 115 and 136 tonnes, and the floor is to be subjected to truck vehicle loading. Parking lots, roadways, and utility bedding schemes are also to be evaluated. The preliminary plan is to support the structure with a deep foundation bearing on bedrock and to support the floor on the landfill.

The preliminary investigation includes six soil borings and 23 test pits through the fill. The locations of these are shown in Figure 1.

The borings were performed with a truck-mounted hollow-stem auger

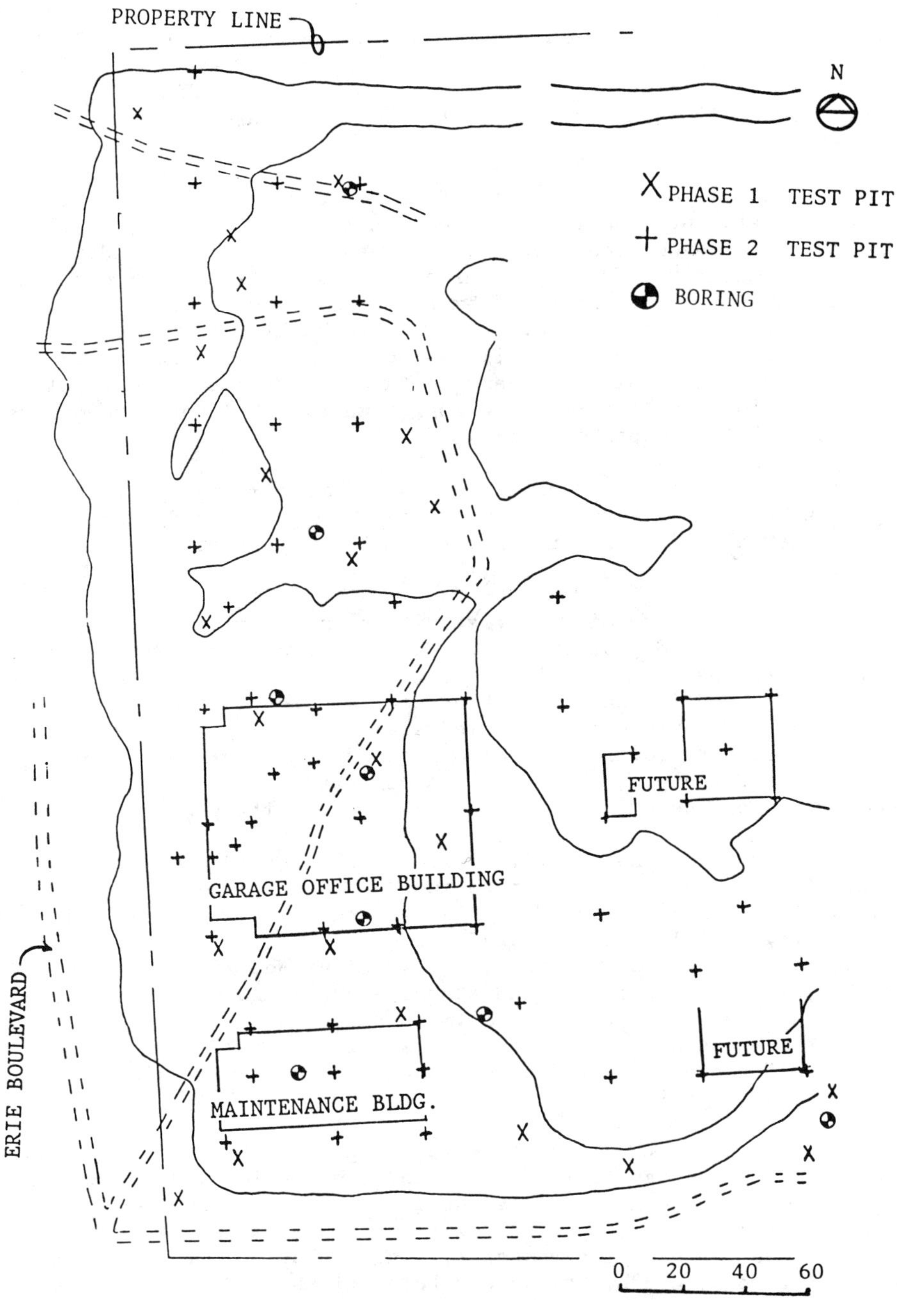

FIG. 1 Site Plan with Test Pit and Boring Locations

drilling rig. A 20cm diameter auger was used to advance the hole through the landfill. In general, the auger was advanced with little difficulty. However, a few localized refusals were encountered and the boring relocated. Once virgin material was encountered, standard split spoon sampling was performed until bedrock was encountered. Virgin material was found to be glacial lake Albany varved silt and clay and it varied in thickness between 4 and 5m. Shale bedrock was found at depths between 12 and 14m. The surface of the bedrock dips gently to the east.

The landfill material was found to be no more than 8.5m thick. Test pits yield more continuous visual information on the landfill material than conventional borings. It was, therefore, decided to perform the majority of the preliminary investigation with test pits. A track-mounted backhoe with a 7.5m reach was used to excavate and backfill the test pits. All test pits were backfilled with the excavated material. The backfill was not compacted. Excavating the test pits was relatively easy, and most test pits stayed vertically open during logging. It required approximately 30 minutes to excavate each test pit to a depth of 7.5m.

Most of the elongated particles in a landfill end up with their long axis horizontal resulting in a laterally reinforced arrangement of the material [1]. This reinforcement helps test pit walls stand vertically.

During excavation of each test pit the material encountered was carefully logged and photographed. Particular attention was paid to distinguishing between material originating from municipal solid waste and from construction and demolition debris. Several test pits encountered immovable concrete objects at or near original grade. These were of particular importance because of the planned deep foundation and the probable existence of substantial concrete structures from the previous water treatment facility.

The portion of the landfill which originated from construction and demolition debris is herein termed construction debris. The construction debris consisted mainly of sandy silt and clay soil. The soil was layered with an estimated 10 to 20% man-made debris, consisting of chunks of concrete, brick, tires, decomposing wood (decking, beams, tree limbs, and stumps), and various metal pieces. The layers of debris varied up to 1m in thickness. Soil layers of 0.2 to 0.4m thickness and more were used as daily covers. The random placement of the pieces of debris results in localized voids and an overall loose deposit.

The portion of the landfill which originated from municipal solid waste is herein termed municipal refuse. The municipal refuse consisted mainly of silt and clay soil, although not as much as in the construction debris. Interbedded with the soil was an estimated 20 to 40% of varied waste including paper, glass, plastic, cloth, ashes, decomposing wood, brush, brick, concrete, tires, and metals. Even though putrescible refuse such as food waste was not present, the municipal refuse had a much stronger odor than the construction debris and in some instances was warm.

Figures 2a and 2b are photos of test pit spoil piles from municipal refuse and construction debris, respectively. The sides of the test pits were littered with a variety of waste. Some waste was cut cleanly with the bucket, but often the waste protruded from the wall. When viewed from the edge of the test pit, the waste protruding from the wall concealed the wall beneath. This can result in overestimating the quantity of waste. The mixing of waste and soil during excavation allows for a more realistic estimate of quantities. In fact, during backfilling, the waste (especially decomposing wood) disintegrates and mixes with soil and results in the mixture having the appearance of contaminated soil.

The landfill had been capped with 0.6 to 0.9m of clayey silt and sand. People familiar with the site indicate that sinkholes were commonly observed over the past years and that soil was used to fill these sinkholes. The cap was noted to be a meter or more thick in these localized areas. The sinkholes were the result of raveling or collapse of hollow structures not compacted during placement of the fill.

In general, the amount of municipal refuse increased to the north and east. Subsurface profiles were prepared and studied to select the better areas for the building and the less desirable for parking and green areas. The southwest quadrant of the landfill consisted mainly of construction debris. The water table was near original grade, and perched water tables existed within the landfill. It was concluded that this area was best suited for building. Some of the fill in this area had been in place for seven to eight years and some for decades. The debris consisted mainly of soil mixed with partially decomposed wood, metals, concrete, brick and tires.

At the completion of the preliminary investigation phase the following tentative conclusions were drawn:

1. It is feasible to build on the NADL and the building complex would best be located within the southwest section of the landfill.
2. Feasible foundation types include piles or caissons to bedrock or shallow spread or combined footings. A shallow foundation would require extensive subgrade improvements.
3. The floors could be designed as slabs on grade on an extensively improved subgrade, provided the owner was willing to accept some differential settlement after several years of service.
4. The settlement of slabs could be halted and the slabs releveled by pressure grouting under the slabs at certain locations.
5. Dynamic compaction (DC) could be used to improve bearing, reduce settlement, and improve stability at locations of structures, pavements, and utilities.

On the basis of these tentative conclusions, the location of the complex was decided and a site plan was developed with the guidance of the geotechnical engineer. Also, the scope of work was increased from one to two buildings after it had been concluded that building on the landfill was indeed feasible. The owner decided to found the building

FIG. 2a Photo of Municipal Refuse Spoil Pile

FIG. 2b Photo of Construction Debris Spoil Pile

structures on piles or caissons bearing on bedrock. However, he did not want to support the floor or parking areas on piles and agreed consequently to accept greater than normal settlements.

Main Investigation

The main investigation phase was subsequently planned to examine the subsurface in the vicinity of the proposed foundations and utilities in the southwest quadrant of the landfill. Final design recommendations for building foundations, floors, parking lots, and utility bedding were made at the conclusion of this investigation.

Two soil borings were performed and 54 test pits were excavated. One boring was advanced through the fill with a 15cm diameter hollow-stem auger and the other with a 10cm diameter rotary wash-drilling operation. The wash fluid was pressurized air.

The test pits were located at all column locations and other areas of interest. The test pits were excavated with a track-mounted backhoe and carefully logged and photographed. Bag samples were obtained for organic and moisture content tests. The organic contents varied between 5 and 12% and the moisture contents between 14 and 68%.

It is not practical to test laboratory samples in conventional devices, therefore, a soils engineer should log test pit. A universally accepted classification system should be adopted for landfill materials such as suggested by Landva and Clark [2].

In the vicinity of the proposed construction the existing landfill material was, in general, found to be construction debris. A limited area at the northwest corner of the garage/office facility contains municipal refuse. The settlement caused by decay of the municipal refuse is expected to be relatively large. Therefore, it was decided that the floor system in that area should be supported by deep foundations.

The floor plan was designed so that the lightly loaded office section of the floor was supported by the deep foundations. The remainder of the floor in the garage section of the building would be supported on the landfill. A control joint in the floor at the transition had to be incorporated.

At the completion of the main investigation, the following design recommendations were made:

1. Perform dynamic compaction wherever it is desirable to minimize post-construction settlement, for example, under floor slabs, pavements, and utilities.
2. Support the building structure and one portion of the floor on deep foundations to ensure normal settlement tolerances.
3. Support floor slabs, driveways, and parking lots on the dynamically compacted landfill.
4. Incorporate a slab support system which allows releveling should settlement magnitudes warrant.

These recommendations are discussed in more detail below.

DESIGN RECOMMENDATIONS

Dynamic Compaction (DC)

The purpose of dynamic compaction is to densify the landfill, thereby decreasing post-construction settlement. Since future expansion at the site is planned, the compaction should be planned so that future DC will not cause excessive vibrations of the initially constructed buildings.

Densification of the entire depth of landfill is needed. Not only should deep-seated settlement be avoided, but adequate bearing capacity at the subgrade is needed to spread traffic and floor loading.

Based on engineering judgement, it was decided to spread a 0.6m thick layer of a granular soil over the entire area to be subjected to DC. The layer would provide a working mat, thereby improving access of construction vehicles. This layer will provide good bearing and promote drainage of the pavement subgrade. Also, the layer would provide some confinement of the landfill upon DC impact and lessen heave.

It was recommended that the compacted fill be demonstrated to have a minimum coefficient of subgrade reaction (Ks) under building areas of 270kPa/cm at design subgrade and at 1.5m below design subgrade; and under pavement areas of 400kPa/cm at design subgrade and 270kPa/cm at 1.5m below design subgrade. These magnitudes were back-calculated from subgrade moduli recommended by the Asphalt Institute of America and the Portland Cement Association.

Plate bearing tests performed in accordance with ASTM Specification D1194 using a circular plate with a minimum diameter of 0.8m will be utilized to verify the criteria. Ks will be calculated from the load versus settlement curve generated during each test. These tests will be performed at both the subgrade level and in test pits 1.5m below the subgrade level. This quality assurance will be required throughout the entire project. A demonstration test area of at least 300m^2 will be closely scrutinized as adjustments are made to the DC procedure.

Peculiar weight behavior upon impact is common in landfill deposits. Springy weight behavior is an indication of concentrations of large wood or steel. These pieces should be removed, because they inhibit effective densification of adjacent areas.

The DC may also reveal soft spongy areas. These areas will either be removed and replaced with more suitable material, or coarse gravel with cobbles will be pounded into the spongy areas.

Records will be kept of the crater sizes and of the net change of volume after each pass, since the change of volume is a measure of the effectiveness of compaction. A larger crater volume may indicate compressible material in a certain location, and these locations will require more compactive energy.

Lukas [3] and Welsh [4] indicate that the net settlement of a landfill subjected to DC will vary between 5 and 25%. The low end of this range represents old landfills and fills with low percentages of refuse. The high end represents recent landfills with high percentages of refuse. It was estimated that the NADL would lose 10% of its volume during DC and settle approximately 0.6m. This estimate was made from the literature review and discussions with DC specialty contractors.

Building Foundations

Since the landfill is more compressible and less uniform than inorganic soils, settlement of a shallow foundation supported on the landfill could be large and differential. A well performed DC program would minimize settlement. However, large voids could remain after DC, especially at the location of concrete vaults, tanks, and tunnels at the previous water treatment facility. A flexible structure with provisions for jacking and grouting column bases could be used. Figure 3 illustrates how adjustable column bases could be used to relevel a light building.

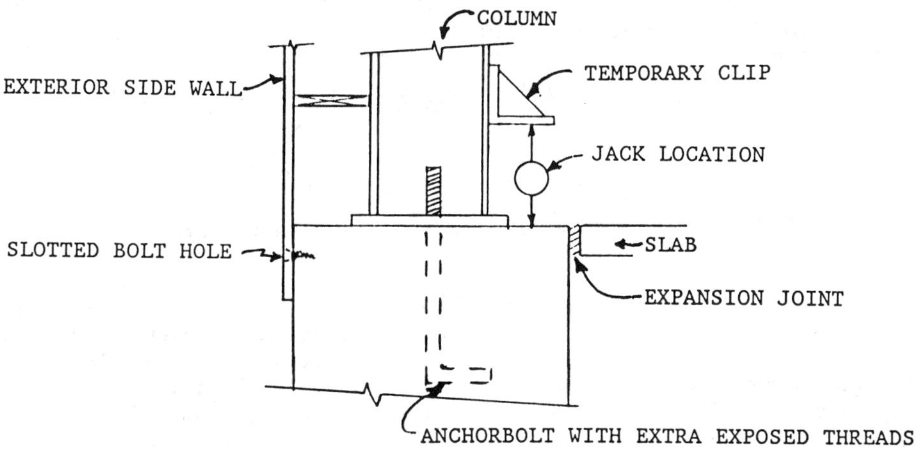

FIG. 3 Illustration of Adjustable Column Base

Deep foundations are an alternative, since bedrock is relatively shallow (approximately 9 meters). Obstructions such as concrete rubble and steel debris are present in the landfill. Driving piling may be difficult and result in damage of the piles. Some areas may require excavation and removal or breaking of the obstruction. Rebar, conduit, wire or other obstructions may make augering or percussion drilling difficult locally.

A negative skin friction load of 15 to 20% of the pile or caisson design capacity should be included because of the high compressibility of the landfill. These percentages were determined by estimating the shear strength of the refuse and soil mixture and its adhesion to the pile or caisson.

The corrosivity of the landfill is expected to be greater than that of inorganic soil because of the harsh chemical and biological environment. Its effect on the piles, caissons, or caps must be addressed.

Floor Slabs and Pavements

Figure 4 gives a section view of the slab support system which can be used on a compacted landfill material.

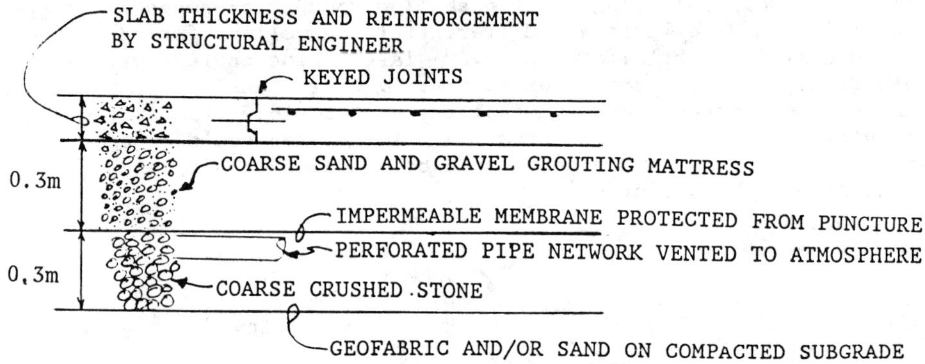

FIG. 4 Slab Releveling Mattress and Vent Systems

A subgrade vent system is required for removal of gases emitted from the landfill. This system was designed by others and envisioned as a layer of coarse crushed stone separated from the supporting soil by a geofabric. A network of perforated pipe will be placed at the upper limit of the layer of crushed stone. The layer will be covered with an impermeable membrane to trap the gas. This membrane may be placed on another layer of geofabric and/or filter sand to prevent puncture.

This membrane also serves as the lower bound of a foundation mattress. This mattress will be a 0.3m thick layer of compacted sand and gravel. The membrane should be protected from puncture. This mattress was designed to support the floor slab and allow releveling of the slab by selectively pressure injecting grout beneath the slab.

Articulated and well-keyed reinforced slabs may be used to localize differential settlement and limit faulting of the slabs at the joints. Separation is needed to prevent the slabs from hanging up on pile caps and grade beams.

A less expensive option is to pave the floor with a more flexible asphalt concrete or stabilized granular surface. Provisions would have to be made to allow future shimming and repaving as necessary.

The driveways and parking areas were designed according to recommendations of the Asphalt Institute of America. The design included a geofabric laid on the subgrade coupled with either a 45cm layer of compacted run-of-bank gravel, or a 35cm layer of compacted crushed stone base. This design would bridge voids in the subbase adequately at a reasonable cost and provide a substantial life before repaving.

Utility Foundations

Four fuel tanks (total capacity of 120 cubic meters), water, sewer, and power lines will be placed on a dynamically compacted imported granular fill after removal of the landfill material to 0.6m below these utilities. The gas line will be placed on the same fill after removal of all the landfill material to original grade.

Post-Construction Settlement

Natural settlement of landfills occurs by several mechanisms: consolidation or decreasing voids; ravelling or washing fine material into large voids; collapse of hollow structures; creep; and chemical and biological decay accompanied by gas production and reduction in solid volume. The first three mechanisms are accelerated by DC. The creep and decay cannot, however, be substantially changed by DC. On the other hand, the rate of anaerobic decay is substantially less than aerobic decay [5], and densification will decrease the amount of oxygen in the landfill resulting in a decreased rate of decay. The planned development will also reduce the quantity of water percolating into the landfill, thereby slowing the rate of decay.

The NADL is many years old, and much of the expected settlement due to consolidation, ravelling and structural collapse will occur during DC.

The settlement due to decay has at least two causes: conversion of material to liquid or gas (that is, disappearance of the material); and decay of hollow structures coupled with collapse and ravelling. The settlement due to decay and collapse of hollow structures coupled with ravelling may result in local sinkholes and differential settlement of structures supported by the landfill. Effective DC and sealing the ground surface will minimize the occurrence of sinkholes.

The settlement due to creep is analogous to secondary consolidation of conventional soil. Creep settlement and conversion to liquid or gas will be more uniform than settlement due to decay and subsequent collapse. Furthermore, the magnitude of long-term settlement due to creep is probably substantially greater than that due to decay, except in the event of a sinkhole caused by structural collapse.

The literature on predicting long-term settlements of landfills is scant. A study of case records reported in the literature was made. The reported settlement was normalized by dividing by the landfill thickness and plotted versus log time in Figure 5. The case records plotted are summarized in Table 1.

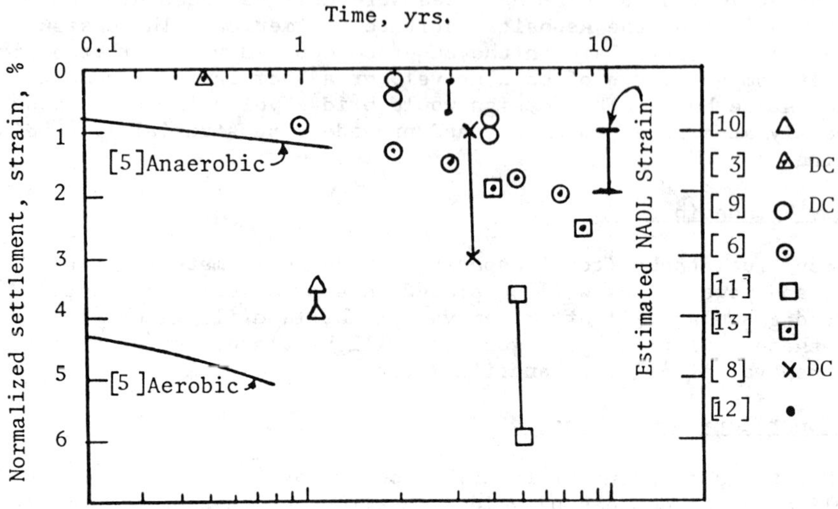

FIG. 5 Normalized Settlement Versus Log Time

Diagrams by Sowers[5] differentiate aerobic and anaerobic decay conditions. His data is plotted as two curves in Figure 5. The landfill material contained garbage refuse and building debris. The stress on the landfill varied between 25 and 50kPa. The age and thickness of the landfill or any treatment are unknown.

Another paper by Sower[6] reports the long-term settlement of a one-story building constructed on a 3m embankment which was placed on an old landfill. The landfill varied in thickness up to 7.5m and received no treatment prior to construction. Settlement of the building was measured annually over a seven year period.

Welsh[4] reports on the DC of a 10m thick municipal landfill which had been closed four years prior to the DC treatment (750 tonne-meters per square meter). A roadway embankment was placed over the site [7] and is reported by Lukas[8] Settlement was measured three and one-half years after treatment.

Charles et al.[9] report on the DC of a 6m thick municipal landfill which was 15 years old when treated (250 tonne-meters per square meter). Settlement was measured at the location of a 3m high embankment two and four years later.

Lukas[3] reports on DC of an 18m thick burned refuse with miscellaneous material landfill which was more than 30 years old when treated (200 tonne-meters per square meter). Settlement of a two-story building was measured 6 months after construction.

Chang and Hannon[10] report on a roller-compacted 5.5 to 6m thick landfill which was 7 to 10 years old when treated. The treatment

Table 1 Summary of Case Records

References	Landfill Thickness	Age	Description	Loading	Treatment
Sower[5]	?	?	Garbage refuse & building	25-50kPa	?
Sowers[6]	7.5^{m}	Old	Sanitary landfill	3m Embankment & building	None
Lukas[8]	10m	?	Municipal	Embankment	$DC750^{Tm}/m^2$
Charles et al.[9]	5-6m	15 years	Refuse	Embankment	$DC250^{Tm}/m^2$
Lukas[3]	18m	>30 years	Burned refuse with misc. material	2 Story bldg	DC
Chang & Hannon[10]	5.5-6m	7-10 years	?	3m Embankment	Roller
Moore & McGrath[11]	1.5-7.5m	4-18 years	Dumpfill	1m Embankment	Roller
Burlingame[12]	2-9m	5-15 years	Residential & Ind. waste underlain by organic silt & peat	Embankment	2m Surcharge
Yen & Scanlon[13]	6-38m	Recent	Residential	Selfweight	Bulldozer compacted in 1.5-6.1m lifts

included 105 passes with a 45 tonne roller. Settlement was measured at the location of a 3m thick embankment thirteen months after construction.

Moore and McGrath[11] report on a roller-compacted 1.5 to 7.5m thick dump fill which was 4 to 18 years old when treated. The treatment included 8 passes with a 27 tonne roller and 38 passes with a 45 tonne roller. Settlement was measured at the location of a 0.6 to 1m thick embankment five years after construction.

Burlingame[12] reports on a surcharge compacted 2 to 9m thick residential and industrial waste landfill which was 5 to 15 years old when treated. This landfill was partly underlain by organic silt and peat. Settlement was measured 3 years after removal of the 2m surcharge.

Yen and Scanlon[13] report on three recent residential refuse landfills which were bulldozer compacted in 1.5 to 6.1m lifts during placement. Settlement was measured under selfweight over a nine year period and average values are shown in the figure.

An examination of Figure 5 and Table 1 indicates that surface roller compaction of an existing landfill is less effective at limiting the magnitude of settlement than dynamic compaction.

The available data shown in Figure 5 are scattered and limited in extent. However, it is useful to predict a range of settlement up to a decade after construction, but more data are needed to conclude that such a figure represents true settlement behavior and to refine the time-settlement relationship.

Using Figure 5 it is estimated that after 10 years the NADL site will experience settlements between one and two percent (8 and 16cm). It is the opinion of the authors that the site will experience settlement near the low end of the range.

This paper was submitted before construction was started. It is expected that a future publication will cover the construction of the facility and the post-construction performance.

CONCLUSIONS

The following conclusions were reached during the investigation at NADL. Much of the information discussed herein can be utilized when planning construction on a landfill.

1. A phased subsurface investigation program should be used when planning construction on a landfill. The results obtained guide the choice for further investigation and foundation type selection.
2. The landfill should be zoned and identified by potential foundation design schemes.
3. A universally accepted classification system for landfill materials should be developed and used by practitioners.
4. The settlement due to decreasing voids, ravelling, and collapse of hollow structures will be much less after a landfill has been

dynamically compacted [Figure 5]. Heavy surface rolling is not as effective at limiting settlement.

5. The magnitude of long-term settlement can be estimated from Figure 5. More data are however needed to refine the time-settlement relationship of landfills. These data should include detailed descriptions of the landfill material, placement method, and any treatment such as grouting, dynamic compaction, and other ground improvement.

ACKNOWLEDGEMENTS

The authors wish to thank the City of Albany, NY, and Smith and Mahoney, PC, of Albany, NY.

The views expressed in this paper are those of the authors.

REFERENCES

[1] Landva, A. O., and LaRochelle, P., "Compressibility and Shear Characteristics of Radforth Peats," Testing of Peats and Organic Soils, ASTM STP 820, Philadelphia, 1983.

[2] Landva, A. O., and Clark, J. I., "Geotechnics of Wate Fills," Geotechnics of Waste Fills - Theory and Practice, ASTM STP 1070, Philadelphia, 1990.

[3] Lukas, R. G., "Dynamic Compaction for Highway Construction Volume I: Design and Construction Guidelines," U.S. Department of Transportation, FHWA/RD-86/133, McLean, Va., 1986.

[4] Welsh, J. P., "Dynamic Deep Compaction of Sanitary Landfill to Support Superhighway," Proceedings of the 8th European Conference on Soil Mechanics and Foundation Engineering, Helsinki, 1983.

[5] Sowers, G. F., "Settlement of Waste Disposal Fills," Proceedings of the 8th International Conference on Soil Mechanics and Foundation Engineering, Moscow, 1973.

[6] Sowers, G. F., "Foundation Problems in Sanitary Landfills," Journal of the Sanitary Engineering Division, ASCE, Vol. 94, No. SA1, Feb. 1968.

[7] Research Section, Materials and Research Division, Arkansas State Highway and Transportation Department, "Dynamic Compaction of a Sanitary Landfill Progress Report," U.S. Department of Transportation, FHWA-TS-85-227, Washington, DC., 1985.

[8] Lukas, R. G., "Densification of Loose Deposits by Pounding," Journal of the Geotechnical Engineering Division, ASCE, Vol. 106, No. GT4, April 1980.

[9] Charles, J. A., Burford, D., and Watts, K. S., "Field Studies of the Effectiveness of Dynamic Consolidation," Proceedings of the 10th International Conference on Soil Mechanics and Foundation Engineering, Stockholm, 1981.

[10] Chang, J. C., and Hannon, J. B., "Settlement Performance of Two Test Highway Embankments on Sanitary Landfills," New Horizons on Construction Material, Envo Publishing Company, Inc., Lehigh Valley, Pa., Vol. 1, 1976.

[11] Moore, L. H., and McGrath, M. E., "Highway Construction on Refuse Landfills," Highway Focus, Vol. 2, No. 5, Dec. 1970.

[12] Burlingame, M. J., "Construction of a Highway on a Sanitary Landfill and Its Long-Term Performance," Transportation Research Record 1031, TRB, National Research Council, Washington, D.C., 1984.

[13] Yen, B. C., and Scanlon, B., "Sanitary Landfill Settlement Rates," Journal of the Geotechnical Engineering Division, ASCE, Vol. 101, No. GT5, May 1975.

Hari D. Sharma[1], Michael T. Dukes[2], Donald M. Olsen[3]

FIELD MEASUREMENTS OF DYNAMIC MODULI AND POISSON'S RATIOS OF REFUSE AND UNDERLYING SOILS AT A LANDFILL SITE

REFERENCE: Sharma, Hari D., Dukes, Michael T., and Olsen, Donald M., "Field Measurements of Dynamic Moduli and Poisson's Ratios of Refuse and Underlying Soils at a Landfill Site", Geotechnics of Waste Fills - Theory and Practice, ASTM STP 1070, A. Landva, G. D. Knowles, editors, American Society for Testing and Materials, Philadelphia, 1990.

ABSTRACT: This paper briefly discusses the applicability of laboratory and field test methods commonly used to determine the dynamic moduli and Poisson's ratio of undisturbed soil and rock and soil-rock fill and refuse fill materials. Down-Hole geophysical methods were used to measure dynamic shear and compression wave velocities of municipal refuse fill and the underlying soils at a sanitary landfill. These data are used to compute the in situ elastic and shear moduli and Poisson's ratio of the materials at the landfill site and are compared with values reported in the literature. Finally, conclusions are drawn regarding the use of dynamic moduli and Poisson's ratios obtained with the Down-Hole test method.

KEYWORDS: Dynamic Shear Moduli, Dynamic Elastic Moduli, Poisson's Ratio, Field Measurements, Down-Hole Test, Municipal Refuse, and Sanitary Landfill.

INTRODUCTION

The dynamic moduli, Poisson's ratio, and strength characteristics of natural materials (i.e., soil and rock) and of man-made materials (i.e., soil-rock fills and refuse fills) must be estimated if their engineering behavior is to be understood when subjected to dynamic loadings, such as, machine vibrations, blasting, and earthquakes.

---

1 Hari D. Sharma, Ph.D., Chief Geotechnical Engineer, EMCON Associates, 1921 Ringwood Ave., San Jose, CA 95131.

2 Michael T. Dukes, Executive Manager, EMCON Associates, 1921 Ringwood Ave., San Jose, CA 95131.

3 Donald M. Olsen, Project Geotechnical Engineer, EMCON Associates, 1921 Ringwood Ave., San Jose, CA 95131.

This paper addresses measurements of the dynamic moduli (Young's elastic modulus and shear modulus) and Poisson's ratios of municipal refuse and of the underlying soils at a landfill site in Richmond, California. These data can be used to evaluate deformations under earthquake loadings for a landfill. The following presents a summary of the various commonly used test methods to determine dynamic properties of materials, the site conditions evaluated, the dynamic test method used, a discussion of the test results, and conclusions.

TESTING METHODS

Dynamic properties of materials are determined by either laboratory or field test methods. Generally, field test methods are preferred because: 1) they measure actual properties over a large area rather than the properties of a discrete sample, and 2) it is difficult to duplicate field conditions in the laboratory. The following presents a brief discussion of several laboratory and field test methods. Determination of the dynamic material properties from the test data are presented at the end of the paper.

Laboratory Test Methods

In general, cyclic triaxial compression, cyclic simple shear, and resonant column laboratory test methods have been used to measure the dynamic properties of materials. A brief discussion of each method is given in the following:

Cyclic Triaxial Compression Test: The test apparatus consists of a standard flexible wall triaxial compression load cell with a cyclic loading mechanism, and an electronic recording system to collect stress-strain and pore pressure data. The major disadvantage of this test is that the field conditions are not adequately simulated. For example, during actual earthquake loading, there is cyclic reorientation of the principal stress direction through some angle relative to its initial position, while in the laboratory triaxial compression test, the major principal stress can only act in either the vertical or horizontal directions. Additionally, the triaxial compression test apparatus can not simulate plane-strain conditions which are believed by the authors to be more representative of actual field conditions.

Cyclic Simple Shear Test: The test apparatus consists of a simple shear box, a cyclic loading mechanism, and an electronic recording system that collects stress-strain and pore pressure data. A typical simple shear box used for both static and dynamic testing contains a square sample specimen with side lengths of 6 centimeters (cm) and a thickness of 2 cm. The box has two fixed side walls and two hinged end walls so that the sample can be subjected to simple shear deformations. The stress-strain conditions developed in the simple shear test are very similar to those of a soil element in the ground when subjected to upwardly propagating shear waves during an earthquake. Thus, the simple shear test apparatus is generally preferred over the triaxial test apparatus.

Resonant Column Test: The test apparatus consists of a flexible wall triaxial compression cell, modified so that the soil specimen can be excited by either longitudinal (compression) or torsional (shear) vibrations. The test is performed by adjusting the excitation frequency until the specimen resonates. The dynamic moduli are computed from the resonant frequency and geometric properties of the specimen according to the theory of wave propagation in prismatic rods. The damping property of the specimen can also be estimated from this test.

One of the drawbacks of all laboratory tests is that representative samples can not always be tested in the laboratory apparatus. This is a major problem with refuse materials which are usually heterogeneous and anisotropic because they may typically consist of a mixture of soil, concrete blocks, metals, plastics, organics, and other materials. Therefore, field tests are the most appropriate methods used for measuring the dynamic properties of refuse fills.

Field Test Methods

Field test methods frequently used are: Cross-Hole, Up- or Down-Hole, and Refraction Wave Propagation Methods. Less commonly used and more complex methods are the Rayleigh and Love Waves Propagation Test, Block Resonance Tests, and Cyclic-Plate-Load Tests. The Cross-Hole, Up- or Down-Hole and Refraction Wave Propagation methods are briefly described in the following:

Cross-Hole Wave Propagation Method: This method measures the velocities of compression and shear waves propagating from a bore hole equipped with a down-hole hammer to at least one other bore hole equipped with directional geophones. The hammer is used to generate both compression and shear waves. The compression and shear wave velocities are measured at specific depths as they propagate away from the source bore hole. To obtain high quality data usually requires a near vertical set of borings with the down hole equipment set at the same elevation. These condition will allow the shortest path of wave propagation to be measured, which is assumed to be the distance between the borings at the surface. This method is described in detail in ASTM [1].

The Cross-Hole method is capable of locating thin, low velocity strata which are of importance to ground settlement studies. The minimum thickness that can be identified depends on the vertical measurement interval used, bore hole separation and the velocities in the surrounding materials.

Up- or Down-Hole Wave Propagation Method: Both the Up- and Down-Hole methods can be used with a single bore hole. The Up-Hole test is performed by placing the geophone at the surface and the wave-generating hammer at various depths in the bore hole. The Down-Hole test is performed with the wave-generating hammer at the surface and the geophone placed at various depths in the bore hole. The Down-Hole test method was used at the Richmond Site. Both methods yield the average vertical compression and shear wave velocities between the hammer and geophone for each measurement location in the bore hole. Average velocities are measured because the propagating waves are affected by both overburden pressure increases and changes in material types with depth.

Interpretation of Up- and Down-Hole velocity data can be very difficult if the surface waves generated by the hammer reach the bore hole casing before the compression and/or shear waves. The surface wave causes the bore hole casing and geophone to vibrate with a large amplitude and low frequency. As a result the arrival of both the compression and shear waves will be difficult to detect. The surface waves that travel down the bore hole casing are called "tube waves."

Refraction Wave Propagation Method: This method measures the gross velocity of materials and allows interpretation of the subsurface structure lying between the point of wave generation on the surface and a series of collinear aligned geophones. Both compression and shear waves are generated at the surface in the same manner as described for the Down-Hole Test Method. The average velocities, thicknesses, and dip angle of each layer overlying a refracting boundary are determined from the travel times of the refracted waves.

The successful use of this method is dependent on the following factors: 1) a distinctive velocity contrast between adjacent material layers, and 2) each layer should be relatively isotropic and homogeneous. A major disadvantage of this method is that a low velocity layer lying below a high velocity layer can not be detected and therefore, will appear as part of the overlying higher velocity layer.

Applicability of Test Methods

Use of the appropriate test method, whether in the laboratory or in situ, is highly dependent on the magnitude of strains expected to occur at the site both during and following construction. However, it is recommended that dynamic moduli be determined for a wide range of strain levels by several methods prior to choosing a value or range of values to use for design purposes. Table 1 summarizes the advantages and disadvantages of the various field methods used to measure dynamic materials properties.

SITE CONDITIONS

The landfill site is located in the City of Richmond, California along the southeastern shores of San Pablo Bay. The active landfill area encompasses about 180 acres of reclaimed marshland. The site has been in continuous operation as a solid waste and liquid waste disposal facility since the early 1950's.

Numerous subsurface investigations have been performed at the site for the purpose of characterizing the vertical and horizontal limits of the refuse fill, geologic and hydrogeologic conditions, and geotechnical engineering properties of the underlying natural soils. These investigations have revealed the following general site conditions. The refuse fill ranges in thickness from 15 to 95 feet (4.6 to 28.9 meters). The landfill is predominantly underlain by silts and clays with discontinuous lenses of sand that are locally known as Bay Mud. Ground water is usually encountered at about elevation 3 feet (0.31 meter) above mean sea level (MSL). The Bay Mud soils are considered to be saturated, while the refuse fill is considered to be saturated below elevation +3 feet (0.31 meter) MSL and unsaturated above elevation +3 feet (0.31 meter) MSL.

TABLE 1 -- Advantages and Disadvantages of Field Methods[1]

FIELD TECHNIQUE AND MEASUREMENTS	ADVANTAGES	DISADVANTAGES
Refraction P-Wave Velocity S-Wave Velocity	Reversible polarity, works from surface, samples large zone, preliminary studies	Misses low velocity zones, low strain amplitudes
Cross-Hole P-Wave Velocity S-Wave Velocity	Known wave path, reversible polarity, works in limited space	Needs two or more holes, need to survey holes for verticality
Down-Hole or Up-Hole P-Wave Velocity S-Wave Velocity	Need only one hole, reversible polarity, works in limited space	Measures average velocities
Surface Vibratory S-Wave Velocity Attenuation of Rayleigh waves	Works from surface	Uncertain about effective depth, need large vibrator

1 After Woods [2], as cited by Prakash [3].

Three borings were drilled for the purpose of measuring dynamic material properties of the refuse fill and underlying natural soils. Each boring was completed by grouting a Slope Indicator Casing with a cement-bentonite mixture, and installing a permanent surface seal with a locking top device. Borings GT-1 and GT-2 were drilled through the eastern perimeter levee and into the underlying Bay Mud soils. GT-3 was drilled in the refuse fill and was terminated above the Bay Mud. Figure 1 shows the location of each boring and a generalized geologic cross-section through the borings.

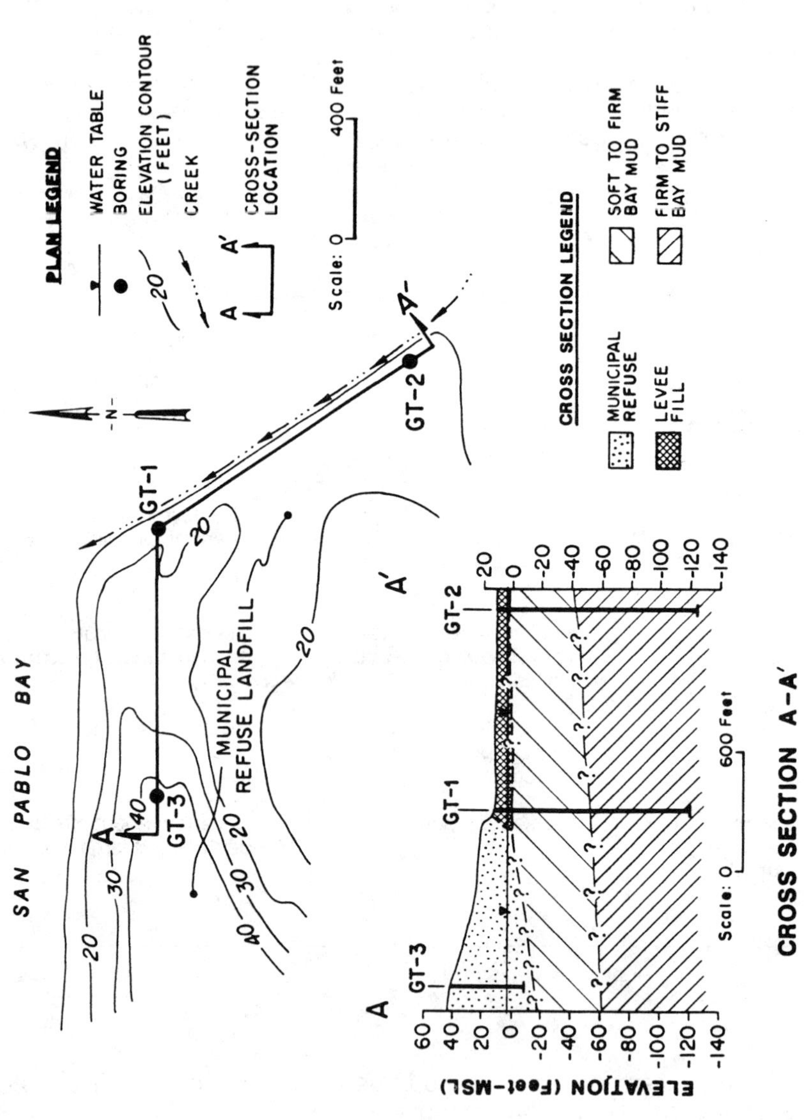

FIGURE 1 -- Site Plan and Generalized Cross-Section

The sediments were logged in accordance with the Unified Soils Classification System (USCS). The Bay Mud lying above elevation minus 60 feet (18.3 meters) MSL generally consist of saturated, dark gray, soft to firm, normally- to over-consolidated, low- to high-plasticity, clayey silts (ML to MH) and silty clays (CL to CH). The Bay Mud lying between elevations minus 60 feet (18.3 meters) and minus 130 feet (39.6 meters) MSL generally consist of saturated, grayish brown to brown, stiff to very stiff, normally- to over-consolidated, low- to high-plasticity, clayey silts (ML to MH) and silty clays (CL to CH). The Bay Mud also contains sand layers that are encountered at various depths.

TEST PROCEDURE AND RESULTS

The Down-Hole test method was used at each boring location to measure the dynamic material properties of the refuse fill and Bay Mud soils. This method was selected on the basis of cost because it requires only one boring to be drilled at each location. This is an important point since the cost of drilling and completing two borings to depths of 130.5 and 136.5 feet (39.8 and 41.6 meters) in the Bay Mud and one boring through 50 feet (15.2 meters) of refuse is high. The compression and shear wave velocities were measured and reported by Redpath Geophysics[4].

Down-Hole Test Procedure

The shear- and compression-wave velocities were determined by measuring the time required for a seismic pulse to travel from the surface to a geophone placed at various depths within the bore hole as shown in Figure 2. Use of casing with tracks to guide the geophones significantly aided in the collection of high quality data. The casing tracks enabled the azimuthal orientation of the geophone to remain fixed (in line with the shear-wave source) while being lowered down the bore hole.

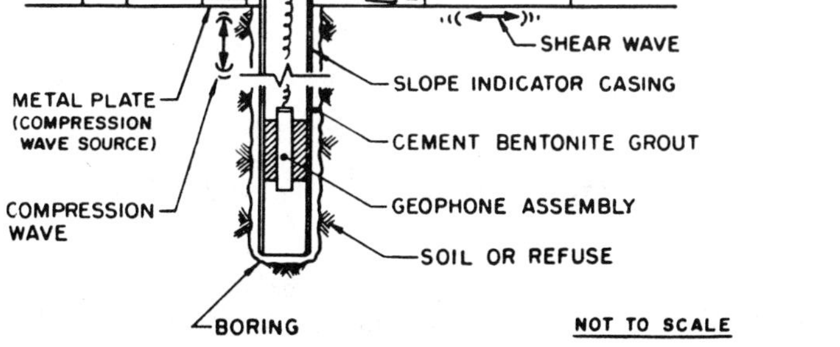

FIGURE 2 -- Down-Hole Test Method Set Up

Shear-wave pulses of opposite polarities were generated by horizontal sledge-hammer blows to opposite ends of a 7-foot long (2.1-meter), 6-in^2 (15.2-cm^2) wood plank, which was held in contact with the ground surface by parking the front wheel of a vehicle on the plank. Compression-wave pulses were generated by vertical sledge-hammer blows to a 1-inch (2.54-cm) thick, 6-in^2 (15.2-cm^2) metal plate placed on the ground surface. In order to improve the signal-to-noise ratio of each travel time record, four hammer blows were recorded and stacked. Seismic records were recorded at 10-foot (3.1-meter) intervals for the full depth of each boring.

Velocity Profile Results

The travel-tlme data recorded at each geophone location were evaluated to determine the first arrival times for both the shear- and compression-waves. Figure 3 shows a plot of the shear- and compression-wave first arrival times in milli-seconds (msec) versus depth in feet (ft) and interpreted velocities for the levee fill and underlying natural soil encountered in boring GT-1 and GT-2. Figure 4 shows similar plots and interpreted velocities for the refuse fill encountered in boring GT-3.

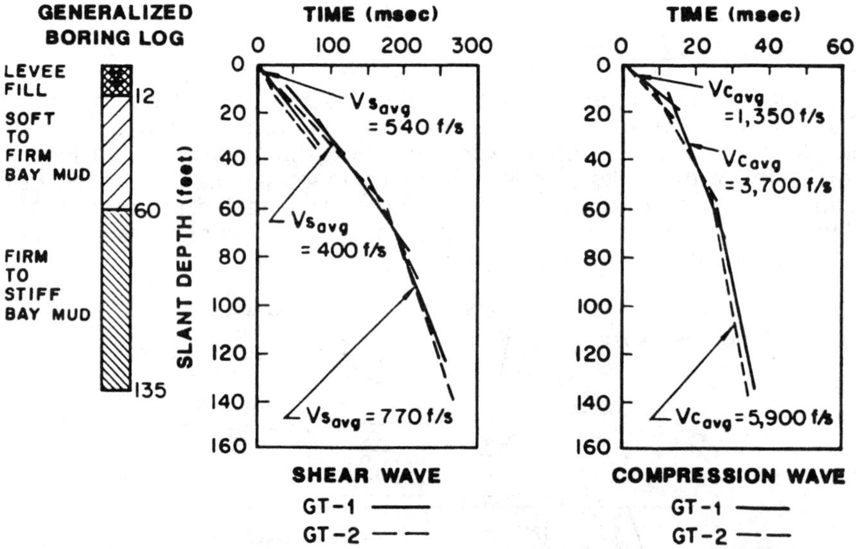

FIGURE 3 -- Shear- and Compression-Wave Velocity Profiles for Borings GT-1 and GT-2

The velocity profiles of borings GT-1 and GT-2 clearly show the levee fill lying between 0 and 12 feet (0 and 3.7 meters) below the surface, the soft to firm Bay Mud lying between 12 and 60 feet (3.7 and 18.3 meters) below the surface, and the firm to stiff Bay Mud lying between 60 and 130 feet (18.3 and 39.6 meters) below the surface. The velocity profile of the refuse encountered in boring GT-3 was also of excellent quality. The average shear- and compression-wave velocities measured for each material encountered in borings GT-1, GT-2 and GT-3 are summarized in Table 2. These velocities were used to estimate the dynamic moduli and Poisson's ratios of the municipal refuse and underlying soils.

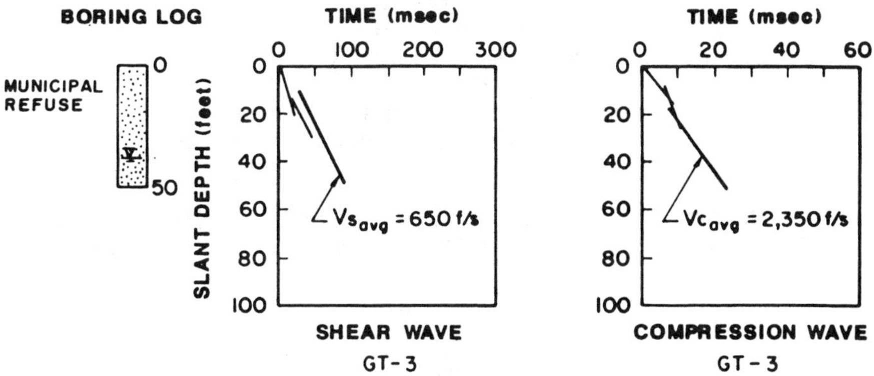

FIGURE 4 -- Shear- and Compression-Wave Velocity Profiles for Boring GT-3

TABLE 2 -- Average Velocity Data

MATERIAL TYPE	DEPTH (meter)	DEPTH (feet)	SHEAR v_s (m/s)	SHEAR v_s (f/s)	COMPRESSION v_c (m/s)	COMPRESSION v_c (f/s)
Refuse Fill	0-15.3	0-50	198.3	650	716.8	2350
Levee Fill	0-3.7	0-12	164.7	540	411.8	1350
Soft Bay Mud	3.7-18.3	12-60	122.0	400	1128.5	3700
Stiff Bay Mud	18.3-39.7	60-130	234.9	770	1799.5	5900

DYNAMIC MODULI AND POISSON'S RATIO

The dynamic shear modulus, modulus of elasticity, and Poisson's ratio of a material can be estimated from its mass density and characteristic shear-wave velocity. The shear modulus can be determined from Equation 1 as follows:

$$G = v_s^2 \rho \tag{1}$$

where

G = shear modulus (F/L^2)
v_s = shear-wave velocity (L/T)
ρ = mass density = γ/g (M/L^3)
γ = unit weight (F/L^3)
g = acceleration due to gravity (L/T^2)
Units: L = length, T = time, M = mass, and F = force

The elastic modulus (Young's Modulus) can be determined from the mass density, shear-wave velocity, and Poisson's ratio of the material as follows:

$$E = 2v_s^2 \rho(1 + \upsilon) \tag{2}$$

where

E = modulus of elasticity (F/L^2)
υ = Poisson's ratio (dimensionless)

The elastic modulus of a material can also be determined from the mass density, compression-wave velocity, and Poisson's ratio as follows:

$$E = v_c^2 \rho(1 + \upsilon)(1-2\upsilon)/(1-\upsilon) \tag{3}$$

where

v_c = compression-wave velocity (L/T)

From Equations 2 and 3 the Poisson's ratio can be expressed as follows:

$$\upsilon = (v_c^2-2v_s^2)/(2(v_c^2-v_s^2)) \tag{4}$$

Mass densities or unit weights of natural soils are readily obtained from relatively undisturbed soil samples. On the other hand mass densities or unit weights of refuse fills are not as easily determined because they must be indirectly estimated from weigh station records and volume changes of the landfill over a given period of time. The volume changes can be estimated from historic topographic relief maps of the landfill surface that are periodically prepared by land surveyors. The average unit weight of the compacted refuse placed at the Richmond landfill was estimated to be 46 pcf (736.9 kg/m^3). Although, the unit weight of refuse fill varies considerably this value is in reasonable agreement with typical values reported by others as summarized in Table 3.

TABLE 3 -- Refuse Fill Average Unit Weights

SOURCE	REFUSE PLACEMENT CONDITIONS	UNIT WEIGHT (kg/m^3)	(pcf)
NAVFAC [5]	Sanitary Landfill		
	a) Not Shredded		
	• Poor Compaction	320	20
	• Good Compaction	641	40
	• Best Compaction	961	60
	b) Shredded	881	55
Sowers [6]	Sanitary Refuse: Depending on Compaction Effort	481-961	30-60
NSWMA [7]	Municipal Refuse:		
	• In a landfill	705-769	44-49
	• After Degradation and Settlement	1009-1121	63-70
Landva and Clark[a] [8]	Refuse Landfill (Refuse to soil cover ratio varied from about 2:1 to 10:1)	913-1346	57-84
EMCON Associates[b] [9]	For 6:1 refuse to daily cover soil	737	46

a These values were obtained from test pit measurements of refuse at eleven municipal landfills located in Canada. Values measured for the Halifax landfill and the August 1983 measurements at the Edmonton and Calgary landfills have not been included, as suggested by the authors.

b Based on tonnage records and areal survey maps recorded during the period from April 1988 through April 1989.

The average unit weights of the levee fill, soft to firm Bay Mud, and firm to stiff Bay Mud materials were estimated to be about 105 pcf (1683 kg/m^3), 100 pcf (1602 kg/m^3), and 105 pcf (1683 kg/m^3), respectively. These average unit weights were determined from relatively undisturbed samples taken during drilling of borings GT-1 and GT-2. The samples were taken with either Osterberg Piston or Pitcher Barrel Samplers equipped with Shelby Tubes.

A review of Table 3 indicates that refuse unit weights can vary significantly. Accordingly, computation of the shear modulus (G) from the unit weight and measured shear-wave velocity (v_s) by Equation [1] will also vary significantly. Thus, site specific measurements of the refuse unit weight should be made. The dynamic moduli and Poisson's ratio of the refuse landfill were estimated by substituting the shear- and compression-wave velocities, and mass densities measured at the site into Equations [1], [2] or [3] and [4]. The results of these computations are summarized in Table 4.

TABLE 4 -- Dynamic Moduli and Poisson's Ratios

MATERIAL TYPE	SHEAR MODULUS G		ELASTIC MODULUS E		POISSON'S RATIO υ
	(MPa)[a]	(ksi)[b]	(MPa)	(ksi)	(dimensionless)
Municipal Refuse	28.9	4.19	84.4	12.24	0.49
Levee Fill	45.5	6.60	127.5	18.49	0.49
Soft to Firm Bay Mud	23.8	3.45	70.9	10.28	0.49
Firm to Stiff Bay Mud	92.6	13.43	275.9	40.01	0.49

[a] MPa = units of megapascals.
[b] ksi = units of kips per square inch.

Table 5 presents a summary of typical shear moduli and Poisson's ratios for soils as reported by Bowles [10]. A comparison of Tables 4 and 5 indicates that field measurements of the shear modulus for soils underlying the landfill are higher than the values summarized by Bowles. The Poisson's ratios in both cases are comparable with saturated clays. This is reasonable because, the Bay Mud soils underlying the landfill site consist predominantly of saturated silts and clays. A comparison of the dynamic shear moduli and Poisson's ratio obtained for the refuse fill was not possible due to a lack of published values.

TABLE 5 -- Typical Shear Modulus and Poisson's Ratio

SHEAR MODULI

Material Type	G, MPa	ksi
Clean dense quartz sand	12.4-20.7	1.8-3.0
Micaceous fine sand	15.9	2.3
Berlin sand	17.2-24.1	2.5-3.5
Loamy sand	10.3	1.5
Dense sand-gravel	69+	10+
Wet soft silty clay	9.0-13.8	1.3-2.0
Dry soft silty clay	17.2-20.7	2.5-3.0
Dry silty clay	27.6-34.5	4.0-5.0
Medium clay	13.8-27.6	2.0-4.0
Sandy clay	13.8-27.6	2.0-4.0

POISSON'S RATIOS

Material Type	υ
Clay, saturated	0.4-0.5
Clay, unsaturated	0.1-0.3
Sandy clay	0.2-0.3
Silt	0.3-0.35
Sand, gravelly sand	0.3-0.4
Rock	0.1-0.4
Loess	0.1-0.3
Ice	0.36
Concrete	0.15

CONCLUSIONS

The following conclusions can be made:

1) The in situ dynamic shear and elastic moduli, and Poisson's ratio estimated for the municipal refuse disposed at the landfill site studied are 4.2 ksi (28.9 MPa), 12.2 ksi (84.4 MPa), and 0.46, respectively. These values were determined from down-hole seismic field test data.

2) Site specific refuse densities must be measured and used to compute dynamic moduli from shear and compression wave data because, the moduli are highly sensitive to changes in material density and because refuse densities are extremely variable. The refuse density at the landfill site studied was estimated to be 46 pcf (737 kg/m^3).

3) The dynamic moduli measured for underlying soils at the site are higher than published values. However, the Poisson's ratios measured at the site are comparable. The dynamic shear and elastic moduli, and Poisson's ratio estimated for the soft to firm Bay Mud are 3.5 ksi (23.8 MPa), 10.3 ksi (70.9 MPa), and 0.49, respectively. The dynamic shear and elastic moduli, and Poisson's ratio estimated for the firm to stiff Bay Mud are 13.4 ksi (92.6 MPa), 40.0 ksi (275.9 MPa), and 0.49, respectively.

REFERENCES

[1] ASTM, "Standard Test Methods for Cross-Hole Seismic Testing," Designation: ASTM D4428/D4428M-84, in American Society for Testing and Materials, Sec. 4, Vol. 04.08, 1989, pp. 661-670.

[2] Woods, R.D.,"Measurements of Dynamic Soil Properties - State of the Art," Proc. ASCE Speciality Conference on Earthquake Engineering and Soil Dynamics, Pasadena, California, June 1978.

[3] Prakash, S., "Dynamic Stress Deformation and Strength Characteristics of Soils," in Soil Dynamics, McGraw-Hill Book Company, 1981, pp. 71-127

[4] Redpath Geophysics, "Down-Hole Velocity Survey, West Contra Costa County Sanitary Landfill, Richmond, California," letter report to EMCON Associates, Sept. 9, 1989, pp. 1-12.

[5] NAVFAC DM 7.3M, "Soil Dynamics, Deep Stabilization, and Special Geotechnical Construction," United States Navy, Naval Facilities Engineering Command, Design Manual 7.3, April 1983, pg. 7.3-78.

[6] Sowers, G.F., "Foundation Problems in Sanitary Landfill," Journal of the Sanitary Engineering Division, American Society of Civil Engineers, Vol. 94, SA1, February 1968, pp. 103-116.

[7] NSWMA, "Basic Data: Solid Waste Amounts, Composition and Management Systems," National Solid Waste Management Association, Technical Bulletin No. 85-6, October 1985, pg. 8.

[8] Landva, A., and Clark, J., "Geotechnical Testing of Wastefill," 39th Canadian Geotechnical Conference, Ottawa, Canada, 1986, pp. 371-385.

[9] EMCON Associates, Personal communications between Mr. Larry Burch and Hari Sharma, July 1989.

[10] Bowles, J.E., Foundation Analysis and Design, McGraw-Hill Book Company, 1988, pg. 100 and 912.

Wei-Hsing Huang* and C. William Lovell*

BOTTOM ASH AS EMBANKMENT MATERIAL

REFERENCE: Huang, W. H. and Lovell, C. W., "Bottom Ash as Embankment Material," Geotechnics of Waste Fills - Theory and Practice, ASTM STP 1070, Arvid Landva, G. David Knowles, editors, American Society for Testing and Materials, Philadelphia, 1990.

ABSTRACT: The physical and chemical characteristics and engineering properties of power plant bottom ash, a byproduct produced from the combustion of coal, are presented. Laboratory studies included evaluation of chemical and index properties, permeability, compaction, shear strength, and compressibility. The results of these tests conducted on Indiana bottom ashes are compared with representative values obtained for granular soils. The leaching behavior of bottom ash and its effects on ground water quality have been studied by laboratory extraction techniques. The findings suggest that untreated bottom ash may be used as a fill material, including backfill for retaining structures, highway embankments, and structural fills.

KEYWORDS: bottom ash, boiler slag, embankment, landfill, waste materials, leaching test, leachates, ash disposal

The utilization of power plant ash as a construction material has received increasing attention because it not only solves a potential solid waste problem but also provides an alternative construction material. Disposal of ash is a serious and expensive problem, especially in urban areas. As supplies of natural construction materials diminish rapidly, obtaining natural material for fills is also becoming more difficult. Thus, it is desired to convert the burdensome waste material into a useful resource; and this can be accomplished at a saving, on a local or regional basis, to both the utility and construction industries.

---

* Research Assistant and Professor, School of Civil Engineering, Purdue University, West Lafayette, Indiana 47907.

The residual materials produced from the combustion of coal in utility power plants are collected in two forms: bottom ash and fly ash. Bottom ash is the ash which builds up on the heat-absorbing surfaces of ·the furnace, and which subsequently falls through the furnace bottom to the ash hopper below. Fly ash is the fine-grained dusty material that is recovered and collected from furnace flue gases by ash precipitators.

While there is considerable information accumulated on the properties of fly ash and its applications, very little has been developed in the constructive use of bottom ash, primarily because of the lack of information on properties of this material. The purpose of this paper is to provide the physical, chemical, and engineering properties that are likely to affect the use of bottom ash as fill material, based on laboratory investigations conducted on Indiana bottom ashes. Also, the potential environmental effects of bottom ash utilization, which are of growing concern to the public, are evaluated.

PRODUCTION AND NATURE OF BOTTOM ASH

Uncombustible, mineral matter accounts for 10-20% by weight of the coal consumed in power plants. Ash characteristics are affected as much by the type of coal burned, as well as by the type of boiler. Today's large utility boilers burn pulverized coal and have dry-bottom furnaces, which produce dry bottom ash. In a dry bottom furnace, bottom ash solidifies and agglomerates into coarse particles and then fall into the ash hopper at the bottom of furnace. The ash particles are vesicular and irregularly shaped, and have a rough, gritty texture. Dry bottom ash represents about 20% by weight of the ash originally in the coal. The remaining 80% is collected as fly ash.

Wet bottom ash, often referred to as boiler slag, is generated from a slag-tap furnace. The word "wet" refers to the molten state of the ash which leaves the furnace as a liquid. The molten ash is then quenched in the water-filled hopper to form boiler slag. As it is solidified from a molten state, boiler slag is a hard, black, glassy, angular material with a smooth surface texture. It can also be vesicular in nature, especially for coarse sizes, if gases are trapped in the molten slag as it is tapped from the furnace. In a wet bottom furnace, much higher percentages of ash are produced in the form of boiler slag. In cases where fly ash is returned to the furnace and recovered as boiler slag, the percentage of slag can be as high as 100 percent [1].

DISPOSAL AND UTILIZATION

Basically, ash handling and disposal is accomplished either by wet or dry methods. Dry disposal implies transport and deposition of dry or moistened ash. This may involve temporary storage of ash in silos, subsequent hauling by trucks, and compacting at a landfill. Most power stations operating in urban areas handle ash by the dry method, due to land limitation. An alternative method of disposal is to add sufficient amount of water to produce a slurry and enable transport of the ash by pipeline to settling ponds or lagoons. This is termed wet disposal, and is more commonly used, because of economy. Ash ponds also minimize dust problems and are simple to operate. Generally, crushing of bottom ash from the hopper is required for both dry and wet disposal methods to facilitate handling.

About 17.5 million tons of bottom ash and boiler slag were produced in the United States during the year 1986. Of this, 13.4 million tons were dry bottom ash and 4.1 million tons were boiler slag. About 27 percent of the dry bottom ash and 51 percent of the boiler slag were utilized. Areas of utilization include fills, embankments, road base, and aggregate in concrete products and bituminous mixtures. On a national scale, ash disposal costs ranged from $5 to $10 per ton and the total cost of ash disposal to the electric utility industry in 1980 ranged from $375 to $740 million [2].

EXPERIMENTAL PROGRAM

Selection of Ash Sources

A total of 11 bottom ashes were collected for study from 10 power stations, with consideration to boiler type, source of coal burned, geographic distribution, and ash disposal method. Of the samples selected for testing, 2 were of the wet ash type, and 9 were of the dry ash type. Only two plants disposed of their ashes by dry method and specimens in this case were collected directly from the hoppers or silos. For the other 9 ashes that were wet-disposed, bottom ashes were collected as grab specimens from ash deposits at the end of sluice pipe. The approximate locations of the selected sources of bottom ash are shown in Figure 1.

In order to study the variability of ash properties, each source was sampled at least twice. All ashes were subjected to a series of chemical and physical characterization tests, and then representative ashes were chosen for detailed testing on the engineering properties and the potential environmental effects of ash materials.

Chemical Analyses

The chemical compositions of bottom ashes were determined by atomic absorption spectrophotometric techniques and the results are shown in Table 1. The principal constituents are silica (SiO_2), alumina (Al_2O_3), and iron oxide (Fe_2O_3). There are smaller quantities of calcium oxide (CaO), magnesium oxide (MgO), potassium oxide (K_2O), sodium oxide (Na_2O), and sulfur trioxide (SO_3), as well as minute traces of other elements. As can be seen from Table 1, the chemical composition of each bottom ash shows a reasonable degree of

FIGURE 1 -- Approximate locations of bottom ash sources in Indiana

uniformity, except those ashes from Perry, Stout, and Richmond. These stations were burning different sources of coal just prior to the dates of sampling, and this is reflected by greater variations in the chemical composition of the bottom ash. The loss on ignition determined at 600°C gives an approximate indication of the unburnt carbon content.

For purpose of comparison the typical range of chemical composition for most ashes [3] are shown at the bottom of Table 1. Bottom ash from Stout shows a rather high content of iron, otherwise the ashes are reasonably typical.

Gradation

Grain size analyses were performed using sieve analysis in accordance with ASTM C136 procedures. Figure 2 shows the range of gradation for the 11 bottom ashes, as well as for fly ash from the same sources, which was determined in a previous study [4]. The fine portions of bottom ash passing the No. 200 sieve are non-plastic and range from 0 to 12% by weight. Particles coarser than 38 mm (1.5 in.) are rarely found.

Among the 11 bottom ashes studied, 10 ashes are classified by the Unified Soil Classification System as sand. The other one is classified as gravel. Ten out of 11 samples have the coefficient of uniformity ranging from 7 to 33, while the most uniform ash has a uniformity coefficient of 3.7. By and large, bottom ash is a relatively well-graded, sand-sized material.

TABLE 1 -- Chemical composition of bottom ashes

Ash source	date sampled	Percent by weight								
		SiO_2	Fe_2O_3	Al_2O_3	CaO	MgO	K_2O	Na_2O	SO_3	loss on ignition
Schahfer unit 14	6-19-87	60.1	5.2	10.4	16.6	5.7	0.9	0.4	0.9	0.3
	5-12-88	53.4	6.0	13.5	18.5	5.7	1.2	0.3	1.0	0.1
Schahfer unit 17	6-19-87	58.1	15.2	12.7	7.0	0.8	1.9	0.3	2.2	0.1
	5-12-88	52.1	23.2	13.2	4.8	0.9	1.4	0.2	1.5	0.8
Gibson	5-18-87	58.7	14.6	14.1	3.1	0.8	2.0	0.4	1.3	0.4
	5-17-88	53.6	20.8	14.8	2.6	1.0	1.9	0.5	1.1	1.0
Gallagher	5-26-87	41.2	28.4	11.2	12.6	0.7	1.6	0.3	1.0	0.9
	5-14-88	49.3	24.2	16.4	3.9	0.9	1.7	0.2	2.6	1.4
Perry[a]	5-19-87	48.9	22.2	13.0	0.8	0.7	2.2	0.3	0.6	7.2
	7-19-88	52.5	6.0	24.3	0.9	0.1	2.3	0.4	0.6	6.2
Mitchell	6-19-87	58.8	6.8	7.8	7.9	2.2	1.4	0.1	3.3	8.1
	5-12-88	51.3	6.5	14.2	8.5	3.0	0.9	0.3	1.0	8.0
Wabash	6-23-87	55.7	21.5	14.3	1.7	0.7	1.9	0.3	0.8	0.2
	4-26-88	51.7	23.0	16.0	1.7	0.9	1.9	0.3	0.6	1.0
Richmond[a]	8-17-87	48.3	33.3	11.9	1.3	0.4	0.9	0.2	1.7	2.2
	5- 5-88	41.6	20.9	18.6	1.3	0.6	1.1	0.1	1.9	14.1
Stout[a]	5-27-87	24.2	42.0	6.9	2.2	0.4	0.6	0.2	0.8	18.4
	6-20-88	54.9	20.2	16.7	1.6	0.9	1.9	0.8	1.8	0.3
Culley	8-21-87	35.6	30.1	11.7	14.6	0.8	1.4	0.3	1.0	0.0
	5-14-88	32.0	31.1	11.8	13.9	1.1	0.5	0.2	0.9	0.3
Brown	8-21-87	48.1	27.6	13.4	3.1	0.8	2.1	0.3	1.7	1.9
	5-17-88	38.5	38.0	12.6	3.8	0.7	1.3	0.2	3.3	1.1
Average		48.6	21.2	13.6	6.0	1.4	1.5	0.3	1.4	3.4
Typical range[3]		20-60	5-35	10-35	1-20	0.3-4	1-4[b]		0-12	...

.[a] Plants were burning different sources of coal prior to the dates of sampling.
.[b] Range for the sum of K_2O and Na_2O

The gradation curves for bottom ashes sampled at different times provide an indication of the potential variability in the gradation. Figure 3 gives typical variations in the gradation of bottom ash. Generally, the ashes with uniform gradations tend to have less variation in the gradation between samplings.

Specific Gravity

The results tabulated in Table 2 show that the specific gravity of bottom ashes, as determined by ASTM D854 procedures, ranges from 1.94 to 3.46. This is a much wider range than for most soils (range from 2.5 to 2.8 [5]). The

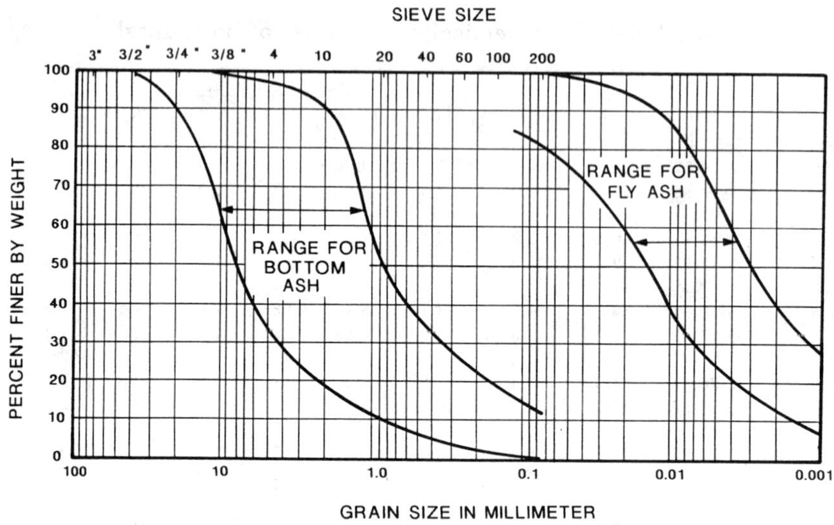

FIGURE 2 -- Ranges of gradation for bottom ash and fly ash

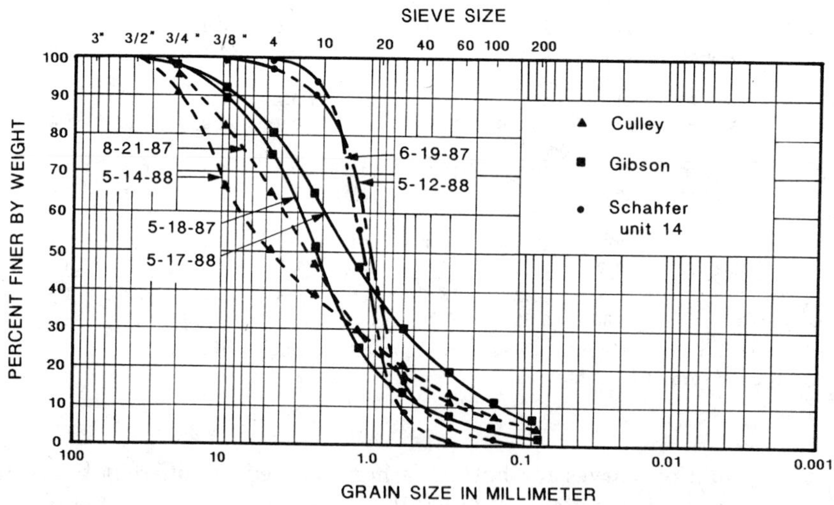

FIGURE 3 -- Typical variations in gradation of bottom ash

specific gravity of the ash is a function of the chemical composition. Obviously, high carbon content will result in a low specific gravity, whereas high iron content will produce high specific gravity. To some degree, specific gravity is an indicator of the quality of bottom ash. Poor ash that contains large amount of porous and popcornlike particles, which are undesirable for engineering purposes, can have a specific gravity as low as **1.6** [**6**].

TABLE 2 -- Specific gravity of bottom ashes

Ash source	type of ash	1st sample	2nd sample
Schahfer			
unit 14	wet	2.82	2.81
unit 17	dry	2.57	2.61
Gibson	dry	2.67	2.56
Gallagher	dry	3.08	2.64
Perry*	dry	2.12	1.94
Mitchell	dry	2.44	2.47
Wabash	dry	2.56	2.48
Richmond*	dry	2.90	2.40
Stout*	wet	3.46	2.45
Culley	dry	3.21	3.23
Brown	dry	2.71	2.97

* Plant burned different sources of coal.

Permeability

The coefficient of permeability of bottom ash was measured by falling head permeability tests. The permeameter had a 10 cm (4 in.) diameter and ash specimens were about 20 cm (8 in.) high. Table 3 gives the results of permeability tests conducted on bottom ashes compacted to 95 percent of the maximum dry density determined using ASTM D698 before testing. The percentage of fines seems to have a predominant effect on the permeability of bottom ash. Generally, the permeability of bottom ashes are comparable to those of soils with similar gradings.

TABLE 3 -- Coefficients of permeability of bottom ashes

Materials	Particle size classification	Permeability (cm/sec)	Percentage of fines
Schahfer			
unit 14	uniform coarse sand	0.101	0
unit 17	well-graded sand	0.034	3
Gibson	well-graded sand with gravel	0.005	6
Gallagher	well-graded sand with gravel	0.002	10
Uniform coarse sand [5]		0.4	0
Well-graded sand and gravel [5]		0.01	0

Moisture-Density Relation

The relationship between the moisture content and unit weight was determined for several selected samples of bottom ash using ASTM D698 method C. The resulting relations are shown in Figure 4. The shape of the compaction curves is typical of that for cohesionless materials [7,8]. These curves are characterized by a fairly high unit weight for the air-dried condition, low unit weight at low water contents, and high unit weight at the high water contents. The variation in dry unit weight for each ash is relatively small with respect to the wide change in moisture content.

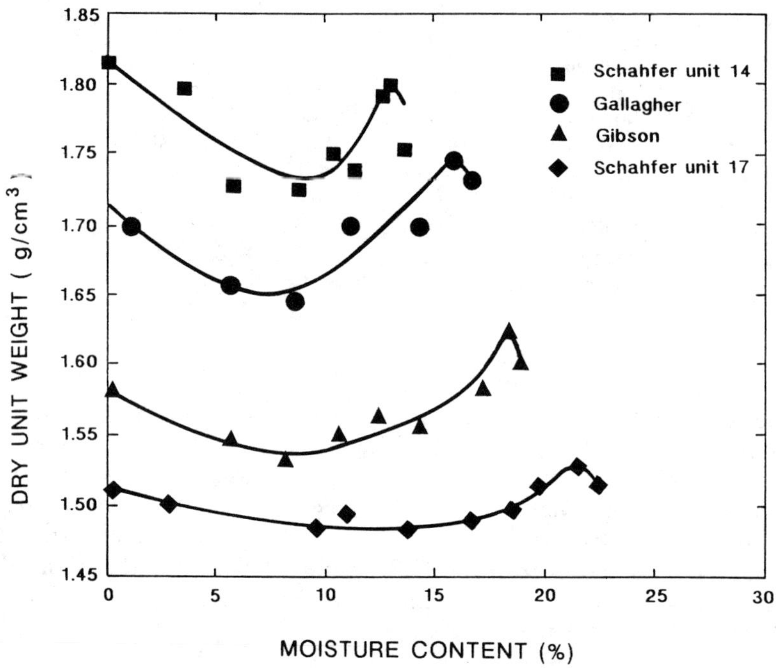

FIGURE 4 -- Compaction curves for bottom ashes

Because the compacted unit weights at the air-dried conditions are comparable to those at optimum moisture contents, it may be beneficial to compact ash air-dried. Thus, much effort and cost in the control of moisture content during compaction can be saved. However, some bottom ashes are reported to lose stability when dry [9]. To date, little data on the relative effects of gradation and water content on the compacted unit weight of bottom ash are available. Therefore, more research on the compaction characteristics of bottom ash, especially field compactions, is highly desired.

Angle of Shearing Resistance

The shear strength of bottom ash is of major importance if it is to be used in structural fills and embankments. A number of direct shear tests were conducted on dry bottom ash. The direct shear box has a diameter of 64 mm (2.5 in.), and only materials finer than 3.6 mm (3/8 in.) were used. The tests were conducted at various relative densities with the normal stress varied from 34 kPa (5 psi) to 240 kPa (35 psi).

Figure 5 shows the angles of friction obtained for bottom ash, along with the range of friction angles generally obtained for various sandy soils [10]. It is found that, at a given initial relative density, the friction angle of bottom ash is higher than that obtained for natural sandy soils. This can be attributed to the rough surface texture and angularity of the bottom ash particles, such that a higher degree of interlocking was developed in the shearing process. If bottom ash is used as an embankment material, the stability of the embankment can be higher than that for natural sandy soils.

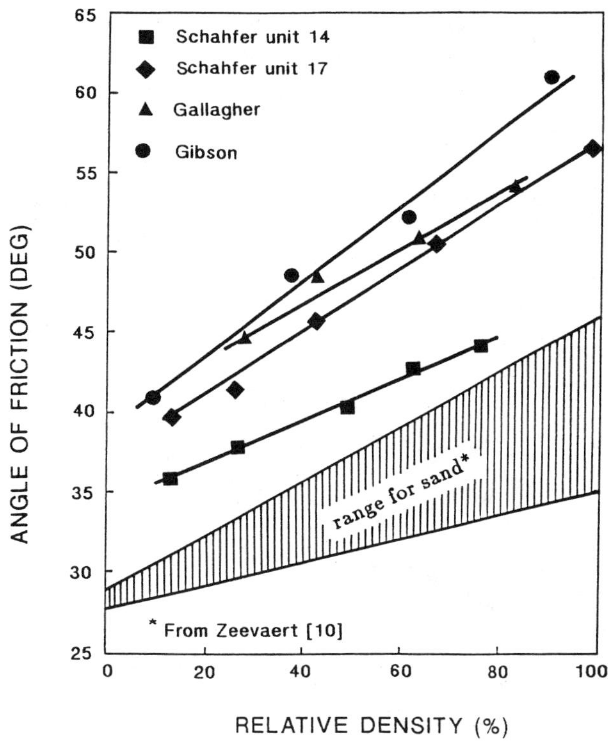

FIGURE 5 -- Angle of friction of bottom ash at varying relative densities

One-Dimensional Compression

One-dimensional compression tests were performed using a consolidometer. Samples were loaded incrementally in an electronically controlled hydraulic loading system. The consolidation ring had a diameter of 102 mm (4 in.) and the bottom ash samples were 38 mm (1.5 in.) high. Initially, the samples were subjected to a seating pressure of 5 kPa (ASTM D2435), and then soaked for at least 24 hours to measure swelling/collapse of the sample. Finally, the samples were maintained under maximum stress until no further deformation was observed. Due to the high permeability of bottom ash, the deformations took place in a very short time. No measurable creep was found. One bottom ash experienced a small swelling of 0.05% and the swelling pressure was later found to be only 30 kPa (4 psi).

The stress-strain relationships for several bottom ashes obtained from one-dimensional compression tests are presented graphically in Figure 6. In order to relate the compressibility of bottom ashes to more familiar soil materials, the stress-strain curve obtained for a uniform medium sand is also shown [11]. It can be seen that bottom ashes are slightly more compressible than the sand. This can be expected because an angular material is known to be more compressible than a well-rounded material [12].

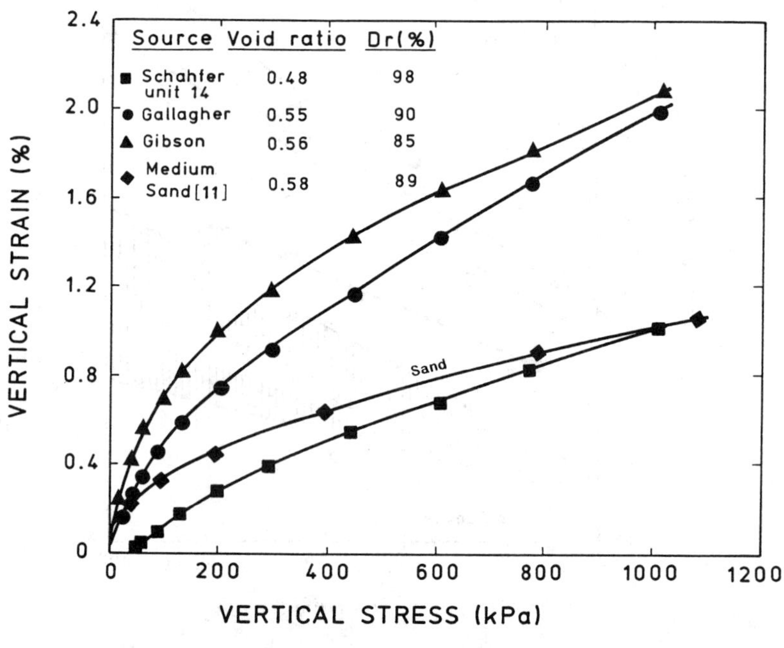

FIGURE 6 -- Stress-strain curves for bottom ashes

If bottom ash is used as a fill material the compression of the bottom ash layer is usually estimated by elastic theory. When vertical loads of large lateral extent are applied to the bottom ash layer, the compression behavior becomes one-dimensional and the parameter used in estimating settlement is the secant constrained modulus.

The secant constrained modulus is the rate of change of vertical stress with respect to the vertical strain under conditions of zero lateral strain, and can be expressed as [13]

$$D = \frac{\Delta\sigma_v}{\Delta\epsilon_v} = \frac{\sigma_{v2} - \sigma_{v1}}{\epsilon_{v2} - \epsilon_{v1}}$$

where D = secant constrained modulus,
ϵ_{v1} = vertical strain at a stress level of σ_{v1}, and
ϵ_{v2} = vertical strain at a stress level of σ_{v2}.

Figure 7 shows the secant constrained modulus calculated from zero stress to various stress levels. A comparison is made between the D values for bottom ash and those for a uniform medium sand [11], and the results are shown in Table 4. It is found that the moduli for one ash (Schahfer unit 14) are comparable to those of the well-graded sand. The values for the other two ashes are somewhat lower than those for sand, especially at the high stress level. The crushing of angular particles at high stress may play an important part in the phenomenon [13].

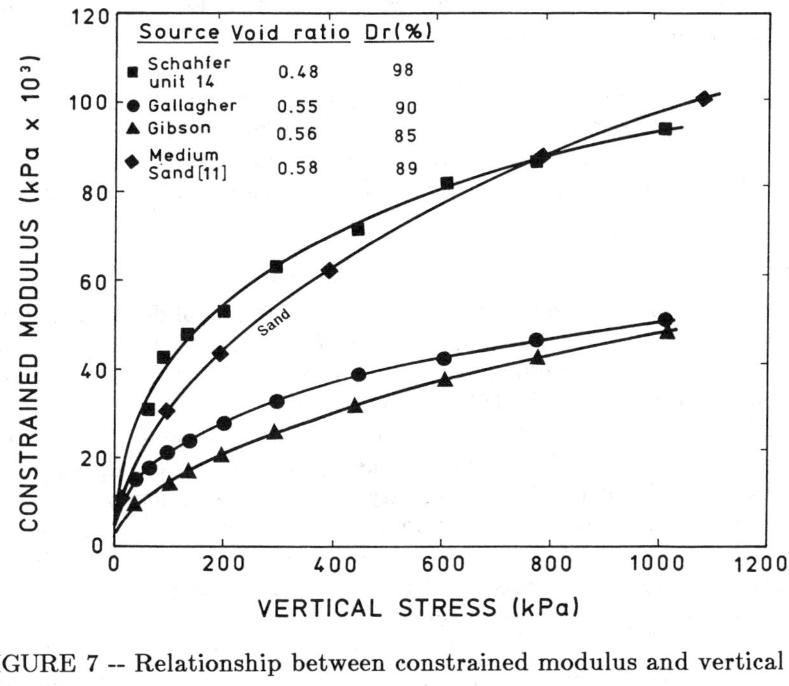

FIGURE 7 -- Relationship between constrained modulus and vertical strain

Environmental Effects

The environmental concerns regarding the use of ash center around possible leaching of heavy metals and soluble salts from ash-constructed embankments. The potential environmental effects are determined by the amount of heavy metals and salts leached from a fill. Small amounts of heavy

TABLE 4 -- Secant constrained modulus of bottom ashes

Materials	Relative density	Modulus (kPa x 10^3) σ_{v1} = 45 kPa, σ_{v2} = 98 kPa	Modulus (kPa x 10^3) σ_{v1} = 770 kPa, σ_{v2} = 1010 kPa
Schahfer unit 14	98	51.5	145.8
Gibson	85	26.5	82.8
Gallagher	90	30.5	73.1
Uniform medium sand[11] (0.1mm<d*<0.6mm)	89	53.9	178.4

* d = particle size

metals released to the environment may constitute a hazard both to environment and health. The high content of salts may adversely affect the quality of ground water, although it does not constitute any danger to human health. These salts are principally calcium and sulfate, but also chloride, sodium, potassium, and magnesium.

The chemical composition of ash is important to the leaching processes but it forms an insufficient basis for an estimate of the leachate composition (contaminants and concentrations). Consequently, the environmental effects resulting from ash fills must be evaluated based on direct analyses of the leachate properties.

The Environmental Protection Agency (EPA) designed an Extraction Procedure (EP) toxicity test to simulate the leaching a solid waste will undergo in a sanitary landfill [14]. In this test a representative sample of a solid waste is extracted with deionized water maintained at a pH of 5 using acetic acid. The maximum contaminant levels (MCL) specified for characterizing hazardous solid wastes are such that they are one hundred times the National Primary Drinking Water Standards. Table 5 summarizes the results from the analysis of bottom ash leachate generated by the EP toxicity test. The concentrations for bottom ash extract are far below the maximum contaminant levels specified by the EPA. Therefore, bottom ashes are characterized by the EP toxicity test as nonhazardous. Moreover, it seems that bottom ash extracts would also satisfy the primary drinking water standards.

The salt content of bottom ash leachate was tested by the leaching method test specified in the Indiana Administrative Code 329 IAC 2-9-3 [15]. The Indiana leaching method test is conducted as specified for the EP toxicity test, except with no addition of acetic acid. Table 6 summarizes the test results and the maximum concentrations specified for restricted waste site in the code, along with the Secondary Drinking Water Standards. Again, the salt concentrations of the bottom ash extracts meet all the requirements.

TABLE 5 -- Results of EP toxicity tests

Contaminant	Concentrations (mg/L) Schahfer unit 17	Gibson	Schahfer unit 14	Perry	EPA MCL*
Mercury	0.0002	0.0001	<0.0001	0.0002	0.2
Silver	0.001	<0.001	<0.001	<0.001	5.0
Cadmium	0.0008	0.025	0.0007	0.0004	1.0
Chromium	0.0009	0.0005	0.0012	0.0009	5.0
Arsenic	0.020	0.010	0.005	0.008	5.0
Selenium	0.005	0.005	0.003	0.004	1.0
Barium	0.098	0.103	0.136	0.108	100.0
Lead	0.007	0.002	<0.001	0.005	5.0

* MCL = maximum contaminant level

TABLE 6 -- Results of Indiana leaching method tests

Contaminant	Concentrations (mg/L) Schahfer unit 17	Gibson	Schahfer unit 14	Perry	Indiana MCL*	Secondary MCL*
Barium	0.098	0.103	0.136	0.108	1	1
Copper	<0.1	<0.1	<0.1	<0.1	0.25	1
Chlorides	<1	<1	<1	1	250	250
Iron	0.1	0.4	0.1	0.1	1.5	0.3
Sodium	0.8	1.0	<0.5	1.5	250	...
Sulfate	31	55	19	26	250	250
Total Dissolved Solids	90	140	10	145	500	...
Calcium	19	24	2	30	...	...
Magnesium	0.7	2.0	0.2	0.1	...	...
Potasium	1.0	0.7	0.1	2.0	...	...
Zinc	0.1	0.3	<0.1	<0.1	2.5	5

* MCL = maximum contaminant level

Due to the nature of the transport system, it was not possible to sample bottom ashes that had not been exposed to some degree of leaching, if the ash was wet-disposed. In this study, only Perry ash could be sampled directly from the hopper; and supposedly, it would have the highest portion of leachable metals and salts. However, no significant difference in the concentrations of leachate was observed between Perry and other ashes.

In order to prevent erosion, soil covering and plant growth on the ash-finished fills and slopes are essential. In addition, vegetation reduces leachate production effectively by increased evaporation [16]. Evaporative transpiration generated by vegetation is estimated to remove more than half of the yearly precipitation and thus limits leachate production significantly.

SUMMARY AND CONCLUSIONS

Laboratory studies on the chemical and physical properties of bottom ash, and potential environmental effects resulting from the use of bottom ash have been presented. Based on the laboratory studies, it can be concluded that the properties of power plant bottom ash compare favorably with those of traditional natural granular soils, and that the material can be successfully utilized as a fill material, e.g., backfills for retaining structures, highway embankments, and structural fills. Before this can be achieved, effective construction techniques need to be developed through trial uses in the field.

Based on the results from leaching tests performed in the laboratory, it is believed that the bottom ashes produced from coal-burning power plants are nonhazardous and that their effects on the quality of ground water are minimal.

It is obvious that utilization of such an extensively produced byproduct of the power industry as a construction material could become more desirable in the future. From a technical standpoint, if the conclusions from the present work hold true for other power plants, bottom ash shows very good promise for use as a fill material.

ACKNOWLEDGMENTS

The authors would like to thank the Indiana Department of Highways and the Federal Highway Administration for support of this work. The invaluable cooperation of the utility companies who provided information and samples is gratefully appreciated. The authors are especially indebted to Mrs. Janet Lovell for her technical advice and assistance in laboratory testing.

REFERENCES

[1] Steam, Its Generation and Use, The Babcock and Wilcox Company, New York, 1972.

[2] Valley Forge Laboratories, "Development of Federal Procurement Guidelines for Use of Power Plant Ash in Transportation Products," Devon, Pennsylvania, August 1983.

[3] Seals, R. K., Moulton, L. K., and Ruth, B. E., "Bottom Ash: An Engineering Material," Journal of the Soil Mechanics and Foundation Engineering Division, ASCE, Vol. 98, No. SM4, April, 1972, pp. 311-325.

[4] Diamond, S., "Selection and Use of Fly Ash for Highway Concrete," Report JHRP-85-8, Purdue University, West Lafayette, Indiana, 1985.

[5] Hough, B. K., Basic Soils Engineering, 2nd ed., The Ronald Press Company, New York, 1969.

[6] Anderson, D. A., Usmen, M., and Moulton, L. K., "Use of Power Plant Aggregate in Bituminous Construction," Transportation Research Record 595, 1976, pp. 18-24.

[7] Foster, C. R., "Field Problems: Compaction," in Foundation Engineering, G. A. Leonards, editor, McGraw-Hill, New York, 1962, pp. 1000-1024.

[8] Lee, P. Y., and Suedkamp, R. J., "Characteristics of Irregularly Shaped Compaction Curves of Soils," Highway Research Record 381, 1972, pp. 1-9.

[9] Moulton, L. K., Seals, R. K., and Anderson, D. A., "Utilization of Ash from Coal-Burning Power Plants in Highway Construction," Highway Research Record 430, pp. 26-39.

[10] Zeevaert, L., Foundation Engineering for Difficult Subsoil Conditions, 2nd ed., Van Nostrand Reinhold Company Inc., New York, 1982.

[11] Schultze, E., and Moussa, A., "Factors Affecting the Compressibility of Sand," Proceedings of the Fifth International Conference on Soil Mechanics and Foundation Engineering, Vol. 1, 1961, pp. 335-340.

[12] Roberts, J. E., and De Souza, J. M., "The Compressibility of Sands," Proceedings of the American Society for Testing and Materials, Vol. 58, 1958, pp. 1269-1277.

[13] Hendron, A. J., Jr., The Behavior of Sand in One-Dimensional Compression, Ph.D. thesis, University of Illinois, Urbana, 1963.

[14] Environmental Protection Agency, "Test Methods for Evaluating Solid Waste, Physical/Chemical Methods," 2nd ed., Report SW-846, Washington, D.C., 1985.

[15] Indiana Register, Volume 12, Number 5, February 1, 1989.

[16] Gidley, J. S., and Sack, W. A., "Environmental Aspects of Waste Utilization in Construction," Journal of Environmental Engineering, ASCE, Vol. 110, No. 6, Dec., 1984.

Arvid O. Landva and Jack I. Clark

GEOTECHNICS OF WASTE FILL

REFERENCE: Landva, A.O. and Clark, J.I. "Geotechnics of Waste Fill" Geotechnics of Waste Fill - Theory and Practice, ASTM STP 1070, Arvid Landva, G. David Knowles, Editors, American Society for Testing and Materials, Philadelphia, 1990.

ABSTRACT: Geotechnical investigations of waste fills are rarely undertaken, and consequently far too little is known by the geotechnical community about the engineering properties of fill, particularly waste fill such as refuse and woodwaste. In many communities waste fills are being used to create recreational areas to elevations significantly above the original terrain. Slope stability then becomes an important consideration. In addition many communities have expanded to encompass waste fill areas once thought to be beyond the limits of development. The engineering properties and long-term behaviour of these fill areas are therefore important to land use considerations.

In 1983 the authors initiated a programme of investigations of the geotechnical properties of waste materials. Particular emphasis was placed on refuse landfills and woodwastes such as barkfill (hogfuel), sludge and ash wastes. Novel equipment and methods had to be introduced both in the field and in the laboratory. The field investigations were carried out in waste fills across Canada. Much of this work had to be concerned with the development of suitable equipment and test procedures.

It is concluded that geotechnical investigations of these unusual and difficult materials are feasible, as long as it is recognized that conventional testing methods and analyses may not apply and that a different approach is required. No direct evidence was found that the shear strength of refuse changes with the degree of decomposition.

KEYWORDS: geotechnical investigations, refuse landfill, waste fill, woodwastes, unconventional equipment and testing methods, placing methods, settlement and stability

Dr. Landva is a Professor of Civil (Geotechnical) Engineering at the University of New Brunswick, Department of Civil Engineering, P.O.Box 4400, Fredericton, New Brunswick, Canada, E3B 5A3. Dr. Clark is the Director of C-CORE, Memorial University of Newfoundland, St.John's, Newfoundland, Canada, A1B 3X5.

INTRODUCTION

Waste fill is a term used here to denote man-made fill consisting of (i) any type of waste material or (ii) mineral fill contaminated with waste materials.

The geotechnical engineer, when called upon to give advice on foundations in waste fill or on the stability of waste fill areas, faces two major problems: firstly, he will be dealing with an unusual soil, one that he is probably not well equipped to deal with, and secondly, he will have to cope with a degradable material whose geotechnical properties may change with time. Geotechnical deterioration as a result of decomposition is a topic that has so far received little attention in the geotechnical world. One reason for this may be that we generally have tended to avoid sites underlain by waste fills, or if they could not be avoided, we may have been able to base our design on a relatively short term only. Expanding communities now frequently encompass waste fill areas once thought to be beyond the limits of development. As a result we have to build more frequently in disposal areas and over wasteland, and we sometimes have to operate with a design life of several generations. Moreover, new waste fill areas are being designed and constructed in more creative ways, the end use of the land being designated at the outset. For example, several waste fill areas in Canada are designated for recreational use, including ski hills. By constructing grades well above the original ground level, a much greater volume of waste can be handled. It is important therefore to be able to quantify the geotechnical properties of the waste and of the often poor foundation soils underneath, and it is necessary to do this for both the short-term and the long-term conditions.

Quantification of geotechnical properties of waste materials can obviously be very difficult, particularly in the case of heterogeneous materials such as urban refuse and contaminated fill. An attempt has been made by the authors to develop testing and sampling methods for these materials that can yield useful geotechnical parameters. A description of these methods is given in the present paper, and some typical results from field and laboratory testing are presented.

FIELD INVESTIGATIONS

Field Reconnaissance

A study of old fills should always be commenced with an examination of maps, plans and aerial photographs. The authors have found that it is not advisable to rely too heavily on estimates based on local memory. For example, estimates of fill thickness or depth are usually too high, sometimes by factors of two or three.

It is often possible to obtain survey records of the fill area for the period before the fill was started, and this will generally yield the necessary information on the thickness and extent of the fill.

Information on the placing of the fill may or may not be on record. Again, it is the authors' experience that descriptions based on memory can lead to conflicting conclusions.

Drilling and Sampling

The type of drilling and sampling used in waste fill depends entirely on the type of fill being investigated. Conventional standard penetration tests are possible in fine-grained fill but cannot be used at all in wastes such as domestic and industrial refuse or in barkfill and other woodwastes.

Sampling was attempted [1] with a "Becker" type of drill in recent fill (active domestic refuse) as well as in old domestic refuse. Practically no material could be sampled, mostly because the drive shoe constantly became clogged with wooden debris or other large particles. Also, progress was quite slow, and drilling came to a complete stop whenever large objects were encountered, which was quite frequent. In such cases drilling could be continued only by relocating the drill hole.

Attempts were also made to use split spoon samplers of various diameters in refuse (domestic) landfill, but again the results were negative. Very little material was recovered in the sampler, and the blow counts were extremely erratic. This was also the case with dynamic cone penetration in a recent domestic refuse landfill [2] .

Auger drilling proved to be the most suitable method for probing and sampling waste fill, both old fill of all kinds and recent domestic refuse. Continuous augers of diameters varying between 100 and 230mm were tried out. The best overall auger with respect to production rate and quality and size of sample was a solid-stem 130mm auger (140mm bit). A reasonably heavy drill rig was generally required in order to penetrate or displace very resistant materials such as wood, tires, rocks, concrete blocks, steel objects etc. In order to obtain samples representative of a certain depth, it was necessary to withdraw the auger completely every 1.5 or 3m depth interval. Sampling was carried out to depths of 30m in both old and active refuse landfills and to 20m in a seafill that had been placed about 30 years ago.

Below the ground water level the augered borehole is likely to collapse if the fill contains much sand or gravel. In typical refuse, however, the auger may simply be advanced to the previous sampling depth whether the borehole collapsed or not, and further auger advance secures a sample of the underlying fill.

An experienced driller is required for auger sampling in waste fill.

Test Pits

Test pits are useful for most fills, although they are generally limited to the upper 4 m depth or so. In general, it is not feasible to excavate below the ground water level.

Sampling and classification: Sampling from test pits in waste fill should be accompanied by classification in situ. The description should be supplemented by colour photographs.

Analyses of numerous types of waste and a comprehensive review of the literature led the Bureau of Solid Waste Management to select the following categories for purposes of classification: (i) food waste, (ii) garden waste, (iii) paper products, (iv) plastic, rubber, leather, (v) textiles, (vi) wood, (vii) metal products, (viii) glass and ceramic products, and (ix) ash, rocks, dirt.

Observing that some waste constituents are readily biodegradable, some are slowly biodegradable, and some are not degradable or very slowly degradable, the following broad classes are suggested for use in engineering applications:

(O) Organic
(OP) Putrescible (monomers and low-resistance polymers, readily biodegradable):
Food waste
Garden waste
Animal waste
Material contaminated by such wastes

(ON) Non-putrescible (highly resistant polymers, slowly biodegradable):
Paper
Wood
Textiles
Leather
Plastic, rubber
Paint, oil, grease, chemicals, organic sludge

(I) Inorganic

(ID) Degradable
Metals (corrodible to varying degrees)

(IN) Non-degradable
Glass, ceramics
Mineral soil, rubble
Tailings, slimes
Ash
Concrete, masonry (construction debris):

The last three groups (ON, ID, IN) may contain numerous void-forming constituents that will affect the geotechnical behaviour of the fill:

Hollow containers - boxes, crates, cans, bottles, jars, drums, barrels, pipe, tubing, etc.
Platy or elongated items - beams, sheets, plates, etc.
Bulky items - furniture, appliances, auto bodies, etc.

Visual examination alone is generally not sufficient for geotechnical classification purposes. The examination should be supplemented with index properties such as water content, organic content, and specific gravity. Another useful classification test is the particle size analysis.

Unit weight in situ: The unit weight of waste is difficult to obtain because of the erratic and often coarse nature of waste materials. A very large sample of material is required. A test pit excavated with a small backhoe is generally suitable and may even be on the small side. Typically, the volume of such a pit is around 10 m^3, and this is generally sufficient to include some of the larger constituents found in refuse, such as tires, beams, planks, drums, construction debris, etc.

The volume measurements of test pits in the waste fills investigated by the authors were carried out with a sliding stick attached to a survey rod. This arrangement served the purpose of measuring the width and the length at different depths and sections (typically at about 50 points) without entering the pit. The generally jagged nature of the walls and the often very considerable variations in width (say between 0.6 and 1,2 m) do, however, render such measurements somewhat imprecise.

The weights of the excavated material and the truck tares are usually obtained at the site by the fill operators, or the trucks are taken to weigh scales somewhere else. We have found, however, that the possibility of weighing errors is fairly high. It seems likely that some confusion may be arising from the interruption of the normal busy work schedule and perhaps also from over-emphasizing the necessity of obtaining accurate weights. Another contribution to inaccurate weight could be a lack of calibration of the scale or simply an inaccurate scale.

On the basis of relatively extensive field experience, the authors would estimate that the maximum volume error is ±15%.

The unit weights of waste as measured by the authors in various test pits in refuse landfills across Canada have been reported in detail in [1]. The refuse in all cases appeared to contain no free water between the various constituents. Under these conditions the unit weights could be expected to be within in a range of about 7 to 14 kN/m^3, as discussed under "Index Properties" below. The unit weights measured in situ varied within a range of 6.8 to 16.2 kN/m^3. Any value greater than 14 kN/m^3 is almost certainly a result of weighing or measuring errors. For example, a volume error of -15% (i.e. the measured volume is 15% smaller than the actual volume) would reduce the maximum measured unit weight from 16.2 to 14.1 kN/m^3. A weighing error of +10% (i.e. the recorded weight is 10% too high) would further reduce the maximum value to 12.8 kN/m^3. The authors have found that such an error in weighing, although generally not likely, could conceivably occur.

The determination of unit weights on the basis of test pits is often complicated by the presence of one or several "dirt" (earth cover) layers. It is generally not possible to predict with any certainty the configuration of the refuse and dirt layers below the surface, because slopes and flats covered with various combinations of dirt, old access roads and construction debris may exist almost anywhere, and as a rule it is not possible to obtain any exact information on the configuration.

Since cover soils exist almost everywhere, either at the surface or at depth, or both, their unit weight should be measured separately. This is most conveniently done with a nuclear density probe. Fig. 1 shows the results of some measurements of the unit weight of earth covers used in Calgary, Edmonton and Mississauga. The water contents

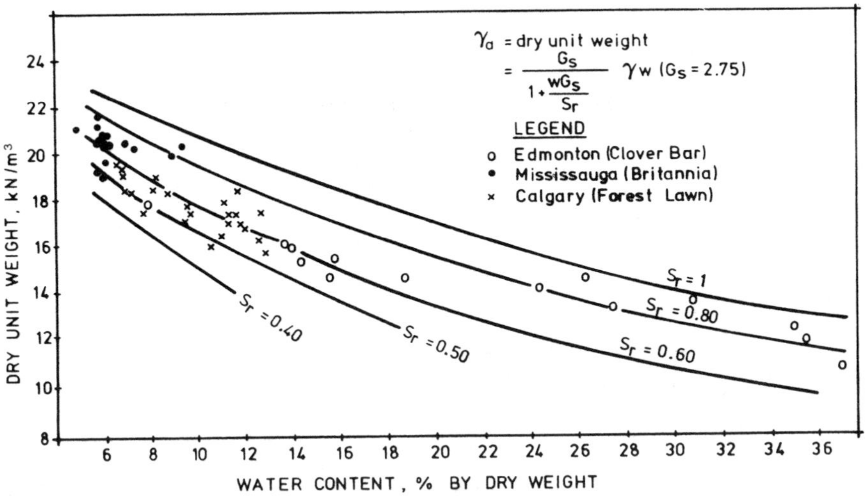

FIG. 1 -- In-service water content and dry unit weight of various earth covers as determined in situ

shown are those measured with the probe, i.e. they are not the as-placed water contents.

The large size of sample obtained from test pits may present a problem with respect to determining water contents and organic contents. However, relatively large samples can be handled in a pottery type of kiln, as long as provisions are made for proper ventilation. The 90cm diameter kiln used at UNB has a temperature control which is sufficiently accurate for both water content (90-100°C) and organic content (450-500°C).

Permeability (hydraulic conductivity): The field permeability of some of the waste fills investigated by the authors was estimated by carrying out large-scale percolation tests in the pits excavated for unit weight measurements. Water supplied by a vacuum truck was dumped into the pit. The permeability was estimated on the basis of the rate of water level recession and on the basis of flow nets applicable to any particular level considered [3].

Permeabilities measured in test pits in Calgary, Edmonton, Mississauga and Waterloo are plotted against unit weights in Fig. 2. The values shown are based on an intermediate stage of water level recession, after the flow had stabilized and before any debris would clog the voids. The measured coefficients of permeability (1x10-5 to 4x10-4 m/s) correspond to those associated with clean sand and gravel.

Plate load tests: Another geotechnical test conducted on the various refuse landfills across Canada was the plate load test. The plate used for these tests had a diameter of 1.13 m (area = 1 m^2). Construction equipment was used as counterweight. Typically, a

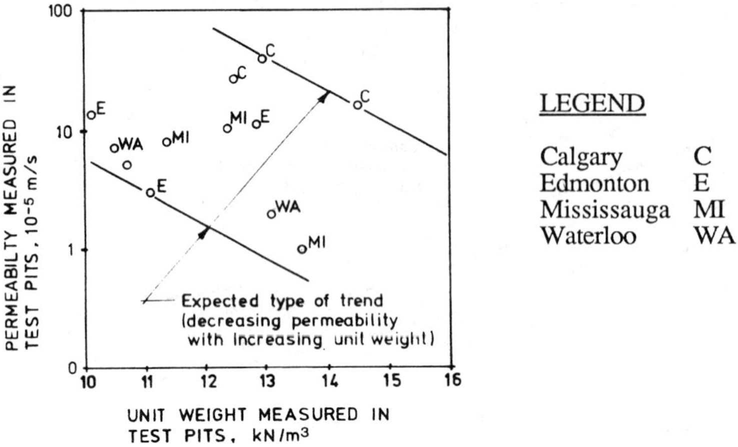

FIG. 2 -- Unit weight and permeability (from percolation) as measured in test pits in situ

Caterpillar D8 dozer or equivalent could supply a reaction of about 20 tonnes (about 200 kPa), while a backhoe or drilling rig would not exceed 2 or 3 tonnes (about 20 or 30 kPa).

Ideally, a load test on waste and other materials exhibiting significant long-term (creep) settlement should last for at least several hours and preferably longer. This is of course not possible when construction equipment is involved, and the test duration therefore had to be limited to 10-15 minutes. These tests consequently do not yield any information on long-term behaviour. However, the plate load tests do reflect in a meaningful way the density or degree of compaction of the fill. The large size of the plate generates a correspondingly large influence zone, say about twice the diameter or about 2 m, which is typical of the thickness of refuse lifts between earth covers. The 1.13 m diameter plate is therefore a reasonable compression test for an entire lift of fill.

The results of plate load tests of various refuse landfills in Canada have been reported in detail in [1]. The values of K_s, which were determined as the ratio of an applied pressure (150 kPa) to the settlement at that pressure after two to five minutes of load application, varied between 1 and 15 MPa/m. Typically, the lower values of K_s (less than 2 MPa/m) are associated with cases of poor compaction, while the higher values (5 MPa/m and higher) correspond to (i) a particularly good compaction or (ii) to a better grade of fill or (iii) to thicker, well compacted earth covers.

In Fig. 3 an attempt has been made to relate the subgrade modulus values to the unit weight of the material. The unit weights were obtained from test pits excavated close to or at the location of the plate load tests. Fig. 3 indicates that there is a tendency for the subgrade modulus to increase with increasing unit weight, although

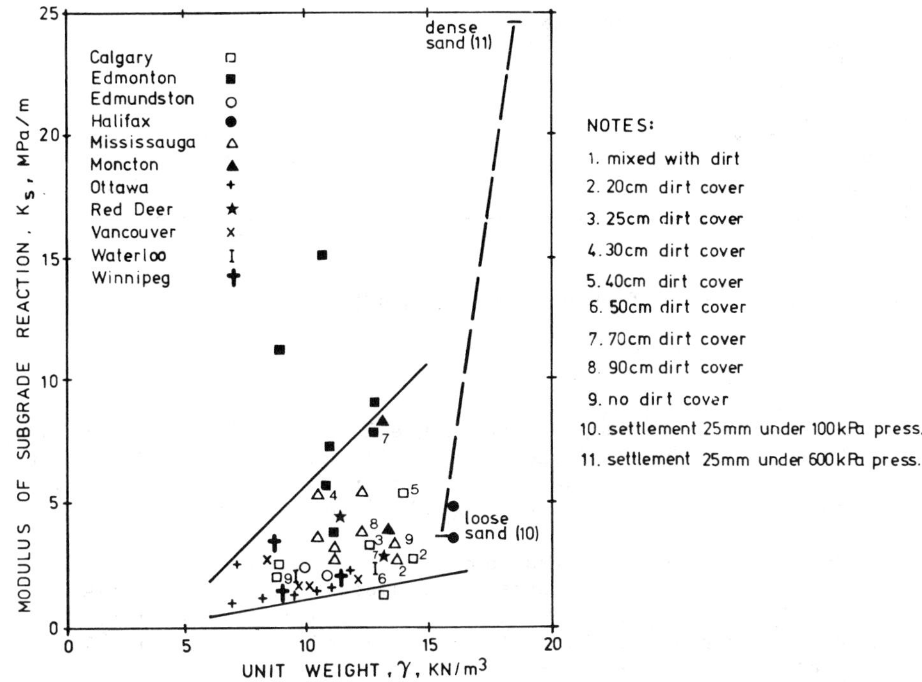

FIG. 3 -- Modulus of subgrade reaction (144 kPa/settlement, 1.13m dia. plate) and unit weight measured in adjacent test pits in various refuse fills. Sand results from conventional relationships of allowable pressure for 25mm settlement (ref. 4)

the range of K_s appears to be rather large. For example, at the typical unit weight of 10 kN/m^3, $K_s \approx 1$ to 5 MPa/m, i.e. the calculated settlement under say 100 kPa applied pressure would range from 2 to 10 cm.

If a comparison is made with sand, it is found, somewhat surprisingly, that the range of K_s-values for loose to dense sand appears to be considerably greater, as calculated on the basis of allowable bearing pressures under footings on sand [4].

In general, since the composition of waste fill and the thickness and type of earth cover tend to be so erratic, it is desirable, after the load test, to excavate a test pit exactly at the location of the test.

Test Fills

The very heterogeneous nature of waste fills and the very wide range of densities to which they have been placed give rise to a wide range of compressibility. Some idea of the magnitude of this range is given by Fig. 4, in which various case records of observed settlements of waste fills have been plotted. A study of these records has not yielded any consistent relationships between the type of material, the magnitude of the loads, and the magnitude and rate

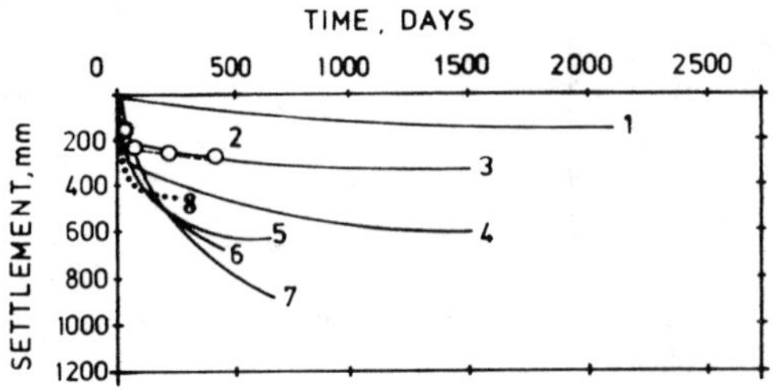

NOTES
1 = ref.6
2 = ref.7
3 = ref.8 (min.)
4 = ref.8 (max.)
5 = ref.9 (eastbound)
6 = ref.10
7 = ref.9 (westbound)
8 = refs.2 and 11

FIG. 4 -- Case records of settlement in refuse landfills

of settlement. For example, the refuse landfill of curve 6 was only 3 m thick, while the Fredericton landfill (curve 8), consisting of the same type of refuse, was about 17 m thick. The applied load was in both cases about 50 kPa, yet the shallower fill settled more.

The Fredericton test fill was about 2.5 m high and had base dimensions of 19 x 16m and crest dimensions of 7.5 x 5m. This corresponds to a significant depth in the order of 20 m, i.e. the entire thickness of the landfill was affected by the applied load. The quantity of fill used was only about 400 m^3. The cost of such test fills may be relatively low if, as in the Fredericton case, the fill used is material hauled in and stockpiled for earth cover.

The authors would strongly recommend that test fills be used for an assessment of the compressibility of waste fill. Such large-scale tests are quite reliable, and their cost can be relatively low. Also, if no further treatment of the waste is planned, the test fill serves the purpose of preloading the waste and thus reduces the settlement of the planned structure. For this reason, and also for the purpose of obtaining reliable parameters for the long-term behaviour of the waste material, the test fill should be planned and placed well ahead of construction, preferably as much as a year ahead.

LABORATORY INVESTIGATIONS

Conventional geotechnical laboratory tests are generally not applicable for waste fill, chiefly because of the coarse and erratic nature of such materials. It is necessary therefore to introduce new and larger testing apparatus that is more suitable for the generally coarse and variable waste materials.

Index Properties

Waste fill generally consists of many different types of constituent, and these constituents are often porous and not fully saturated. The determination of index properties then becomes a formidable task requiring special equipment and procedures. Also, since these materials are often very compressible, properties such as unit weight and water-holding capacity (and also permeability) must be determined as a function of porosity, which again is a function of the method of placing and of the applied pressure.

The determination of the index properties of samples of refuse fill is a long and somewhat complex procedure, but it is possible to determine reliably the following properties: water content, organic content, grain size distribution, average specific gravity for the cases of (i) saturated intraparticle voids and (ii) dry intraparticle voids, and waterholding capacity at different densities.

Since waste fill contains porous constituents, it is necessary to distinguish between intraparticle (within, inside of) and interparticle (between particles) voids. It is entirely possible, for example, and indeed not uncommon, to have saturated or partly saturated intraparticle voids and dry interparticle voids.

An instructive example is an aggregate of metal cans. If made of sheet steel, the unit weight of the solid material would be about 80 kN/m^3, while the unit weight of each empty can could be as low as 2 kN/m^3 (for a porosity of 97.5%) and that of a water-filled can would be about 12 kN/m^3. With an interparticle (i.e. inter-can) porosity of around 40%, the unit weights of the aggregate would be approximately as follows:

Intraparticle voids	Interparticle voids	Unit weight, kN/m^3
Dry	Dry	1
Semi-saturated	Dry	4
Semi-saturated	Semi-saturated	6
Saturated	Dry	7
Saturated	Saturated	11

The average unit weight of the individual constituents of the waste depends on the unit weight of the solid portion of each constituent and on its porosity and degree of saturation. In general, the average unit weight of the constituents is

$$\gamma_c = \frac{1}{\sum_1^n \frac{W_i}{W_c} \times \frac{1}{\gamma_i}} \qquad (1)$$

where W_i/W_c = weight of constituent i as a fraction of the total weight W_c of the constituents,

γ_i = unit weight of constituent i, and

n = number of constituents.

On exposure to water, the unit weight of any constituent absorbing water would increase (e.g. that of food waste, garden refuse, paper, textiles, wood, ash, etc.). The new average unit weight $\gamma'c$ of the constituents could be recalculated from (1), or it could be expressed as

$$\gamma'_c = \gamma_c \left[1 + \sum_1^n \frac{W_i}{W_c} \times \frac{\Delta\gamma_i}{\gamma_i} \right] \qquad (2)$$

TABLE 1. Typical refuse composition (from ref.1)

Category	Percent of total weight	Unit weight, kN/m^3 [1] Dry	Saturated
Food waste	5-42	1.0[2]	1.0[2]
Garden refuse	4-20	0.3[3]	0.6
Paper products	20-55	0.4[4]	1.2[5]
Plastic, rubber	2-15	1.1	1.1
Textiles	0-4	0.3	0.6
Wood	0.4-15	0.45	1.0
Metal products	6-15	6.0	6.0
Glass & ceramics	2-15	2.9	2.9
Ash, rock & dirt	0-15	1.8	2.0

NOTES:

1 With respect to intraparticle voids
2 Assumed value
3 Assume $n = 0.8$ and $G_s = 1.5$
4 Average density of paper is 0.6 g/cm^3 and of cardboard 0.2 g/cm^3
5 $G_s = 1.55$, $n = 0.6$

where W_i/W_c, γ_i and γ_c are the same as above, and $\Delta\gamma_i$ = increase in unit weight of constituent i.

A typical composition of a refuse landfill is shown in Table 1, together with typical unit weights of the constituents in the dry and saturated conditions. These unit weights have either been obtained from handbooks and other literature or they have been measured in the laboratory. The possible range of average unit weights of the constituents can be determined by combining (i) the lightest materials and their dry unit weights and (ii) the heaviest materials and their saturated unit weights, as shown in Table 2. The lightest combination yields an average unit weight of the constituents of 3.8 kN/m^3, and the heaviest combination gives 16.3 kN/m^3. It is again emphasized that these are the average unit weights of the constituent particles and not the unit weight of the waste aggregate. The latter could be calculated only if the interparticle porosity at which the constituents co-exist is known and also if the water content is known (for waste located above the ground water level, the water content will generally be a function of the water-holding capacity).

The refuse landfills investigated by the authors always appeared practically dry with respect to the interparticle voids. Assuming a small amount of moisture ($w = 5\%$) and a range of interparticle porosity of 30 to 60% (depending on the amount of compaction), the lighest combination shown in Table 2 would yield an overall unit weight of 1.6 to 2.8 kN/m^3 and the heaviest would yield 6.8 to 12.0 kN/m^3. The range of overall unit weights measured in situ was 6.8 to 16.2 kN/m^3. The latter value is of course not possible, considering that the highest conceivable average unit weight of

TABLE 2. Lightest and heaviest combinations of refuse constituents (from Table 1)

Category	Lightest combination		Heaviest combination	
	% of total weight	Dry unit weight (kN/m³)	% of total weight	Saturated unit weight (kN/m³)
Food waste	6	1.0	-	-
Garden refuse	20	0.3	-	-
Paper products	55	0.4	40	1.2
Plastic, rubber	-	-	15	1.1
Textiles	4	0.3	-	-
Wood	15	0.45	-	-
Metal products	-	-	15	6.0
Glass & ceramics	-	-	15	2.9
Ash, rock, dirt	-	-	15	2.7

constituent particles is 16.3 kN/m^3. The highest conceivable overall unit weight for, say, a water content of 10% and an interparticle porosity of 20% is 14.3kN/m^3.

All the values of water content and organic content determined on the various refuse samples from across Canada are compiled in Fig. 5. In general, the water content does seem to increase with increasing organic content, as is usually the case.

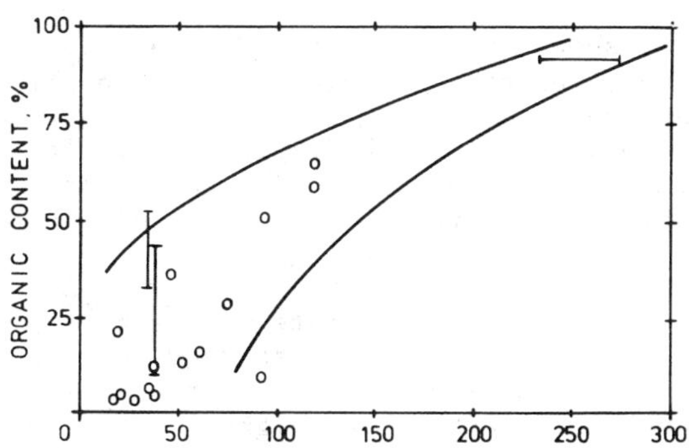

FIG. 5 -- Organic contents and water contents of samples from old fills across Canada

The apparatus shown in Figs. 6(a) and (b) is a combination unit used for specific gravity, saturation, permeability, water-holding capacity, and compression testing with pore pressure measurements. The larger container in Fig. 6(b) is used for saturation of especially coarse samples.

Consolidation and Permeability Tests

Consolidation tests on coarse waste materials may be carried out in the 470mm diameter units shown in Figs. 6(c) and (d) and in the 250mm diameter apparatus in Fig. 6(a). The 500 kN apparatus in (d) has provisions for measuring permeability at constant or falling head up to about 3 m heads. The load plate may be sealed at any vertical compression for measurement of the permeability, or the measurements may be done through overflow without a seal, but still under vertical load. The 1000 kN apparatus in Fig. 6(c) is used for consolidation tests for which measurements of wall friction are desired.

The high compressibility of waste fill is evident from Fig. 7, where consolidation results from five locations are plotted. These tests were all done in the 470mm diameter apparatus. The samples were placed in the container in about 5cm lifts and lightly compacted. The gradient of the log pressure vs strain is the compression number $C'_c = C_c/(1+eo)$, where C_c = compression index (i.e. the gradient of the e-log p curve) and eo = void ratio before the load was increased. The range of C'_c in Fig.7 is 0.2 to 0.5. This is high in comparison with inorganic soils, and this is to be expected.

The coefficient of secondary consolidation c_α (the gradient of the compression vs. log time relationship) was found to be in the range 0.2 to 3.0 percent per log cycle of time, depending on the type of waste involved. Too few tests have been carried out for any firm relationship to be established between the value of c_α and the type of waste, but it does appear that c_α increases with increasing organic content.

Shear Tests

The shear test results shown in Figs. 8 and 9 were obtained with the large direct-shear apparatus shown in Fig. 6(e). The horizontal dimensions of this apparatus are 434 x 287 mm. The vertical load and the shear load are both applied with hydraulic rams. In cases of wet samples, a rubber membrane is installed prior to sample preparation. The membranes are fabricated in the laboratory with a latex product [1].

The shear rate used was about 1.5 mm/minute unless pore pressure measurements indicated that a lower rate was required to maintain drained conditions. This was generally not a problem with the relatively coarse samples tested, but at higher pressures the permeability could decrease sufficiently that the rate had to be lowered.

An examination of the waste materials in both the natural and the dried conditions showed that they possessed a granular and fibrous nature. In the large direct shear box, therefore, it could be

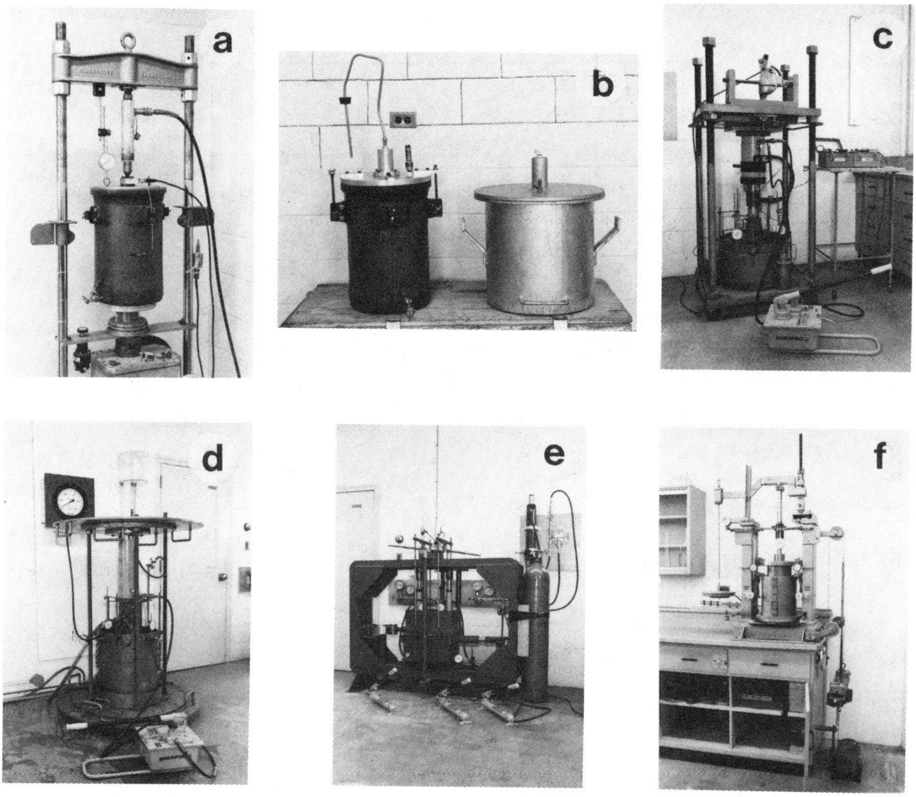

a = 250mm diameter multitester for specific gravity, water-holding capacity, compressibility, and permeability
b = (l) 250mm diameter multitester with cover for application of suction
(r) 450mm diameter container for application of suction to extra large or coarse samples
c = 470mm diameter 1000 kN consolidometer with provisions for wall friction measurements
d = 470mm diamter 500 kN consolidometer and permeameter
e = 434x287mm direct shear apparatus
f = 240x124mm ring-shear apparatus

FIG. 6 -- Laboratory equipment used for testing waste materials

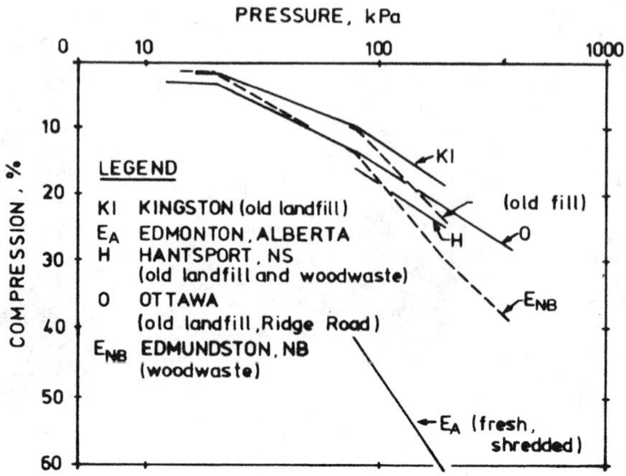

KI	$C_c' = 0.17$ (p = 20-200 kPa)		O	$C_c' = 0.21$ (p = 100-400 kPa)
E_A	$C_c' = 0.35$ (p = 80-200 kPa)		E_{NB}	$C_c' = 0.36$ (p = 100-400 kPa)
H	$C_c' = 0.22$ (p = 80-200 kPa)			

FIG. 7 -- Compressive strain vs log pressure for various fills in Canada (laboratory tests in 450mm dia. consolidometer)

expected that friction parameters similar to granular soils would be obtained. As seen from Figs. 8 and 9, this was indeed the case, the friction angle ϕ varying between 24°and 41°. These materials also had a cohesion parameter c of between zero and 23 kPa.

The fresh shredded refuse from Edmonton consisted of a large amount of plastic sheet waste, and this is no doubt the reason for the low friction angle of 24°. The direction of shear is at right angles to the direction of the principal consolidation stress, and fibrous and elongated particles have been found to tend to align themselves in this direction. Separate direct shear tests on plastic bags stacked horizontally and allowed to slide along the shear plane gave a friction angle of 9°, as shown in Fig. 9.

The shear strength parameters obtained for woodwaste (Fig. 9), viz. c=0 and ϕ=36°, were obtained independently also in ring shear tests in the apparatus shown in Fig. 6(f). The dough-nut-shaped specimen had an outside diameter of 240 mm and an inside diameter of 124 mm.

It is tempting to draw the conclusion from Fig. 8 that aging of refuse reduces its shear strength, in this case from c=19kPa and ϕ=39° to c=16kPa and ϕ=33° for the old Blackfoot (Calgary) refuse. However, the test results shown in Fig. 9 clearly indicate that a large range of shear strength parameters (ϕ=27° to 41°) is possible in waste fill. On the basis of the results shown in Figs. 8 and 9, it must be concluded that the shear strength of waste fill is highly variable, depending on the type of material involved. It should be noted that the same range of shear strength parameters, or even a larger range, is found for conventional soils of different densities

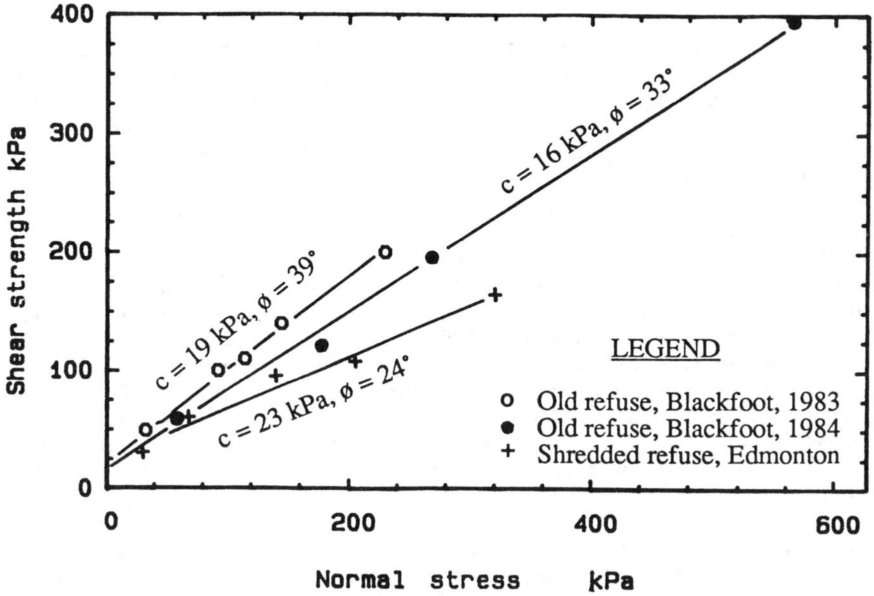

FIG. 8 -- Large direct-shear tests on samples from old fill in Calgary and from fresh shredded fill in Edmonton

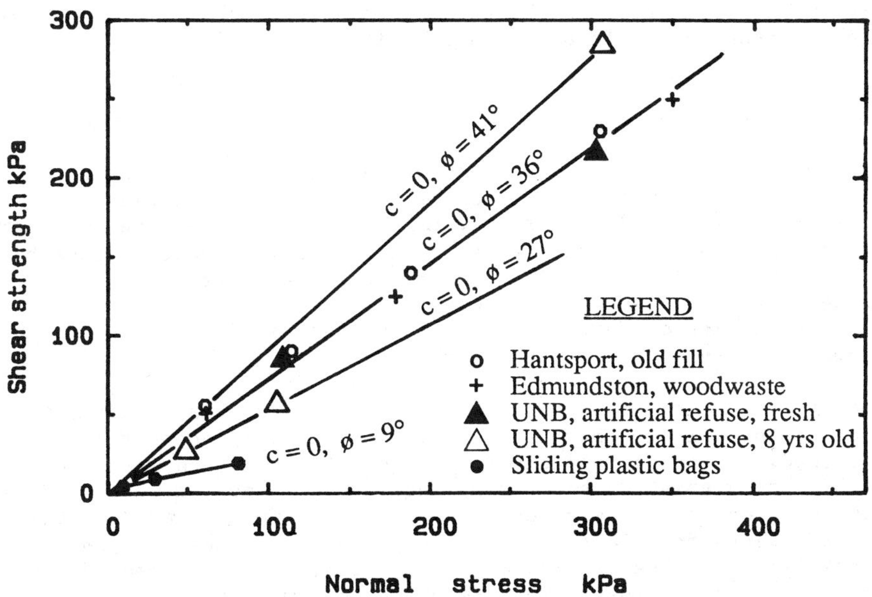

FIG. 9 -- Large direct-shear tests on samples from old woodwaste/refuse fill in Hantsport, N.S., Fraser woodwaste stockpile in Edmundston, N.B., artificial UNB samples, and sliding garbage bags

and mineral compositions.

The concept of a linear relationship between shear strength and normal stress, as applied to conventional soils, may not apply in the case of highly compressible and erratic refuse fill. For example, as the refuse is compressed by higher normal stresses, it becomes denser, and this could lead to higher shear strength parameters if increased interlocking occurs.

If an assessment were to be made of the stability of a landfill on the basis of the results shown in Figs. 8 and 9, the lowest strength would have to be used for the stability today. For the long-term stability, any change in the shear parameters would depend on the nature of the waste fill concerned. For sanitary landfill type of refuse the authors are aware of no direct evidence that the shear strength parameters change significantly with time. The overall shear strength could of course decrease if there is substantial local decomposition which leaves weak zones or cavities, but this type of deterioration could not easily be detected through laboratory shear tests.

Another aspect of landfill stability is the possible existence of local zones of weaker material within the landfill. Large collections of plastic sheet waste, for example, if placed in a potentially critical shear zone, could contribute to an overall low factor of safety. The tests described above for Edmonton suggest that the friction angle could be in the range 9° - 24° in such a zone. Other sources of low strength include that due to seepage along the laminated composite structure made up of horizontal or sloping layers of refuse and less pervious daily covers. Such seepage has also caused internal erosion which eventually led to slope instability.

CONCLUDING REMARKS

The field and laboratory work with waste fill as described herein has shown that it is in general feasible to conduct geotechnical investigations of fill materials, regardless of its nature and degree of decomposition. Undisturbed sampling is not, of course, possible, and field determinations of unit weight and strength are limited to the upper zone. Resort may, however, be had to the use of test fills (e.g. stockpiles of earth cover) for an assessment of the compressibility of the entire depth of fill.

Laboratory testing of waste materials requires special equipment, as described briefly in this paper. All samples are necessarily disturbed, in fact completely destructured as compared with the state in which they existed before sampling. This is not, however, a major drawback, because the samples, upon replacing in the large containers used for waste fill in the laboratory, can be recompacted and reconsolidated. The density of the samples, after reconsolidation to the field value at the depth from which they were taken, is not likely to be significantly different from the field value.

The laboratory investigations reported in this paper were carried out on samples reconsolidated to pressures between 20 and 400 kPa. This range of pressure corresponds to that between depths of 2 and 40 m in a typical refuse landfill.

The waste materials investigated were extremely variable and erratic. For example, their organic content varied within the wide range of 5 to 92 percent. Yet their compression and shear behaviour was not significantly more erratic and variable than that of inorganic soils. The reason for this somewhat surprising conclusion seems clear: waste materials are particulate materials just like inorganic soils, and the same general laws of compression and shear behaviour can therefore be expected to apply, regardless of the nature of the constituents.

The geotechnical behaviour of waste fill can be compared to that of a loose soil containing a variety of fibres. In general, the shear strength may be expected to be at its minimum on planes of failure parallel to the alignment of the fibres. In refuse and other waste landfills, this alignment tends to be parallel to the compacted layers, that is, generally horizontal or sloping up to about 10° from the horizontal.

REFERENCES

1. LANDVA, A.O. and CLARK, J.I. 1986. Geotechnical testing of waste fill. Proc. 39th Canadian Geotechnical Conference, Ottawa, Ontario, pp. 371-385.
2. LANDVA, J. 1982. Investigation of the geotechnical properties of sanitary landfills. Senior Report, Department of Civil Engineering, University of New Brunswick.
3. HEALY, K.A. and LAAK, R. 1973. Factors affecting the percolation test. Journal WPCF, Vol.45, No.7, July, pp.1508-1516.
4. TERZAGHI, K. and PECK, R.B. 1967. Soil mechanics in engineering practice. Wiley, 729 pp.
5. LANDVA, A.O. and LA ROCHELLE, P. 1982. Compressibility and shear characteristics of Radforth peats. ASTM STP 820. Testing of Peats and Organic Soils, P. Jarrett, ed.
6. SOWERS, G.F. 1968. Foundation problems in sanitary landfills. ASCE Journ. Sanitary Eng., 94, SA1, February, pp. 103-182.
7. CHANG, J.C. and HANNON, J.B. 1976. Settlement performance of two test highway embankments on sanitary landfill. New Horizons in Construction Materials, Vol. 1, Int. Symposium, November 1-3, Lehigh University, Envo Publishing Co., Pennsylvania, pp.139-157.
8. KEENE, P. 1977. Sanitary landfill treatment, Interstate Highway 84. Proc. Conf. Geotechnical Practice for Disposal of Solid Waste Materials. Univ.of Michigan, Ann Arbor, June 13-15, ASCE, New York, pp.632-644.
9. SHEURS, R.E. and KHERA, R.P. 1980. Stabilization of a sanitary landfill to support a highway. Transp. Research Board, Transp. Research Record No. 754, pp. 46-53.
10. RAO, S.K., MOULTON, L.K. and SEALS, R.K. 1977. Settlement of refuse landfills. Proc. Conf. Geotechnical Practice for Disposal of Solid Waste Materials. Univ.of Michigan, Ann Arbor, June 13-15, ASCE, New York, pp.574-598.
11. KWONG, A. 1982. Geotechnical investigation of sanitary landfills. M.Sc.Eng. thesis, Department of Civil Engineering, University of New Brunswick.

Stabilization, Compaction, and Consolidation

J.-L. Briaud, M.-L. Liu, and Ph. Lepert

THE WAK TEST TO CHECK THE INCREASE IN SOIL STIFFNESS DUE TO DYNAMIC COMPACTION

REFERENCE: Briaud, J.-L., Liu, M.-L., Lepert, Ph., "The WAK Test to Check the Increase in Soil Stiffness due to Dynamic Compaction," Geotechnics of Waste Fills Theory and Practice, ASTM STP 1070, Arvid Landva, G. David Knowles, Eds., American Society for Testing and Materials, Philadelphia 1990.

ABSTRACT: A test is proposed to check the increase in soil stiffness brought about by dynamic compaction. It consists of hitting the dynamic compaction weight, after it has been dropped by the crane and while it is resting on the ground, with an instrumented sledge hammer, recording the response of the weight through geophones simply placed on the weight. The analysis of the recorded signals gives the stiffness of the soil under the weight. The test is of very short duration and can be used after the weight has been dropped a number of times with the crane at one location to decide whether the soil has become stiff enough or if further compaction is necessary. The test is particularly useful for waste fills where other more conventional tests to check the increase in soil stiffness are very difficult if not impossible to perform.

KEYWORDS: impact test, clay, sand, wastes, dynamic compaction, stiffness, load test, soil dynamic.

INTRODUCTION

In the case of Dynamic Compaction of Waste Fills it is very difficult to obtain a reasonable evaluation of the stiffness of the fill: indeed it is very difficult to obtain the modulus of deformation of a rusted refrigerator or of an old tire. Common sampling techniques, and in situ techniques all fall short of being as helpful as they are for classical soils. This is why the following method was developed by the authors.

Dr. Briaud is head of the Geotechnical Engineering and Surveying Division in the Department of Civil Engineering at Texas A&M University. College Station, Texas, 77843-3136, USA. Mr. Liu is a graduate student working with Dr. Briaud. Mr. Lepert is an engineer at the Laboratoire Central des Ponts et Chaussees, B.P. No. 19, Bonguenais, 44340, France.

Dynamic compaction consists of dropping a weight from a certain height with a crane in order to increase the stiffness of the waste fill. Once the weight is dropped with the crane and while it is resting on the ground, the WAK test can be performed by using an instrumented sledge hammer to impact the weight (Figure 1). The impact is created by a person swinging the hammer on the weight. The force-time signal of the impact is recorded and the response of the soil-weight assembly is recorded by using two geophones placed on the weight. These geophones measure the vertical velocity with which the weight vibrates. The force-time signal from the dynamic load cell on the hammer and the velocity time signal from the geophones on the weight are analyzed using Fast Fourier Transforms and the stiffness K of the soil mass under the weight is extracted from the transfer function of the system. The test is nicknamed the WAK test: Wave Activated Stiffness (K) test (Figure 1) and allows to measure the stiffness of the waste fill within the zone of influence under the weight.

MODELLING PROCESS

The footing is modelled as a single degree of freedom system (Figure 2). The stiffness of the soil underlying the footing is K, the internal and geometrical damping is C, and the mass of the footing plus the added mass of soil, if any, is M. The equilibrium equation leads to [1]:

$$M\ddot{x} + C\dot{x} + Kx = F(t) \tag{1}$$

where x is the vertical displacement of the footing subjected to a dynamic force $F(t)$. For a harmonic force $F(t) = F_o e^{j\omega t}$ and after a few moments the solution to equation 1 reduces to [2]:

$$x(t) = \frac{F_o}{((K - M\omega^2)^2 + C^2\omega^2)^{0.5}} \; e^{j(\omega t - b)} \tag{2}$$

with $\tan b = C\omega/(k - M\omega^2)$. Therefore the response to a harmonic force excitation is also harmonic after a certain time. The velocity is then:

$$v(t) = \frac{F_o\omega}{((K - M\omega^2)^2 + C^2\omega^2)^{0.5}} \; e^{j(\omega t - b + 90^\circ)} \tag{3}$$

The transfer function is defined here as the ratio of the velocity over the force:

$$\frac{v}{F}(\omega) = \frac{\omega}{((K - M\omega^2)^2 + C^2\omega^2)^{0.5}} \; e^{j(-b + 90^\circ)} \tag{4}$$

The modulus of the transfer function is a function of ω (Figure 3):

$$\left|\frac{v}{F}(\omega)\right| = \frac{\omega}{((K - M\omega^2)^2 + C^2\omega^2)^{0.5}} \tag{5}$$

The maximum value of the modulus is obtained when the denominator is minimum. This occurs when ω is equal to the damped natural frequency ω_d :

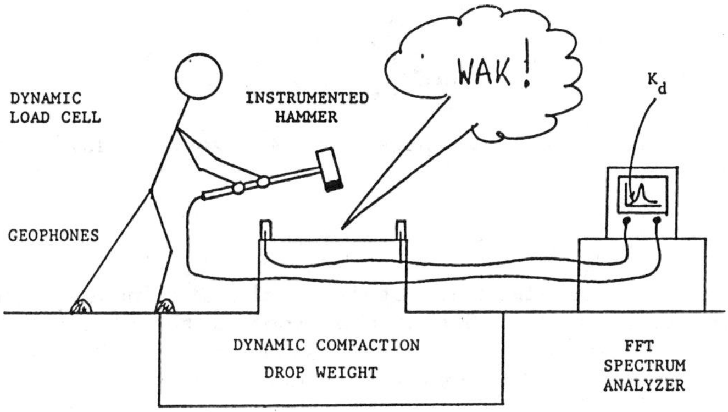

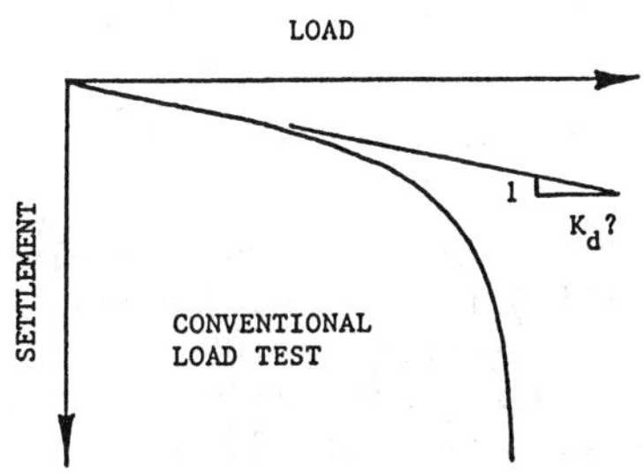

Fig.1 - The WAK test: Wave Activated Stiffness (K) test.

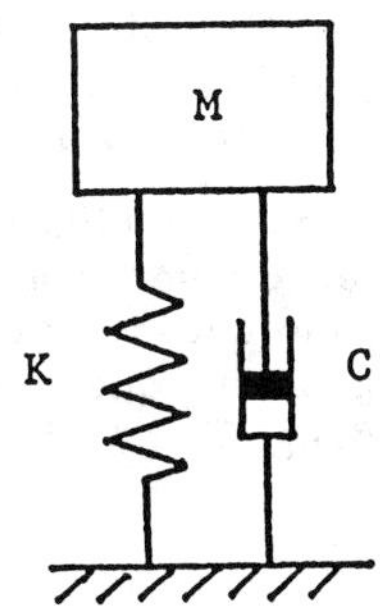

Fig.2 - The footing /soil model.

$$\omega_d = \left(\frac{K}{M}\right)^{0.5} \left(1 - \frac{C^2}{4MK}\right)^{0.5} \tag{6}$$

The maximum value of $|\frac{v}{F}(\omega)|$ is then:

$$|\frac{v}{F}|_{max} = \frac{1}{C}\left(\frac{16KM - 4C^2}{16KM - 3C^2}\right)^{0.5} \tag{7}$$

Since C is usually small compared to KM a good estimate of C is:

$$C \sim \frac{1}{|\frac{v}{F}|_{max}} \tag{8}$$

The WAK test is aimed at obtaining the stiffness of the soil. It can be seen from equation 5 that the slope of the tangent to the origin of the $|v/F|$ versus ω curve is the stiffness K. Therefore once this curve is obtained experimentally one can obtain the soil stiffness K.

$$K = \left(\frac{\Delta\omega}{\Delta|\frac{v}{F}|}\right)_{\text{at the origin}} = 2\pi \left(\frac{\Delta f}{\Delta|\frac{v}{F}|}\right)_{\text{at the origin}} \tag{9}$$

OBTAINING THE $|v/F|$ VERSUS ω CURVE

Experimentally this process is relatively simple. It consist of using a 1500 dollar (1988) instrumented sledge hammer (most of the cost is in the dynamic load cell), two 100 dollar (1988) geophones and a microcomputer with a data acquisition board and a Fast Fourier Transform software package for 5000 dollars (1988).

The geophones are placed symmetrically with respect to the center of gravity of the weight. The sledge hammer is used to hit the weight close to its center of gravity. The force-time signal of the dynamic load cell on the hammer and the velocity-time signal given by the geophones are recorded.

The signals recorded for the velocity and the force are a function of time. Yet the modulus of the transfer function $|v/F|$ is a function of ω. The Fourier Transform technique [3] is used to go from the time domain to the frequency domain and allows to obtain the transfer function. This technique is described in some details in [3].

The stiffness of the soil can be obtained from the tangent to the origin of the transfer function curve as mentioned earlier (Figure 3). However a more reliable technique consists of curve fitting the transfer function. This curve fitting technique allows to obtain not only the stiffness K but also the damping C and the mass M. The curve fitting technique is described later.

EXPERIMENTAL DATA

A first experiment was performed while the first author was on sabbatical leave at the Laboratoire Central des Ponts et Chaussees in France. The details are presented in [3] and are summarized here. A 0.3 m x 0.3 m x 0.3 m concrete block was placed on a 0.55 m thick bed of loose dry sand. Geophones were placed on the weight (concrete block)

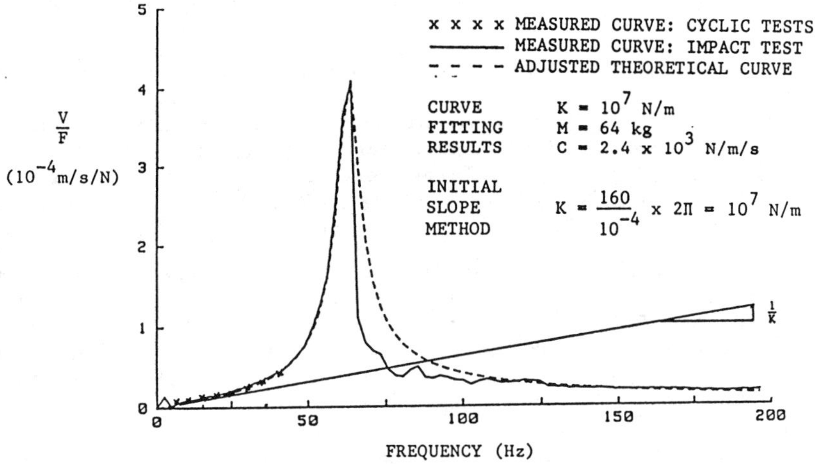

Fig.3 - Measured Transfer Function and Stiffness Calculation.

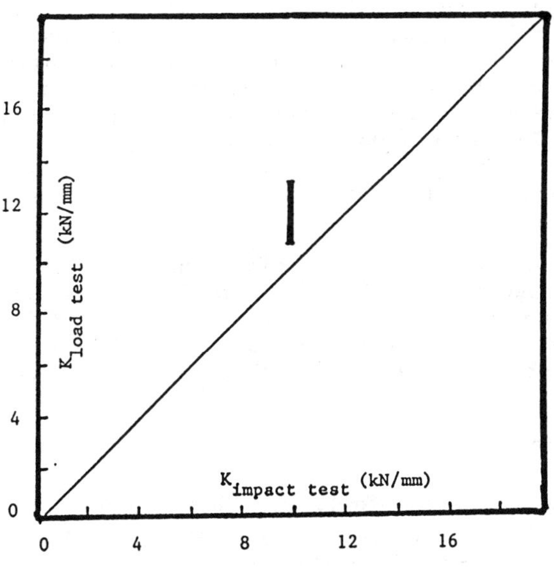

Fig.4 - Comparision of the Stiffness obtained with the Impact Tests and with the Monotonic Test.

which was hit with an instrumented mini sledge hammer. The signals were recorded, then analyzed and gave the soil stiffness as predicted by the impact test, $K_{impact\ test}$. A static load test was also performed on the weight and gave the soil stiffness as predicted by the static load test, $K_{load\ test}$. Figure 4 shows the comparison; the two stiffnesses match reasonably well.

A second set of experiments were carried out at Texas A&M University. Whereas the first experiment was carried out in the laboratory, the second set of experiments were carried out in the field with more rudimentary equipment. Two weights were tested; the smaller one was 1.5 m x 1 m x 0.3 m and the larger one was 3.3 m x 1.5 m x 1.1 m (approximate dimensions). They will be referred to as the small mass and the large mass in the rest of this article.

Figure 5 shows the force-time signal due to the sledge hammer impact. The two signals are plotted on the same graph for convenience. Note that with a 53 N sledge hammer one can generate a peak dynamic force of 23000 N. In this case the mean deceleration of the hammer head during the impact is 430 g, where g is the acceleration due to gravity. Note also that the sledge hammer is simply swung by a person. Such a procedure does not generate a standard dynamic force. This is of no consequence however because the geophones record the velocity response to whatever the input force is and because it is the ratio of the velocity over the force which leads to the stiffness.

Figure 6 shows the velocity-time signal for the two masses. The response is very damped with very little vibrations; this is probably due to the high internal damping on the stiff clay and the very high geometric damping since all the input energy radiates in the half space. A much smaller damping was observed in the laboratory because of the low internal damping of the dry sand and the low geometric damping due to the fact that the sand was in a finite container (1.5 m x 1.5 m x 0.55 m). Indeed in the case of the laboratory experiment the waves will bounce back on the wall and the energy will not dissipate as readily. The other important observation is the noise which is obvious on the velocity signal. Indeed this signal is not as smooth as it could be.

Figures 7 and 8 show the transfer functions for the small and the large mass respectively. These transfer functions which are obtained from field experiments are not as smooth as the transfer function which was obtained in the laboratory experiment [2]. The following parametric study was undertaken to study the reasons for the difference.

PARAMETRIC ANALYSIS

The influence of the following factors was studied numerically: stiffness K, mass M, damping C, noise content. The reference parameters were $K = 10^7 N/m, M = 64 kg, C = 2400 N/m/s$. The impact used had a peak force of 2200 N and lasted 0.02 seconds.

Figure 9 shows the response to two impacts: one for the reference stiffness of $10^7 N/m$ and one for a stiffness of $10^8 N/m$. As can be seen on Figure 10 the transfer function is very sensitive to the stiffness K. The ratio of the initial tangent of the transfer function for the first case to the one for the second case is 10 which matches

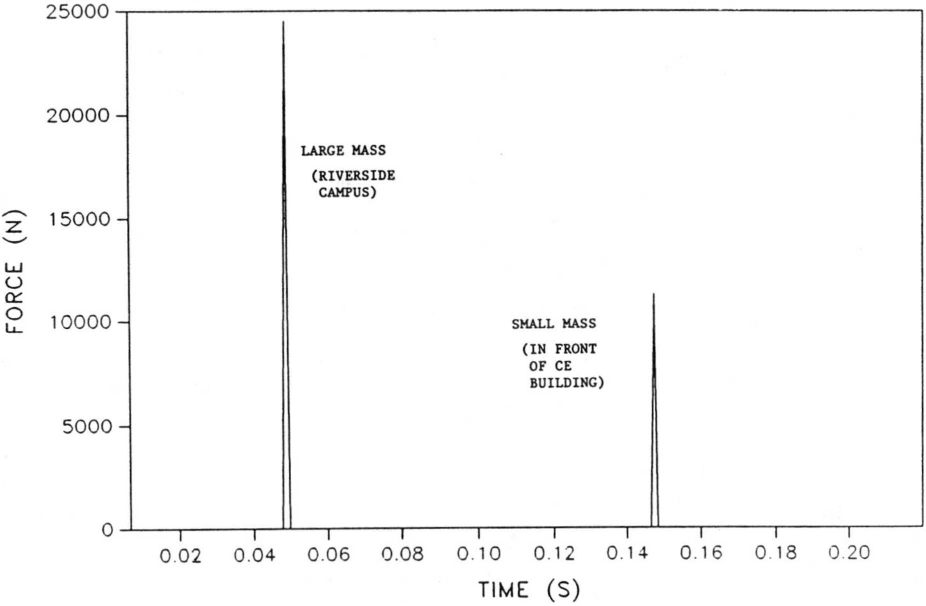

Fig.5 - Force - Time Signals.

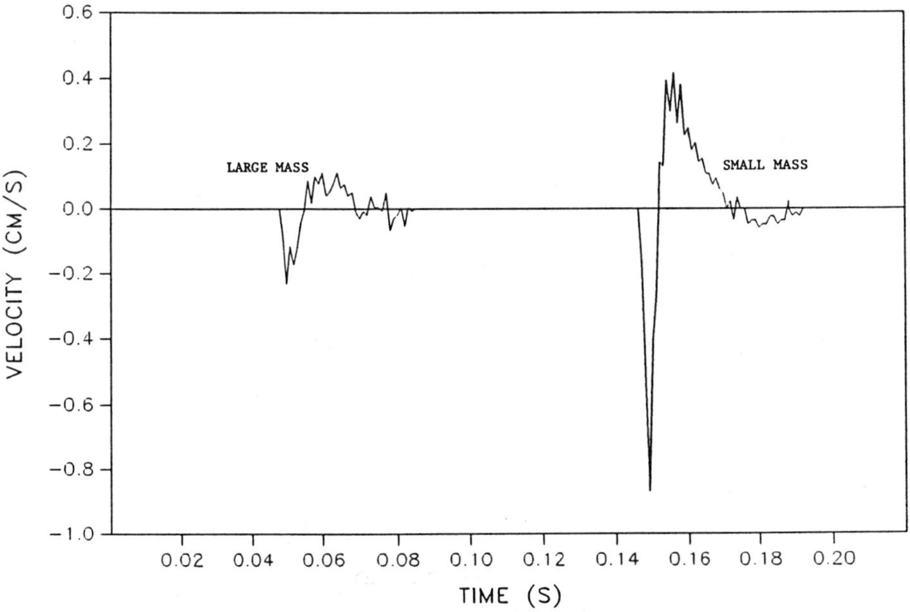

Fig.6 - Velocity - Time Signals.

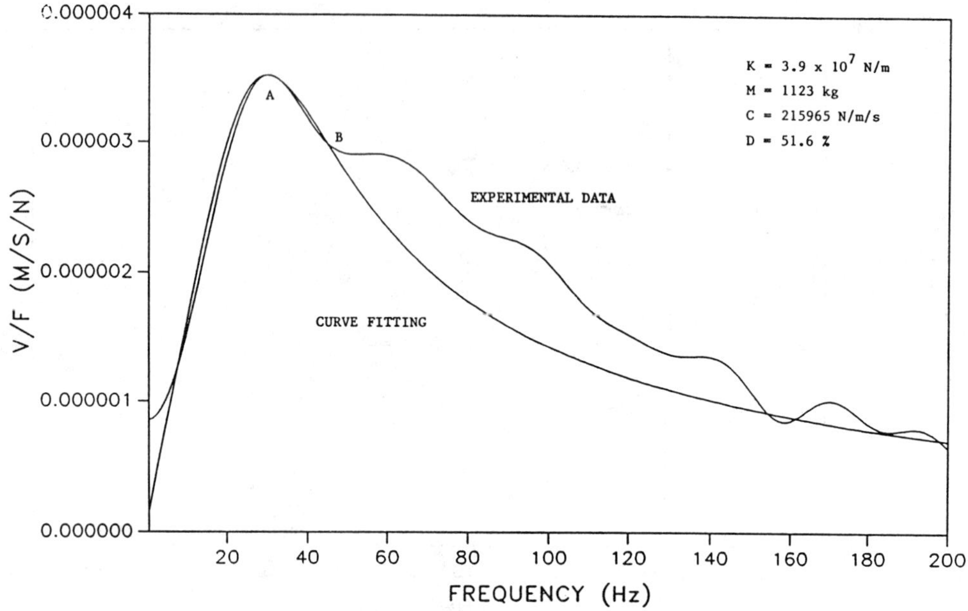

Fig.7 - Transfer Function for the Small Mass.

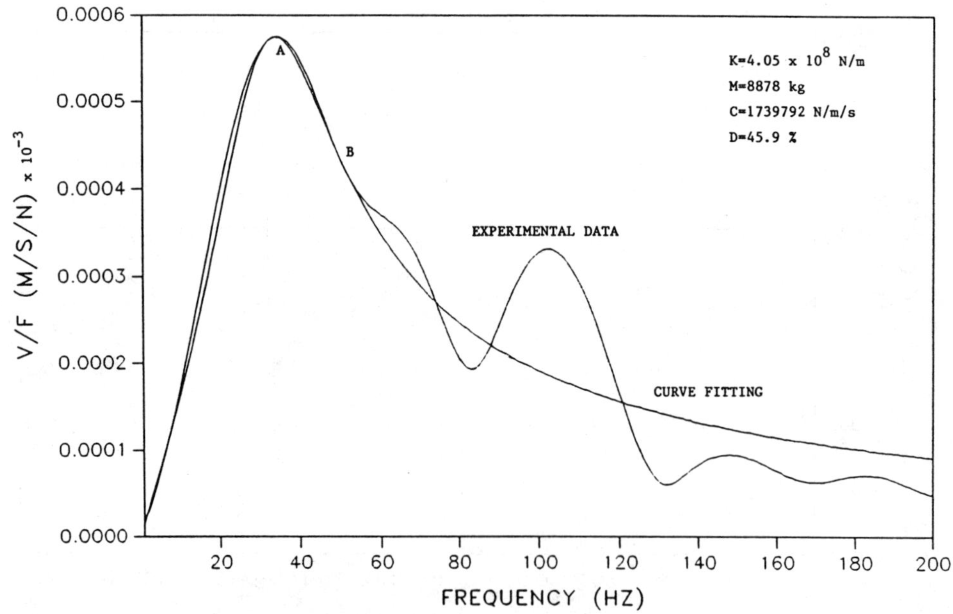

Fig.8 - Transfer Function for the Large Mass.

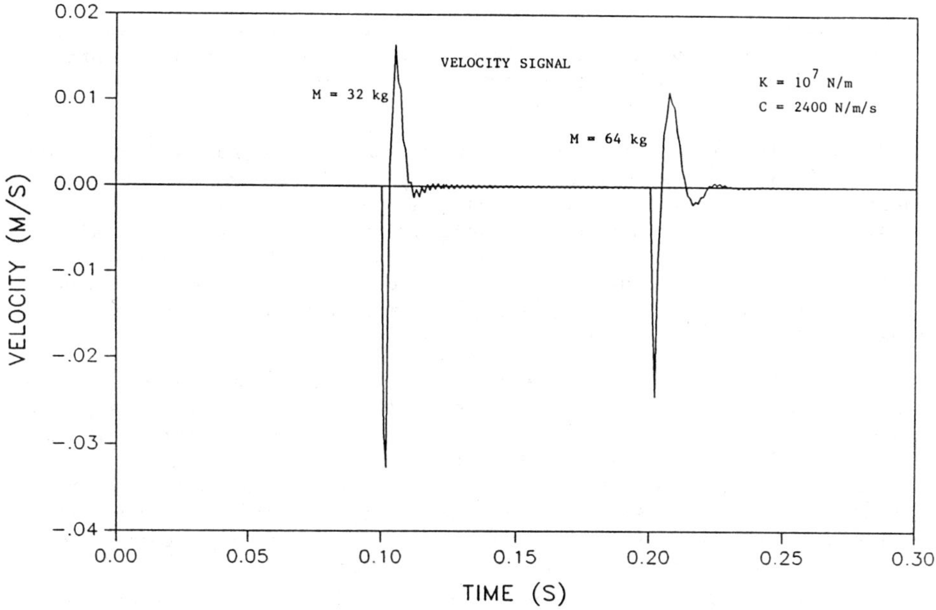

Fig.9 - Influence of the Stiffness on the Results.

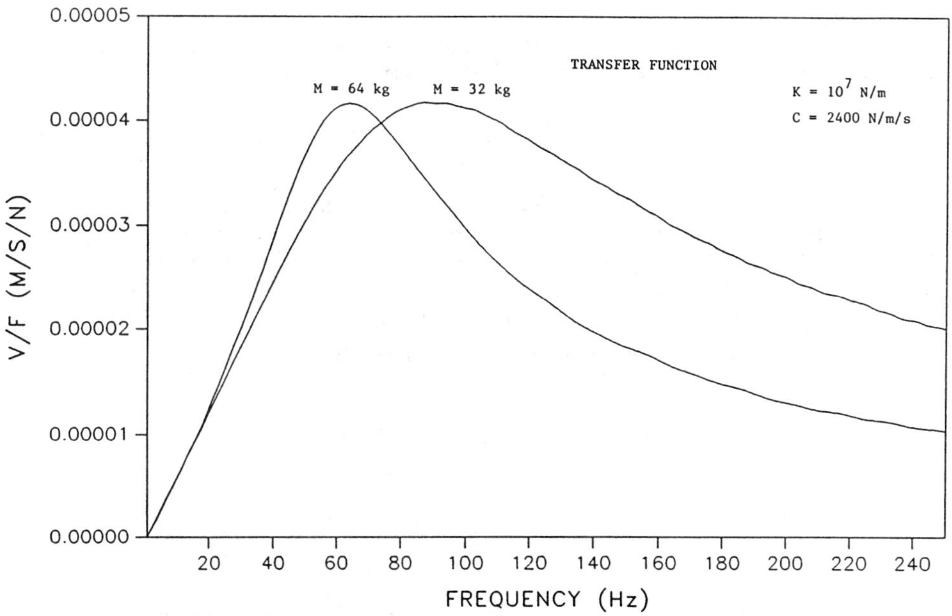

Fig.10 - Influence of the Stiffness on the Results.

the ratio of the stiffnesses. This is due to the fact that the slope of the initial tangent is 1/K as shown earlier (Equation 9). Note that on Figure 10 the frequency scale represents f and not $\omega (\omega = 2\pi f)$.

Figure 11 shows the response to two impacts: one for the reference mass of 64 kg, one for a mass of 32 kg. As can be seen on the transfer functions (Figure 12) the larger mass leads to a lower resonant frequency and lower content in high frequencies.

Figure 13 shows the response to three impacts: one for the reference damping of 2400 N/m/s, one for a damping of 1200 N/m/s and one for a damping of 24000 N/m/s. The velocity signal shows that an increase in damping reduces the number of oscillations. Very few oscillations and therefore high damping was observed in the field tests. The transfer functions (Figure 14) show that an increased damping decreases the peak value without changing the resonant frequency by much. Equation 8 can be verified on this example.

Figure 15 shows the response to two impacts: one with noise in the velocity signal and one without noise in the velocity signal. As can be seen on the transfer function (Figure 16) the effect of the artificially created noise is twofold: 1) it makes the transfer function start from a nonzero ordinate and 2) it creates some post-peak undulations. Another numerical simulation showed that such post-peak undulations are generated when the recording of the velocity signal is stopped before the end of the free vibrations. Indeed in this case some frequencies are missing in the response.

This parametric analysis has allowed to explain the features of the transfer functions obtained in the field and leads to the following comments. The noise level is an important factor and must be decreased to a minimum in order to obtain quality data in the field. The damping in the field experiment is high. Obtaining the stiffness K from the initial tangent to the transfer function is not a reliable means of obtaining K and a curve fitting technique must be used.

PROPOSED CURVE FITTING TECHNIQUE AND STIFFNESS DETERMINATION

It is observed from the transfer functions obtained in the field and in the parametric analysis that the portion of the curve which is the least affected by the ambient noise is around the peak value and in particular the post-peak part of the curve close to the peak. Use is made of this part of the curve as follows.

First, the peak point is used to obtain the damping C (Equation 7):

$$C = \frac{1}{\left|\frac{v}{F}\right|_{max}} \tag{10}$$

This requires that (Equation 7),

$$C \ll KM \tag{11}$$

and this condition must be verified. Second, the peak point is also used to obtain a relation between K and M (Equation 6):

$$\omega_{peak} = \left(\frac{K}{M}\right)^{0.5} \tag{12}$$

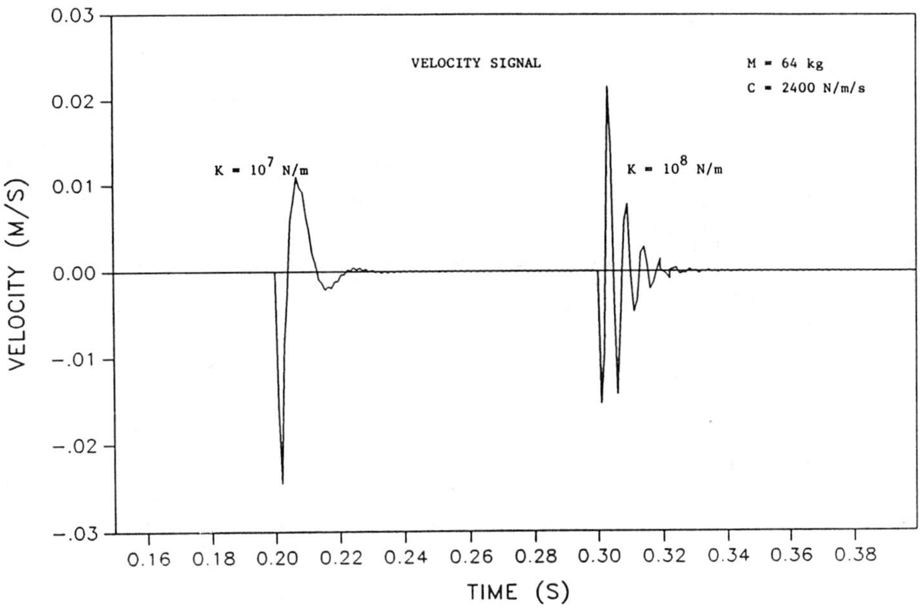

Fig.11 - Influence of the Mass on the Results

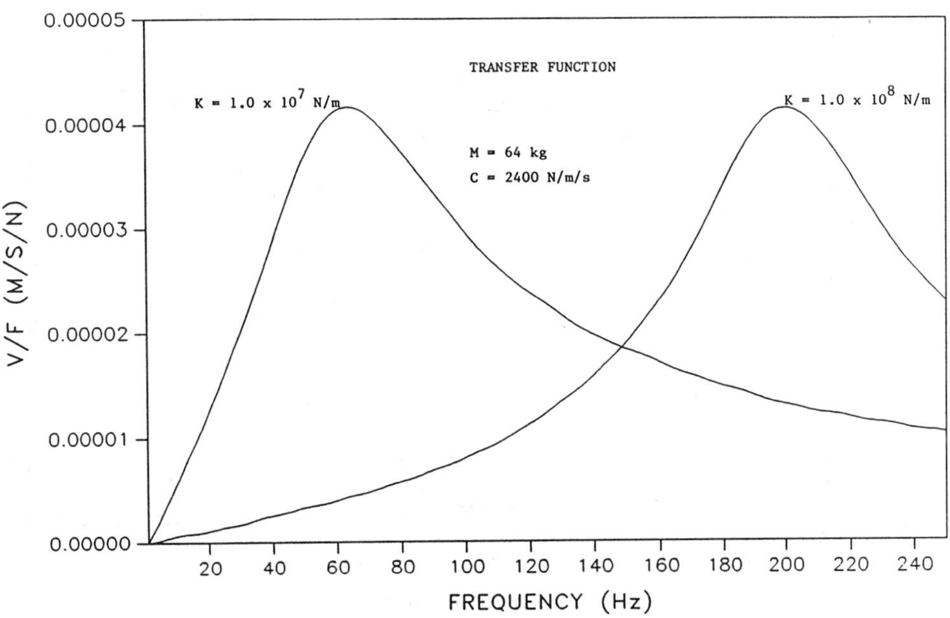

Fig.12 - Influence of the Mass on the Results.

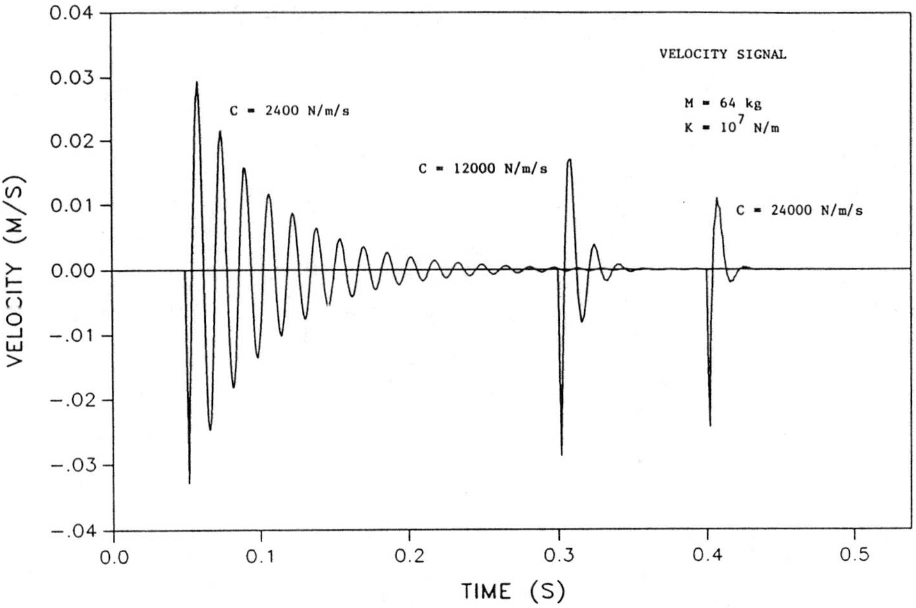

Fig.13 - Influence of the Damping on the Results.

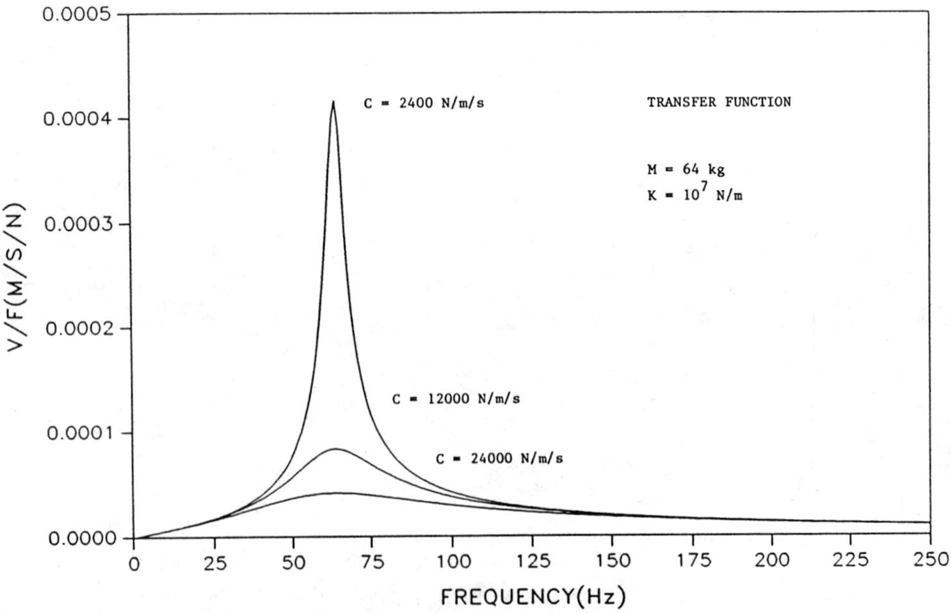

Fig.14 - Influence of the Damping on the Results.

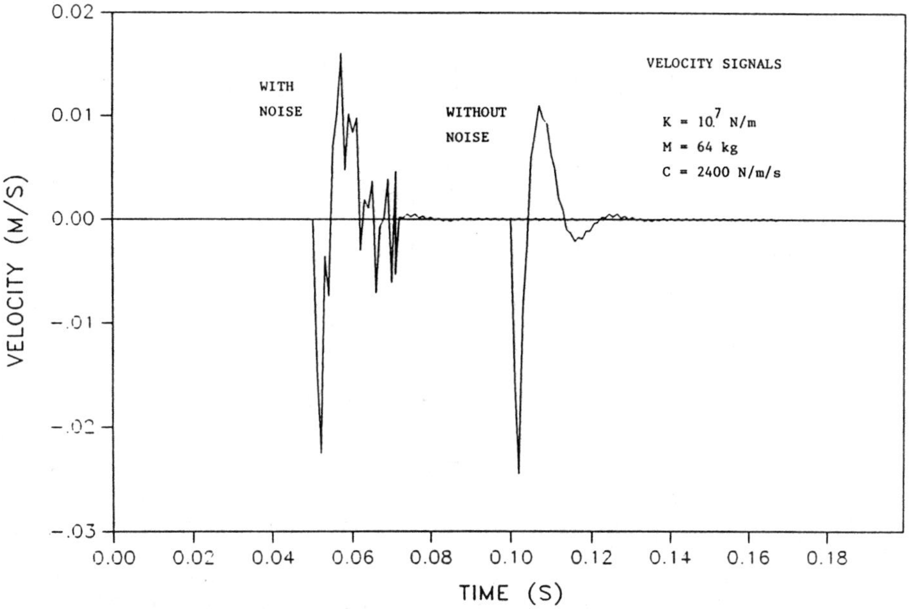

Fig.15 - Influence of Noise on the Results.

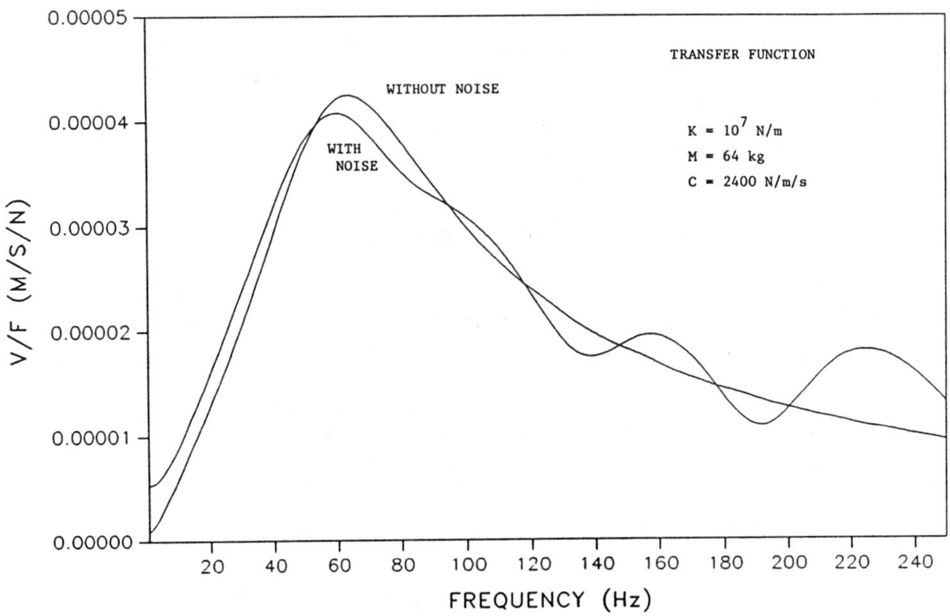

Fig.16 - Influence of Noise on the Results.

This requires that (Equation 6)

$$\left(\frac{16KM - 4C^2}{16KM - 3C^2}\right)^{0.5} \sim 1 \qquad (13)$$

and this condition must be verified. Third, a point on the post-peak part of the curve but fairly close to the peak is selected. For this point B (Equation 5):

$$\left|\frac{v}{F}\right|_B = \frac{\omega_B}{((K - M\omega_B^2)^2 + C^2\omega_B^2)^{0.5}} \qquad (14)$$

Combining Equation 12 and Equation 14 leads to:

$$M = \left(\left(\frac{\omega_B}{\left|\frac{v}{F}\right|_B}\right)^2 - C^2\omega_B^2\right)^{0.5} x \; \frac{1}{\omega_{peak}^2 - \omega_B^2} \qquad (15)$$

then Equation 12 gives K. The damping ratio is then given by:

$$D = \frac{C}{\sqrt{KM}} \qquad (16)$$

Once K, M, and C are known the curve corresponding to equation 5 represents the regression curve for that test. These curves are shown on Figures 7 and 8 together with the experimental transfer function and the backfigured values of K, M, C, and D for the two field tests which were performed.

The following is an example of how to calculate the stiffness by hand. Note that this technique is not recommended because the precision required for the coordinates of the points on the transfer function can only be obtained electronically. Indeed slight errors in the coordinates can change the calculated stiffness drastically. Consider the large mass of Figures 5, 6, and 8. In order to use Equation 10, the coordinates of the peak point A on the transfer function curve are determined electronically (Figure 8).

$$f_A = 34Hz; \left|\frac{v}{F}\right|_A = 5.747x10^{-7}\frac{m/s}{N}$$

A second point B is selected as shown on Figure 8. The coordinates are

$$f_B = 49Hz; \left|\frac{v}{F}\right| = 4.456x10^{-7}\frac{m/s}{N}$$

Therefore Equation 10 gives:

$$C = \frac{1}{\left|\frac{v}{F}\right|} = 1739829\frac{N}{m/s}$$

The circular frequencies are then obtained:

$$W_A = 34x2\pi = 213.63 \quad rd/s$$

$$W_B = 49x2\pi = 307.87 \quad rd/s$$

Equation 15 leads to:

$$M = \left(\left(\frac{307.87}{4.456x10^{-7}}\right)^2 - (1739829)^2 x (307.87)^2\right)^{0.5} x \frac{1}{(307.87)^2 - (213.63)^2}$$

$$M = 8879.67 kg$$

Then Equation 12 gives the stiffness K:

$$K = 8879.67x(213.63)^2 = 4.05x10^8 N/m$$

Equations 11 and 13 need to be verified:

$$C \ll KM; 1739829 \ll 4.05x10^8 x8879.67 \qquad OK$$

$$\left(\frac{16KM - 4C^2}{16KM - 3C^2}\right)^{0.5} \sim 1; \left(\frac{5.754x10^{13} - 1.211x10^{13}}{5.754x10^{13} - 9.081x10^{12}}\right)^{0.5} = 0.968 \qquad OK$$

CONCLUSIONS

A test called the WAK test is proposed in order to check the increase in soil stiffness brought about by dynamic compaction (D.C.). The WAK test consists of placing two geophones on the D.C. mass after it has been dropped by the crane and hitting the mass with an instrumented sledge hammer. The force-time and velocity-time signals are recorded. The Fourier Transform analysis gives the stiffness of the soil under the mass. The test is very fast and can be used as a quality control test on dynamic compaction jobs.

The small scale laboratory experiment showed that the stiffness obtained by the WAK test and the stiffness obtained by the static load test matched reasonably well. The large scale field experiment showed that the noise in the signal influences the results and that the damping in the field is relatively high. A curve fitting technique is proposed to circumvent the influence of the noise. All measured data was found to be consistent with the trends indicated by a parametric analysis.

This WAK test can be used on any soil, fill or even rock. The depth of influence for the WAK test is argued to be similar to the one involved in the dynamic compaction process simply because the two processes are dynamic tests with one generating much smaller strains than the other. As such the WAK test can be used to check the stiffness improvement brought about by dynamic compaction.

ACKNOWLEDGMENTS

This study was started while the first author was on sabbatical leave at the LCPC in France. The results of this follow-up work were sponsored by the Texas Engineering Experiment Station and by AMOCO. Dr. Richardson of TEES and Mr. Funegard of AMOCO are thanked for their support.

REFERENCES

[1] Clough, R.W. and Penzien, J., Dynamics of Structures, McGraw-Hill Book Company, 1975.

[2] Briaud, J.-L., Liu, M.-L., and Lepert, Ph., "Impact Method to Find Soil Stiffness Under a Rigid Mass," Research Report to AMOCO, Civil Engineering, Texas A&M University, 1990.

[3] Brigham, E.O., The Fast Fourier Transform, Prentice Hall, 1974.

Yalcin B. Acar, Roger K. Seals, and Anand J. Puppala

ENGINEERING AND COMPACTION CHARACTERISTICS OF BOILER SLAG

REFERENCE: Acar, Y. B., Seals, R. K., and Puppala, A. J., "Engineering and Compaction Characteristics of Boiler Slag,"Geotechnics of Waste Fills - Theory and Practice, ASTM, STP 1070, Arvid Landva, G. David Knowles, Editors, American Society for Testing and Materials, Philadelphia, 1990.

ABSTRACT: Engineering and compaction characteristics of boiler slag are presented. Laboratory and field compaction tests are conducted in assessment of these characteristics. Field tests are conducted on a slag fill of 0.92 m (3 ft) in thickness. A smooth drum vibratory roller 56 kN (12.5 kips) is used for compaction. The densification is evaluated after 2, 4, 6, 10 and 16 passes of this roller. A methodology is presented for estimating the lift thickness and the number of passes required for a desired level of densification.

KEYWORDS: compaction, slag, lift thickness, roller pass, friction angle, cohesion intercept, relative density, and specific gravity.

INTRODUCTION

Many regions in the United States and the world are facing a severe shortage of natural aggregates. The significant increase in the cost of aggregate processing and handling operations necessitates studies to locate and evaluate the engineering characteristics of various types of synthetic aggregates such as flyash, slag and boiler slag [1].

A by-product of the power generation industry, boiler slag, has been used in a variety of applications in West Virginia, Maryland and Florida, as filter material, structural fill, roofing shingles, roadbeds, embankments and sandblasting grit. In other applications, when mixed with sand, it has been reported to produce a road surface with improved skid resistance [2,3].

This paper presents the results of laboratory studies and field compaction tests conducted to evaluate the engineering and field compaction characteristics of boiler slag

Drs. Acar and Seals and Mr. Puppala are, respectively, Associate Professor, Professor, and Graduate Research Assistant, Department of Civil Engineering, Louisiana State University, Baton Rouge, LA 70803.

produced as a by-product of the synthetic gas generation by the Dow Chemical Company in Plaquemine, Louisiana.

NATURE AND CHEMICAL COMPOSITION

Boiler slag is one portion of the inorganic residue that is produced when coal is burned in a furnace at temperatures between 2500°F to 3100°F (1301°C to 1704°C). It is produced either in wet bottom (slag tap) or cyclone boilers. The fundamental difference between the two boilers is that crushed coal is burned in the cyclone boiler, while pulverized coal is burned in the wet bottom boiler. The slag constitutes 70 to 80 percent of the total ash produced in a cyclone boiler, while this quantity decreases to 40 to 50 percent in a wet bottom boiler. The molten slag flows out from the furnace into cold water and crystallizes, solidifies and forms angular, black, glassy particles. Plate 1 presents a view of the slag used in this study. The needle-like long (l/4 in. to 3/8 in.) crystallized particles are common to boiler slags.

It is estimated that 25 percent of all power plant ash produced in the USA is boiler slag, 65 percent is flyash and 10 percent is dry bottom ash and cinders [2]. In recent years, its use is getting more common as construction material. However, national utilization still lags production rate.

The typical chemical analysis of the Dow gasifier slag is presented in Table 1. The principal constituent is silica as in sand. The U.S. Environmental Protection Agency, RCRA toxicity tests on evaluation of Dow gasifier slag and other slags produced from Western (Peabody), Ohio and Kentucky coals indicate that barium is the only heavy metal detectable and only at levels (0.02 to 0.5 ppm) much below the specified standard (100 ppm) [2]. As a consequence, the Louisiana Department of Environmental Quality has declassified this material from the list of potentially hazardous wastes. In addition, the State of Maryland has classified it as a natural resource. Therefore, this material should be considered as a valuable resource rather than a waste. However, the effects of post-construction changes in the environment should be carefully evaluated.

PHYSICAL AND ENGINEERING CHARACTERISTICS

Representative samples each weighing approximately 27 kg (60 lb) were collected from five different stockpiles at the Plaquemine facility of Dow Chemical Company. These samples were tagged as 1, 3, 4, 18, 19 in sequence with their production cycles.

Tests were conducted to assess the physical and engineering characteristics of the material and provide values that could be used in evaluating the factors which influence construction and performance of a structural fill. All tests were conducted in accordance with applicable ASTM standards.

Specific Gravity

Table 2 compares the specific gravity values determined for the selected samples of each pile with values reported in previous studies for other slags. The samples from pile 1 and pile 3 had relatively low values of specific gravity. However, such variations are expected due to variability in chemical composition and friable material content.

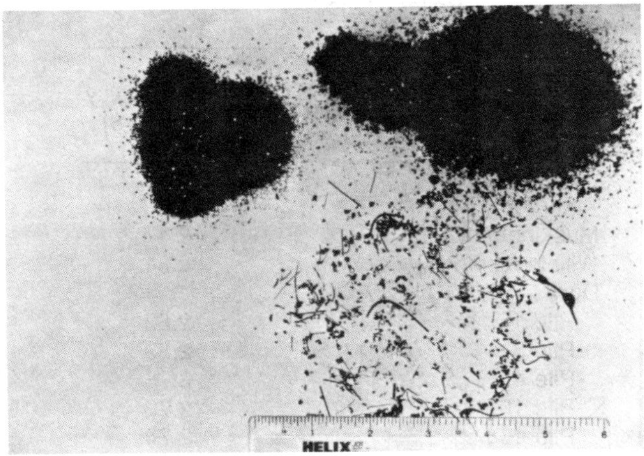

PLATE 1 -- General view of slag used in the study.

TABLE 1 -- Chemical Constituents of Dow Gasifier Slag

Constituent	Amount (%) Dow Gasifer Slag	Amount (%) Power Plant Bottom Ash Slag
Silica, SiO_2	48.8	46 - 54
Alumina, Al_2O_3	20.2	22 - 28
Ferric Oxide, Fe_2O_3	9.6	6 - 14
Calcium Oxide, CaO	15.8	0.4 - 1.4
Magnesium Oxide, MgO	3.9	4 - 5
Sulphur Trioxide, SO_3	0.8	N.R.*
Titanium Dioxide, TiO_2	0.8	N.R.
Sodium Oxide, Na_2O	0.1	0.7 - 1.2

*N.R. - not reported.

It has been reported that the specific gravity of solids decreases with an increase in friable particle content [1]. The production process is also expected to affect the specific gravity.

TABLE 2 -- Specific Gravity of Solids for Selected Boiler Slags

Source	Specific Gravity of Solids (GS)
Kammer [1]	2.72
Muskingham [1]	2.47
Willow Island [1]	2.61
Dow	
Pile 1	2.63
Pile 3	2.53
Pile 4	2.76
Pile 18	2.73
Pile 19	2.74

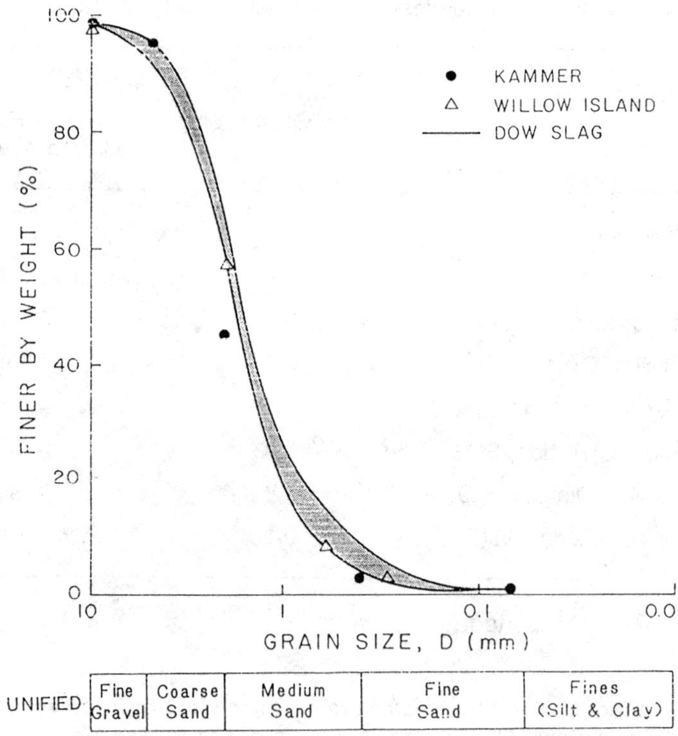

FIG. 1 -- Grain size distribution of slag.

Grain Size Distribution

The quantitative determination of distribution of particle sizes of the slag was conducted separately using dry sieving procedures [4]. The variation of grain size distribution for the Dow slag samples are compared with other slags in Figure 1. The boiler slag is uniform and most of the material is retained on No. 10 (2 mm) to No. 20 (0.85 mm) sieves. The uniformity coefficient, C_u, and concavity coefficient, C_c, values were found to vary between 2.8 to 3.5 and 1.7 to 1.8, respectively. The slag can therefore be described as a poorly graded medium to fine granular material with angular, glassy particles and is classified as SP according to the Unified Soil Classification System. The grain size distribution for the Dow slag is within the ranges reported for other electric generating coal boiler slags [2,5].

Proctor Compaction Tests

The relationship between the moisture content and dry density of soils was determined on samples collected from pile 19, using standard and modified Proctor compaction tests (ASTM D-698 and D-1557, respectively) [6,7].

The moisture content-dry density relationship of the slag is shown in Figure 2. Over the range of moisture contents considered, the highest densities were obtained when the water content was very high (approaching saturation).

Table 3 compares the maximum dry density and optimum moisture content for this boiler slag with others. It is noted that the Dow slag has a higher maximum dry density possibly due to variations in grain size distributions. High energy input may lead to particle breakdown or degradation in the slag. Thus, sieve analyses were performed both before and after the standard and modified Proctor compaction tests to describe the possible effect of degradation. The degradation of the slag was insignificant in case of the standard Proctor compaction test. However, it was significant when the compactive effort was increased beyond that of the standard Proctor test [8]. It is then envisioned that higher compactive efforts in the field may change the grain size distribution, maximum dry density and hence the maximum and minimum void ratios.

Maximum and Minimum Void Ratios

1. Maximum Void Ratio - Minimum Index Dry Density

All boiler slag particles pass a 3/4 inch (19 mm) sieve. Therefore, ASTM D4254-83 [9] Method B was selected to determine the maximum void ratio. The minimum index dry density and maximum void ratio values corresponding to the loosest possible condition were 14.2 kN/m^3 (88.6 pcf) and 0.93, respectively.

2. Minimum Void Ratio - Maximum Index Dry Density

The maximum index dry density may be described as a densified state in which the soil mass has been densified so as to occupy the least possible volume. The determination of the maximum possible index dry density and the corresponding value of the minimum void ratio is often a difficult task. However, it is known that if granular soils are subjected to continuous induced vehicular vibration, they will approach such

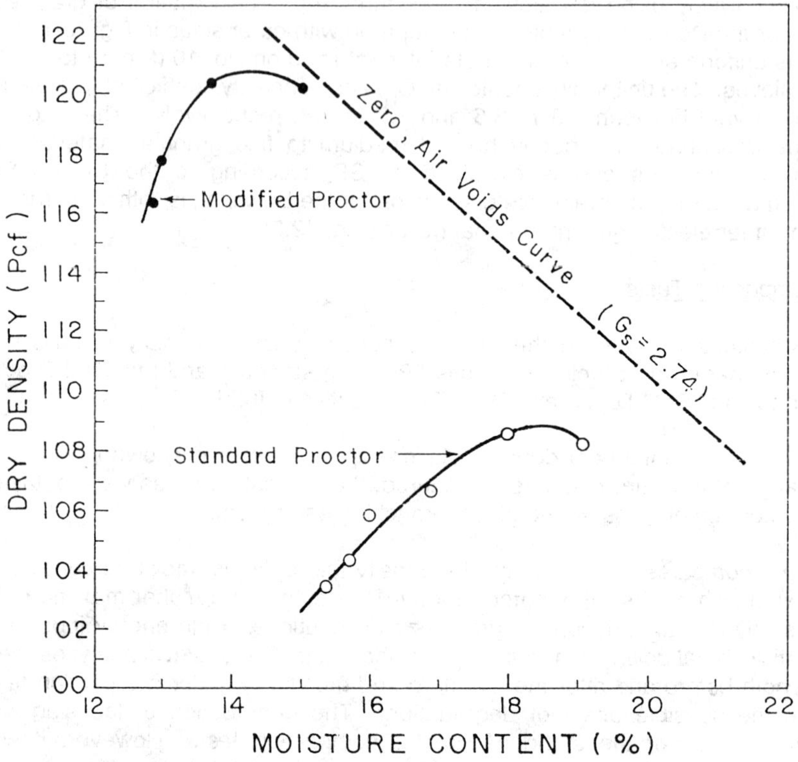

FIG. 2 -- Dry density-moisture content relationships.

TABLE 3 -- Standard Proctor Compaction Results for Boiler Slag

Source	Maximum Dry Density (pcf)	Optimum Moisture Content (%)
Kammer	102.0	13.8
Muskingham	91.1	22.0
Willow Island	92.4	21.0
Dow	109.0	18.5

a state. Impact compaction may be used as an indirect way to evaluate the maximum index dry density for some soils [10]. This method was used for the slag.

The maximum index dry density was determined to be 17.5 kN/m^3 (109.2 pcf) from the standard Proctor test. The corresponding void ratio for this dry density value is 0.57. The maximum dry density and the corresponding value of the minimum void ratio were determined to be 19.4 kN/m^3 and 0.41, respectively, with the modified Proctor test. It is noted that this 19.4 kN/m^3 (121.1 pcf) may be taken as the maximum density only if there were no particle degradations during compaction.

The sieve analysis conducted before and after the standard and modified Proctor compaction tests indicated that the higher energy resulted in a significant amount of degradation in the slag when the energy is increased beyond that of standard Proctor [5].

Therefore, the actual value of minimum void ratio, e_{min}, should be lower than what is determined from the standard Proctor test, but be somewhat higher than that obtained from the modified Proctor test. In order to establish e_{min}, a study was conducted where the energy of compaction was varied. First, the hammer used in standard Proctor test was dropped from 2 in. and the number of layers was varied from one to six. Figure 3(a) indicates that 4 lifts would provide the optimum lift height for e_{min} in the 1/30 ft^3 standard Proctor mold. Subsequently, keeping the number of lifts as 4, the number of blows were increased. Figure 3(b) indicates that e_{min} decreases to 0.51 with further increase in the number of blows.

The density value of 18.1 kN/m^3 (113.2 pcf) and 0.51 may be interpreted to be the values of maximum index dry density and the minimum void ratio of the undergraded boiler slag.

Compressibility

One-dimensional compression tests were carried out as per the procedures of ASTM D2435-80 [11] with normal loads ranging from 50 to 400 kPa on specimens of 8 cm (3 in.) diameter and 1.9 cm (0.75 in.) thick. Specimen was prepared at a relative density of 50 percent. The compression index, C_c, for the Dow slag was calculated to be 0.107. This value is considerably higher than the 0.031 value reported for the Kramer slag [8]. The difference may be attributed to the procedures used in determining the compressibility. The Dow slag was tested saturated, while the Kramer slag was reported to be tested under dry conditions.

Shear Strength

The effective angle of internal friction for the boiler slag at different relative densities was determined by displacement-controlled direct shear tests. Both tangential force and vertical deformation value were recorded as a function of horizontal displacement up to a total displacement of 5 mm. Tests were conducted at vertical stresses of 20, 40, 62 and 82 kPa. The tests were performed on specimens with a relative density values varying from 0 to 66 percent. The density was directly controlled in the shear box by tamping. Figure 4 presents the failure envelopes and Figure 5 presents the change in friction angle with relative density. It is interesting to note that the slag displays both a cohesion intercept and an angle of internal friction. In addition, a substantial internal friction angle is mobilized at relatively low densities. This may be

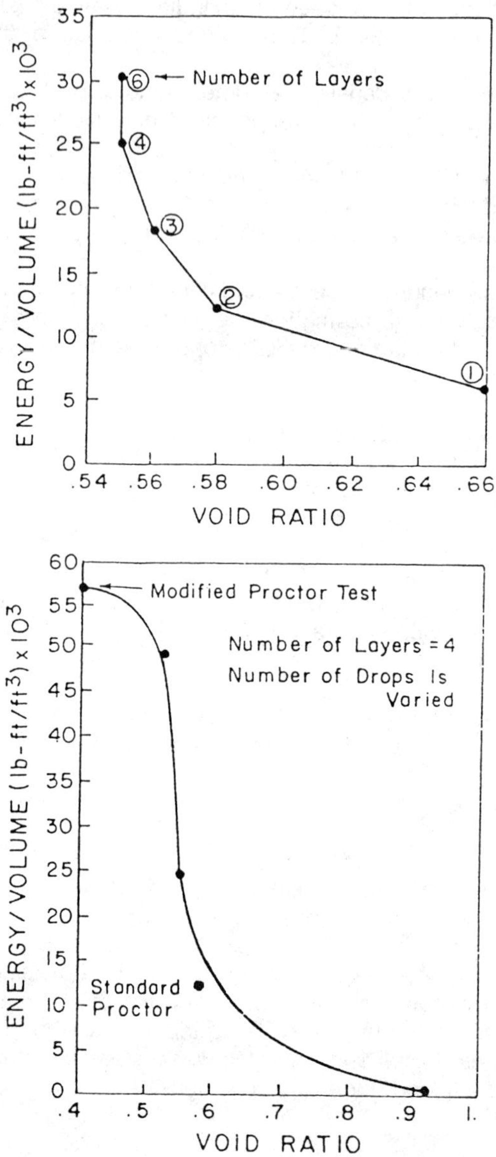

FIG. 3 -- The effect of compaction effort on void ratio.

attributed to the angular characteristics and the interlocking capability of this material. For comparison, the values of internal friction angle for other boiler slags are also summarized in Table 4.

It is quite surprising that a cohesion intercept is indicated in Figure 4. The slag, being granular and noncohesive, would not be envisioned to display such a characteristic. It is then hypothesized that this display of cohesion intercept may have been due to:

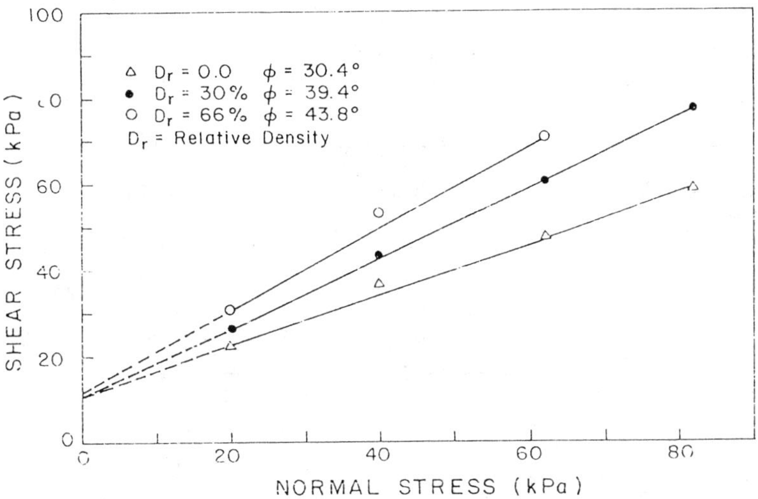

FIG. 4 -- Failure envelopes.

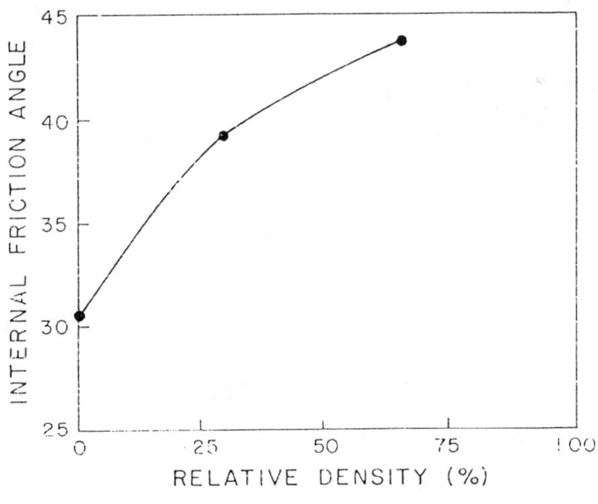

FIG. 5 -- Variation of friction angle with relative density.

1. Any moisture in the tested sample leading to suction and a confinement which is greater than that applied vertically,

2. The reinforcement effect due to the short (2 mm to 2 cm) needle-shaped, fibrous particles in the slag (Plate 1).

TABLE 4 -- Angles of Internal Friction for Boiler Slag

Source	Average Void Ratio	Angle of Internal Friction (°)
Kammer	0.88	41.0
Muskingham	1.33	40.0
Willow Island	1.32	42.0
Dow	0.93	30.4
	0.80	39.4
	0.65	43.8

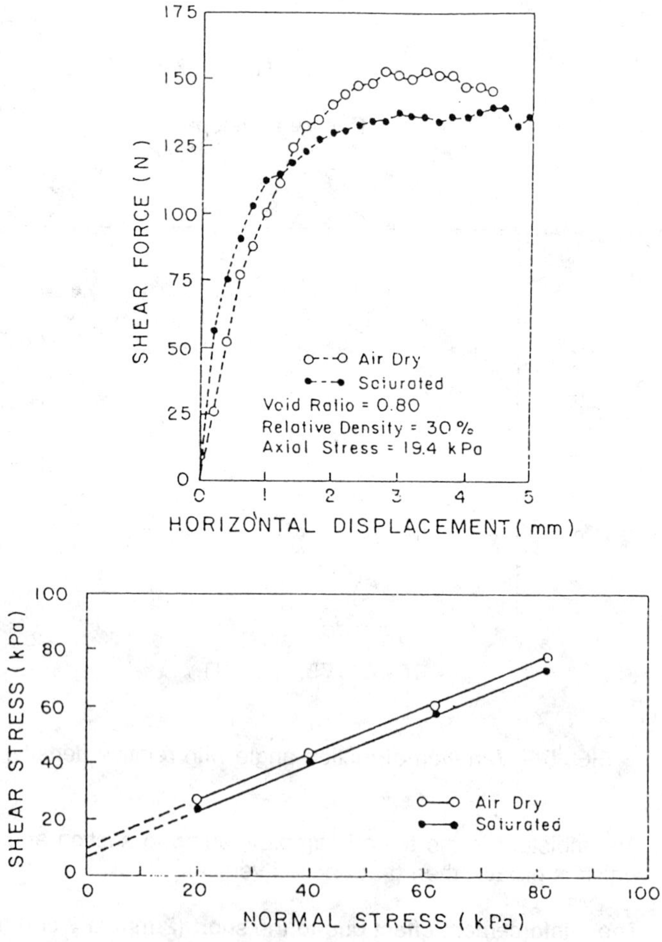

FIG. 6 -- Influence of saturation effect on shear force-displacement behavior and failure envelope.

In order to test these hypotheses, a series of tests were conducted:

1. samples were first saturated and then tested and

2. tests were conducted after removing the fibers longer than 1 cm.

Figure 6 shows the effect of saturation on the shear force-displacement behavior and the failure envelope for the slag. The test was performed on slag samples at a relative density of 30 percent. It is noted that the cohesion intercept was somewhat reduced.

Figure 7 presents the influence of the fibrous particles on the shear force-displacement behavior and the failure envelope. This test was performed at a relative density of 66%. The figure depicts the decrease in strength due to removal of the fibers. When the slag material is sheared in the direct shear box, it tends to dilate (the amount of dilation is dependent upon its relative density and confining stress). The friction mobilized at the slag-fiber interfaces restrains any relative displacement between the slag particles and the fibers. Consequently, tensional forces are developed on the fibers. These forces increase the average normal stress on the failure surface and reduce the average shear stress carried by the soil. Thus, it is believed that the fibers are the main sources of the cohesion intercept and as a consequence, the higher shear resistance of the slag. Parallel displacement of shear envelopes suggests that the fibers do not affect the frictional properties of the slag.

The existence of the fibrous particles will improve the characteristics of the slag. In general, the fibrous particles restrain the volume expansion and provide an interlocking effect to the slag mass. This, in return, leads to the emergence of the cohesion intercept. The critical state diagram obtained from the volume changes in the direct shear test is presented in Figure 8. The slag displays similar characteristics as that of coarse, angular sands.

FIELD TESTS

Procedure

A field study was conducted to assess the field compaction characteristics. Since the gradation of the slag is similar to that of granular soils, a smooth drum vibratory roller (Galion Model 490012) was selected for compaction. The dead weight of this roller was 56 kN (12.4 kips). The roller can be operated at a frequency of 1800 vib/min and the additional equivalent static weight at this frequency was 27 kN (6.1 kips). The following procedure was used:

1. The slag was spread in an area 30.5 m (100 ft) by 6.1 m (20 ft). A 0.61 m (2 ft) thick base layer was compacted by ten passes of the roller over the existing vegetation. No stripping was done since the vegetation would help confine the slag and facilitate compaction. The subgrade consists of clay, silty clay and some silt.

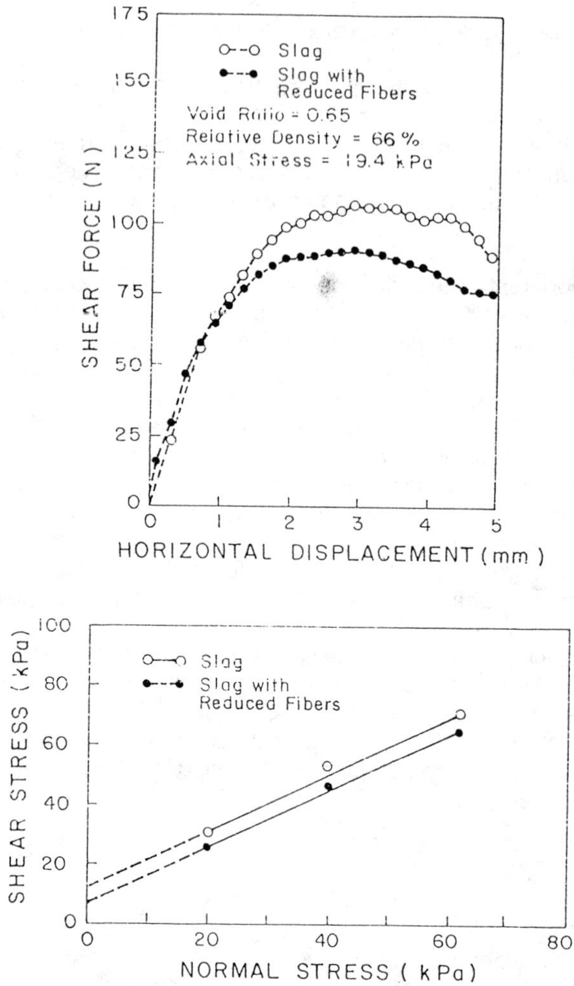

FIG. 7 -- The effect of fibrous particles on force-displacement behavior and failure envelope.

2. A 0.92 m (3 ft) thick layer of slag was placed over the base layer with a dozer. In-situ density tests were conducted in this fill in a zone excluding the outer 1.53 m (5 ft). This eliminates any boundary effects due to the sloping sides.

3. The fill was wetted with water to achieve as high a water content as permitted but below the saturation level and was compacted with two passes of the roller. Density tests were conducted at randomly selected locations by using a random number generator in a hand-held calculator.

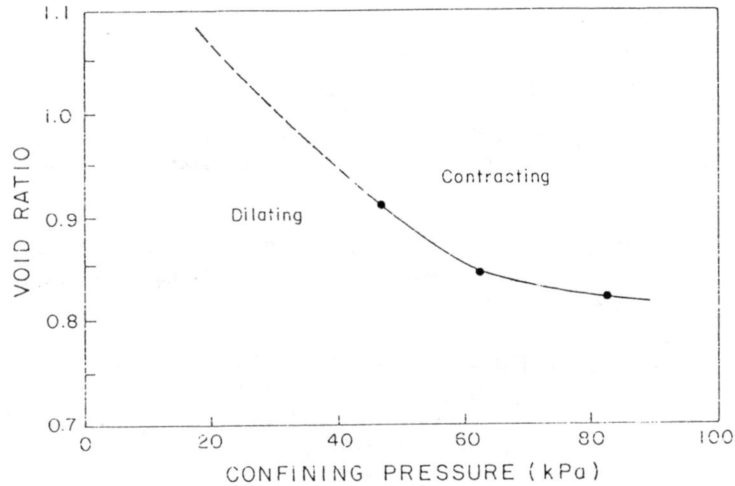

FIG. 8 -- Critical state diagram for slag.

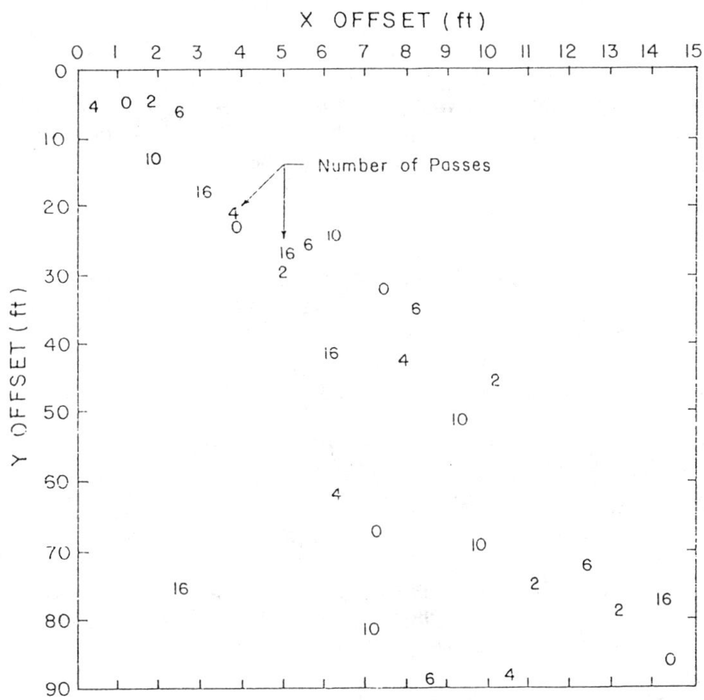

FIG 9 -- Location of tests.

4. The fill was sequentially recompacted by additional number of passes. The total roller passes were 2, 4, 6, 10 and 16. The density tests were repeated after each phase as described above.

The locations of all tests and a schematic view of the test pad are presented in Figure 9. It is noted that the tests scatter diagonally across the fill. This was due to using the same random number in scaling the test location in both horizontal and vertical directions.

The weather conditions were dry during the time when the first 10 passes were completed. However, there was a heavy rain after the 10th pass. Rolling for the last phase (16 passes) was done immediately after this rain.

The sand-cone method [12] was used to determine the insitu density. Tests were conducted at depths of 10 cm (3.9 in.), 25 cm (9.8 in.) and 45 cm (17.7 in.). When the total number of passes were 10, the last test depth was changed to 50 cm (19.6 in.) and another was added at 70 cm (27.6 in.).

Grain size distributions [13] and minimum density tests [9] were conducted on samples of slag collected from the fill after 16 passes of the roller. Grain size distributions before and after compaction did not indicate any difference. This implies that there was not a significant crushing of the particles due to compaction, as was observed in the Standard Proctor laboratory compaction tests.

The minimum density tests indicated a density of 14.3 to 14.4 kN/m^3 (89.1 pcf to 89.6 pcf) as compared with 14.2 kN/m^3 (88.9 pcf) in the uncompacted specimens. These results compared favorably with previous test results [5,14]. It was therefore concluded that compaction with 16 passes of the roller did not result in a change in the grain size distribution and minimum density of the slag.

Analysis

Field dry density values are plotted along the depth of the fill for each compaction sequence. A smooth curve is passed through the mean values. Figure 10 compares this line passing through mean values of field density after each phase of compaction with the range of initial field densities. As expected, the number of passes increases the compaction results in densification at increasingly greater depths.

The effective depth of densification, h_e, is defined as that depth below which a specified relative density cannot be achieved. For example, for a relative density of 50 percent, the effective densification depth increased from 12 cm (5 in.) to almost 55 cm (22 in.) with an increase in the number of passes from 2 to 16.

The relative density, D_r, of granular soils such as sand is the principal parameter that controls the angle of internal friction, ϕ', and, therefore, the shearing resistance. Such soils are conventionally classified as [15]:

1. Very loose, $D_r \leq 15\%$, $\phi' \leq 28°$

2. Loose, $15\% < D_r \leq 35\%$, $28° < \phi' \leq 30°$

3. Medium, $35\% < D_r \leq 65\%$, $30° < \phi' \leq 36°$

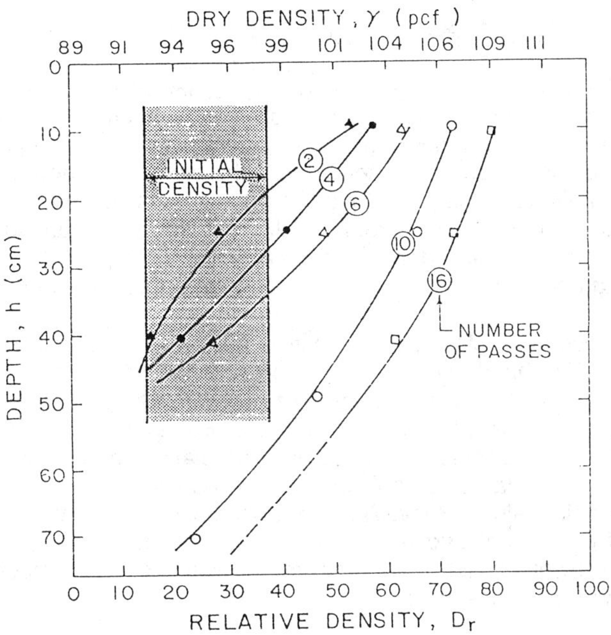

FIG. 10 -- Dry density versus depth in field tests.

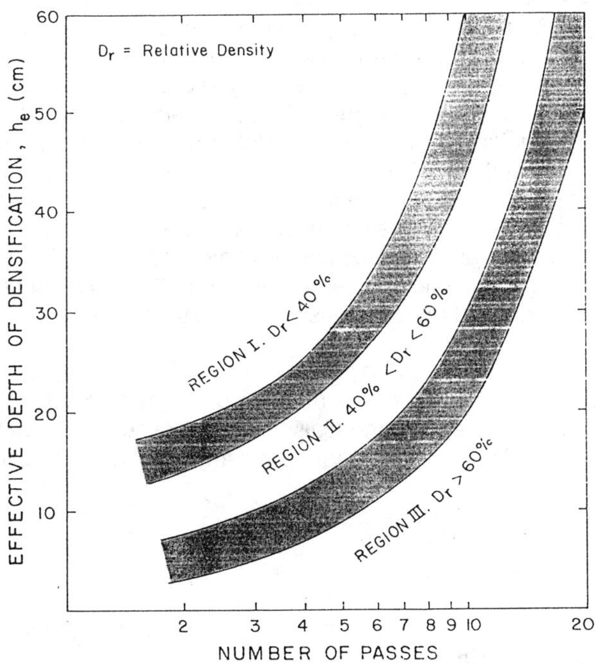

FIG. 11 -- The influence of number of passes on the effective depth of densification.

4. Dense, $65\% < D_r \leq 85\%$, $36° < \phi' \leq 41°$

5. Very dense, $D_r > 85\%$, $\phi' > 41°$

The internal friction angle for the slag is 30° in its loosest state ($D_r = 0\%$) [5]. Therefore, even in its loosest condition, slag has a higher internal friction angle than typical sands because of its interlocking capability. This loose state corresponds to medium dense conditions in sands. Even though the slag would generally exhibit a higher angle of internal friction than sand at a comparative relative density value, it may require more shear and volumetric strain to fully mobilize the shearing resistance. Thus, it is better to require relative density values that will minimize the volume change potential of the slag and, correspondingly, produce high angles of internal friction.

Figure 11 presents a plot of the effective depth of densification versus the number of passes. The bands in this figure represent the data within one standard deviation around the mean. Regions I, II and III defined in the figure represent relative densities less than 40 percent, between 40 and 60 percent, and greater than 60 percent, respectively. It should be pointed out that relative density values of 40 and 60 percent correspond to relative compaction values of 86 and 90 percent, respectively. The highest dry density value achieved in the field test, 109 pcf, corresponded to a relative density of 86 percent and a relative compaction of 96 percent. For most geotechnical and highway applications, a maximum relative compaction of 98 percent (D_r = 93 percent) would be specified.

It should be noted that relatively high friction angles are readily achieved during the initial placement of the slag. However, in this state the slag is very nonhomogeneous. This inhomogeneity may lead to differential settlements of structures or roads founded upon such uncompacted fills. Therefore, it is recommended that the slag be densified to relative densities exceeding 60-70 percent (or relative compactions of 90-93 percent) when used as a base for structural or pavement loads. The chart given in Figure 11 could readily be used to estimate the number of passes and the lift thicknesses required to achieve an appropriate relative density (or relative compaction). This is based on the premise that a roller of similar design and size would be used. Increased densities could be achieved at greater depth and with fewer passes using heavier rollers. The effective depth of densification in this figure can be taken as the maximum lift thickness for the desired level of densification. For the specific roller utilized in this study, the approximate number of passes to achieve the indicated relative densities for lift thicknesses of 1 ft and 2 ft are reported in Table 5.

Since Figure 11 indicates that the number of passes required to achieve a desirable densification increases exponentially, lift thicknesses of greater than 60 cm (2 ft) are not recommended unless additional tests verify that adequate compaction can be achieved.

SUMMARY AND CONCLUSIONS

Boiler slag is a valuable resource. In the last decade, its use as a construction material has become more common.

TABLE 5 -- Number of Passes Needed for Different Lift Thicknesses

Lift Thickness (ft)	Number of Passes	
	$40\% < D_r < 60\%$ (86% < R.C. < 90%)	$D_r > 60\%$ (R.C. > 90%)
1.0	6	12
2.0	10	20*

*Based on extrapolation of test data.

Dow slag studied in this paper is a poorly graded, medium to fine granular material. The slag displays slight crushing and degradation characteristics in the modified Proctor test. Compactions with a smooth drum vibratory roller in the field did not result in any particle degradation in the slag.

Due to its angularity and interlocking capability, the internal friction angle and the shearing resistance of the slag are higher than expected in similar granular material. A cohesion intercept is displayed due to the presence of long fibrous slag particles. Based upon the field compaction studies, the following recommendations are made for the optimum design and construction procedures for a slag fill:

1. A base should be prepared in construction of a controlled fill using the slag. This base provides the necessary confinement for proper compaction of the overlaid slag. A base course of about 30 cm (1 ft) should be laid over the ground surface and compacted with 16 passes of the roller,

2. Subsequently, a lift thickness should be selected. Figure 11 should be used for this purpose. For compactors essentially equivalent to that used in the field study, a maximum initial lift thickness of 30 cm (12 in.) is recommended. For this thickness, adequate densification can be achieved in 8 to 12 passes,

3. The fill should be wetted for efficient densification. This could be accomplished by wetting the fill with water before rolling. The fill should be wetted above the optimum water content since bulking at lower water contents will result in reduced dry densities. However, subgrade softening associated with such an operation should be considered and adequate underdrainage should be provided,

4. For temporary roads, multi-purpose fills, and fills with light structural loads (generally less than 48 kPa), a relative density in the range of 40-60 percent (relative compaction of 92-96 percent) is recommended,

5. For permanent facilities and heavier structural loads, a relative density of 60 percent (relative compaction of 96 percent) is recommended. If these facilities involve foundations for reciprocating equipment, a relative density exceeding 70 percent (relative compaction of 98 percent) should be specified,

6. The performance of a fill or a structure placed on a fill will be dependent on the properties of the soils underlying the fill as well as the fill itself. Thus, an adequate subsurface exploration and laboratory testing program should be considered an integral part of a slag fill project.

ACKNOWLEDGEMENTS

This study was funded by Dow Chemical Co., Plaquemine, Louisiana. The funds provided by this company are gratefully acknowledged. Messrs. Laique Haider and Halis Ider assisted in the laboratory studies pertaining to the engineering characteristics of the slag. Messrs. Pradeep Kurup and Rainer Echle assisted in the field compaction study.

REFERENCES

[1] Seals, R. K. and Moulton, L. K., "Bottom Ash - An Engineering Material," Journal of Soil Mechanics and Foundation Division, ASCE, No. SM4, 1972, pp. 8817.

[2] Dow Chemical Company and Louisiana Gasification Technology, Inc., "Petition for Exemption of Slag from Section 4.2 of Louisiana Solid Waste Rules and Regulations," August 1985.

[3] Moulton, L. K., "Utilization of Ash from Coal-Burning Power Plants in Highway Construction," Highway Research Board 430, Utilization of Waste Materials and Upgrading of High Quality Aggregates, National Academy of Sciences, 1973.

[4] ASTM D422-63, D2487-85, "Standard Test Method for Classification of Soils for Engineering Purposes," Annual Book of ASTM Standards, Vol. 4, Construction, 1989, pp. 288-298.

[5] Seals, R. K., Ider, H. M. and Acar, Y. B., "Assessment of Optimum Design Considerations for the Slag Fill Project - Part I. Laboratory Investigation," Report No. GE86-04, Department of Civil Engineering, Louisiana State University, 1986, 18 p.

[6] ASTM D698-78, "Test Methods for Moisture-Density Relations of Soils and Soil-Aggregate Mixtures Using 5.5 lb (2.49 kg) Rammer and 12-in (305-mm) Drop," Annual Book of ASTM Standards, Vol. 4, Construction, 1988, pp. 149-153.

[7] ASTM D1557-78, "Test Methods for Moisture-Density Relations of Soils and Soil-Aggregate Mixtures Using 10 lb (4.54 kg) Rammer and 18-in (457-mm) Drop," Annual Book of ASTM Standards, Vol. 4, Construction, 1988, pp. 206-210.

[8] Acar, Y. B., Ider, H. and Seals, R. K., "Assessment of Optimum Design Considerations for the Slag Fill Project - Part 2. Site Reconnaissance Study and Preliminary Compaction Specifications for the Slag Fill Project," Report No. GE87-01, Department of Civil Engineering, Louisiana State University, 1987, 23 p.

[9] ASTM D4254-83, "Minimum Density of Soils and Calculation of Relative Density," Annual Book of ASTM Standards, Vol. 4, Construction, 1989, pp. 572-578.

[10] ASTM D4253-83, "Test Methods for Maximum Index Density of Soils Using a Vibratory Table," Annual Book of ASTM Standards, Vol. 4, Construction, 1988, pp. 554-565.

[11] ASTM D2435-80, "Test Method for One-Dimensional Consolidation Properties of Soils," Annual Book of ASTM Standards, Vol. 4, Construction, 1988, pp. 278-282.

[12] ASTM D1556-82, "Density of Soil in Place by the Sand-Cone Method," Annual Book of ASTM Standards, Vol. 4, Construction, 1989, pp. 206-210.

[13] ASTM D2487-85, "Classification of Soils for Engineering Purposes," Annual Book of ASTM Standards, Vol. 4, Construction, 1988, pp. 283-292.

[14] Seals, R. K., Acar, Y. B. and Puppala, A., "Assessment of Optimum Design Considerations for the Slag Fill Project - Part 3. Laboratory Calibration Studies with the Dynamic Penetrometer, Department of Civil Engineering, Louisiana State University, 1988, 21 p.

[15] Gibbs, H. J. and Holtz, W. G., "Research on Determining the Density of Sands by Spoon Penetration Testing," 4th ICSMFE, Vol. 1, 1957, p. 35.

Michael C.R. Davies

THE GEOTECHNICAL PROPERTIES OF CEMENTED COLLIERY WASTE FOR USE IN LAND FILL

REFERENCE: Davies, M.C.R., "The Geotechnical Properties of Cemented Colliery Waste for Use in Land fill". Geotechnics of Waste fills - Theory and Practice, ASTM STP 1070, Arvid Landva, G. David Knowles, Ed., American Society for Testing and Materials, Philadelphia, 1990.

ABSTRACT: The reject resulting from the reworking of colliery waste tips may be utilised for landfill. When the coarser reject is mixed with the tailings from the washing process the mixture presents problems for compaction, which may be alleviated by the addition of cement. A test program was conducted to characterise the geotechnical properties of the cemented material. The properties of the tailings were observed to be governed by the moisture content at which cement was added and the level of effective stress. The properties of the mixed spoil varied according to the proportions of coarse material to tailings. As a result of the testing program conclusions regarding the acceptability of cemented colliery waste for use as a fill material can be made.

KEYWORDS: waste fills, colliery discard, geotechnical properties, cement stabilization, tailings disposal.

INTRODUCTION

Land reclamation in coal mining areas of the U.K. often entails the removal of colliery waste tips. These tips, which may be up to and in some circumstances in excess of a century old, contain quantities of coal which may be commercially recovered. The attractiveness of the benefits from coal recovery means that this process frequently accompanies land reclamation schemes. The unrecovered material from the tip being used for landfill in the program of reclamation.

The process of "washing" - used in the mining industry for maximising coal recovery - is used in the "reworking" of colliery waste tips. This process involves the use of water, vast quantities of which, polluted with fine particles of coal (smaller than

Dr. Davies is lecturer in Soil Mechanics at University of Wales, College of Cardiff, P.O. Box 917, Cardiff, CF2 1XH, U.K.

approximately 1.0mm), has to be treated prior to disposal. Environmental considerations require that this water cannot be disposed of without treatment. It is also necessary to treat the plant water prior to recycling. Hence techniques have been devised to separate as much of the water as possible from the waste material - which is then tipped. Methods of dewatering include centrifuges, filter presses and deep cone thickeners. The result of treating contaminated water with such a process results in the production of a tailings material, the moisture content of which varies very much with the process adopted and its method of operation.

The tailings produced from such processes are mixed with the coarser discard prior to placement as land fill. Despite dewatering the tailings still have a relatively high moisture content; which can be in excess of the liquid limit. When mixed in quantity with coarse discard this results in a very wet material which proves difficult to handle. The technique of increasing the strength of the tailings by the addition of small quantities of cement - typically between 2% and 4% by wet weight - prior to mixing with the coarser material has been adopted in some instances to alleviate the problems of handling these mixtures [1].

In order to design slopes, highways, excavations and foundations in the fill material it is necessary to know the effect of the cement addition on the Geotechnical properties of the discard. This paper describes a program of laboratory tests to investigate consolidation, strength and stress-strain characteristics of specimens of cemented tailings and mixed spoil, in which the specimens were constitute to represent the typical extremes of field conditions experienced in the United Kingdom.

COMPOSITION OF REJECT

In the process of coal washing from tips the ratio of tailings to the coarser material in the reject can be highly variable. Of the material leaving the washery (i.e. coal product and waste materials) typically some 10% is recoverable coal product; the remainder of the material leaving the plant being discard. The discard is of two types (i) coarse discard - which can constitute between 60% to 80% of all the material leaving the plant and (ii) tailings - constituting between 10% and 30% of the total. Therefore, the ratio of tailings to coarse material in the reject typically varies between 11:89 and 33:67. Hence the "as tipped" mixture of coarse reject and tailings can be highly variable.

The coarse discard is separated from the plant water in a screening process; the smallest particle size being retained on the screens being 1.0mm. Sieve analysis of the coarse material from a number of plants indicated that 95% of the course material passes the 20mm sieve - the largest particle size is typically 100mm - and the material may be classified as a well graded gravel, GW [2]. The tailings constitutes all the reject material passing the 1mm screen. Depending on the dewatering technique adopted, the moisture content (weight of water/weight of dry material), w, of the tailings can vary between 25% and 66.6%. Classification tests indicate that typically the LL = 37 and the PL = 18; the material being a sandy lean clay, CL [2]. When cement is added to the tailings this is generally mixed with the wet tailings in ratios of between 2:100 and 4:100 (weight of cement:weight of wet tailings) prior to being mixed with the coarse reject.

TEST PROGRAM

In order to investigate the large variations in grading and moisture content of colliery reject, two series of tests were conducted. In the first the effect on the cemented tailings of variation of the moisture content was investigated, by conducting consolidation and triaxial tests on specimens of tailings mixed with cement at the highest and lowest moisture contents encountered, i.e. 66.6% and 25% respectively, with a control series of similar experiments conducted on specimens of uncemented tailings. The second series involved triaxial testing of four different mixtures of tailings and coarse material. The program entailed testing specimens which reflected the extremes of moisture content and mixture of tailings to coarse reject encountered, as shown in Tables 1 and 2.

TABLE 1 -- Tailings tests - specimen identification

Test I.D. (s)	Moisture content when mixed, w_i, %	Cement	Test Type
COND	N/A	NO	C
CONW	N/A	NO	C
CONCD	25.0	YES	C
CONCW	66.6	YES	C
N1, N2, N3[a]	N/A	NO	T
D1, D2, D3[a]	25.0	YES	T
W1, W2, W3[a]	66.6	YES	T

c = 1-D consolidation, T = 3 No. Triaxial tests
[a] Confining pressures 1 - 230 kPa, 2 - 400 kPa and 3 - 600 kPa

TABLE 2 -- Mixture tests - specimen identification

Test I.D. (s)	Moisture content when mixed, wi, %	Coarse %	Tailings %
AD1, AD2	25.0	89	11
AW1, AW2	26.6	89	11
BD1, BD2	25.0	67	33
BW1, BW2	66.6	67	33

SAMPLE PREPARATION

All specimens were prepared from oven dried material (obtained from a spoil tip) which had previously been subjected to a coal recovery process. Comparison of the results of tests conducted on plant tailings (i.e. undried) and reconstituted dried tailings have indicated differences in measured engineering properties (such as strength parameters), but not in the general pattern of material behaviour. Since the study described herein was fundamental rather than specific, the effects of drying were not detrimental to the observations. The tailings specimens were reconstituted at the required moisture content and cement added (when specified) at 4% by wet weight. In an industrial context the moisture content would have been obtained by moisture content reduction using one of the methods mentioned above. When using a cone thickener or centrifuge quantities of flocculant are added to the slurry to accelerate the settle process. Studies have shown [3] that the presence of flocculant in the doses commonly found in U.K. coal washeries (typically up to 500 g/tonne) has no measurable effect on the geotechnical properties of the tailings and consequently none was added to the powered fines and water in the preparation of the tailings.

The tailings to be mixed with the coarse material for the mixture tests were prepared using the method described above prior to mixing with the coarser fraction of the spoil. Specimens were prepared by compacting the materials in 3 layers, with 27 blows per layer from a standard 2.5 kg compaction hammer falling through 0.3 m [4], into a specially prepared mould 200mm high and 100mm in diameter. Following preparation the specimens containing cement were immersed in a tank of water and permitted to cure for at least 28 days prior to testing.

EXPERIMENTAL RESULTS

Tailings Tests

Consolidation tests: Standard oedometer tests were conducted following the procedures of B.S. 1377 [4]. The specimens were 100mm diameter and with an initial height of 19mm. The plots of vertical effective stress v. voids ratio for the tests, Fig. 1, clearly show, in agreement with other studies of cemented tailings [3,5], that the consolidation behaviour of cemented tailings is dominated by the moisture content, w_i. Comparison of the curve for the cemented specimen CONCW (w_i = 66.6%) with the results of the uncemented specimens CONW and COND indicates that during the initial stages of the tests (i.e. at low stress levels) the change in voids ratio with increasing vertical stress was lower in the cemented specimen. Following this initial stage, as the vertical stress was increased the cemented specimen displayed larger changes in voids ratio than the uncemented specimens. The curve for the cemented tailings mixed at the lower moisture content , CONCD (w_i = 25%), displays similar behaviour at low stress levels; the volume change during consolidation of CONCD was lower than that for CONCW as a result of the lower value of w_i. Results for the uncemented specimens were very typical of natural soils. The plot for CONW was linear whilst the plot for COND (mixed at the lower moisture content) tended to the plot of the initially wetter material with increase in effective vertical stress.

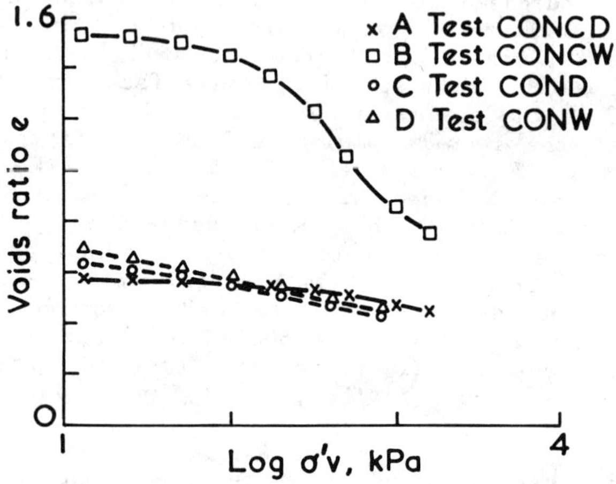

FIG. 1 -- Consolidation tests - cemented and uncemented tailings

The results of these one-dimensional consolidation tests indicate that the presence of cement allows the tailings to sustain a higher voids ratio than the uncemented material in the range of stress tested, a condition known as "meta-stability". This phenomenon has been observed in other artificially bonded and natural residual soils [6]. Studies of meta-stable soils have indicated that as consolidation stresses are increased so the compression curve of the meta-stable soil converges on that of the completely de-structured, i.e. remoulded, material [6]. Figure 1 shows the relative positions of the compression curves for the cemented and uncemented specimens; from which it can be seen that the compression curve for the cemented material (CONCW) is converging on the normal consolidation line for the uncemented tailings.

Triaxial tests: The results of the triaxial tests, summarized in Table 3, indicated that the effective stress parameters are strongly affected by the presence of cement and by the moisture content at which the cement was added to the tailings. The values of the effective strength parameters for the W series show a greater similarity to those of the uncemented tailings than do the parameters for the D series. Comparison of the values of the apparent cohesion measured in each series of tests (Table 3) reveals that at low levels of effective stress the cemented tailings have a greater strength than uncemented tailings; the tailings mixed with cement at low moisture content having a greater strength than those mixed at high moisture content.

TABLE 3 -- Results of triaxial tests on tailings

Test I.D.	Moisture content when mixed, %	Friction angle, ϕ'	Apparent Cohesion, c'
N	no cement	37.04	0
W	66.6	38.14	25kPa
D	25.0	45.00	50kPa

A greater understanding of the relationship between strength and consolidation pressure for the tailings may be obtained by plotting for each specimen the undrained shear strength against the consolidation pressure, Fig.2. Results of the tests on the uncemented material indicate that the uncemented tailings behave like a normally consolidated frictional soil. It is well established that for most normally consolidated soils the undrained shear strength is directly proportional to the effective confining pressure [7]. Hence, a straight line may be plotted through the points obtained from the specimens of uncemented tailings in Fig.2. For the W series of tests the points plotted for the lowest confining stress lie above the uncemented strength line. However the other two tests indicate that the undrained shear strength of the W series cemented material was of very similar value as the uncemented material; being consistently slightly lower in strength. By contrast the results of the D series of tests were significantly greater than the N and W tests at all stresses.

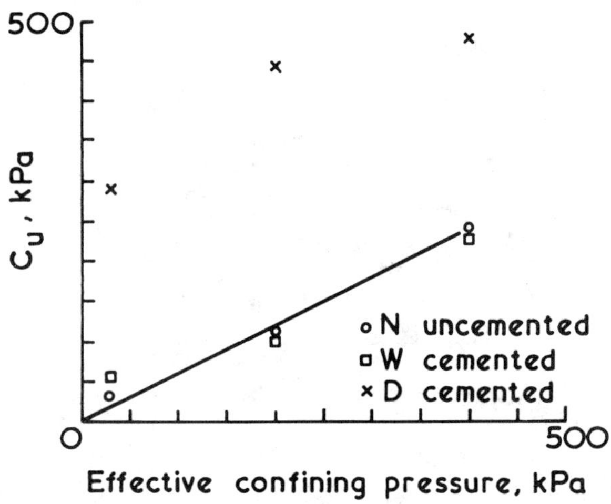

FIG. 2 -- Tailings tests undrained shear strength v. consolidation pressure

The consolidation behaviour of cemented tailings, discussed above, has indicated that the voids ratio of specimens depends on the moisture content of the material when mixed. Hence, at the same effective confining pressure the voids ratio of specimens in the three series of tests would not have been the same. In naturally occurring soils the undrained shear strength is a function of the voids ratio (or if fully drained the moisture content) hence the interpretation of the triaxial test data may be best undertaken by considering the moisture content of each of the specimens following isotropic consolidation.

For naturally occurring, normally consolidated, fully saturated soils it may be shown [8] that at failure there is a linear relationship between the moisture content and the natural log of the mean effective stress. In addition, as has been shown above, a linear relationship exists between the undrained shear strength and the effective confining pressure (i.e. mean effective stress) of the uncemented tailings. Hence, if the natural log of undrained shear strength of the uncemented tailings is plotted v. the moisture content the result will be a linear plot; as displayed in Fig.3. The Figure shows that at the same values of moisture content the undrained shear strength of the cemented specimens is consistently higher than that of the uncemented material. As the moisture content decreases (i.e. as has been seen from the results of the consolidation tests, the voids ratio of the cemented material tends to the voids ratio of the uncemented tailings) the results would imply that the undrained shear strength of the cemented tailings might be approaching that of the uncemented tailings.

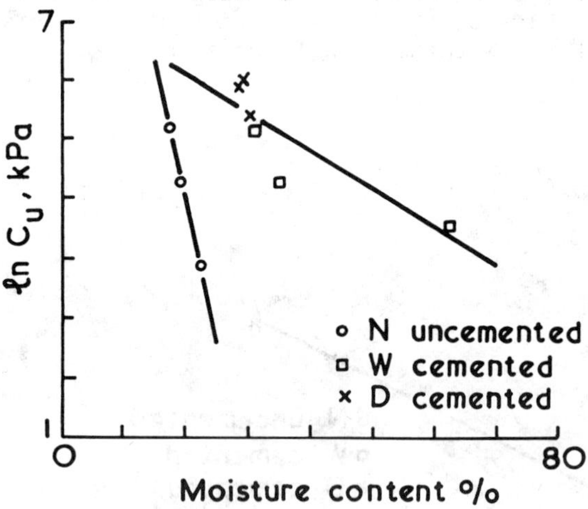

FIG. 3 -- Tailings tests - natural log of undrained shear strength v. moisture content

The results indicate, therefore, that there is some consistency in the total stress parameters for the cemented tailings (i.e. the undrained shear strength is a function of moisture content (or voids ratio). During consolidation the contribution to the strength of the tailings of the cemented matrix reduces and at the same time, due to the increase in effective stress, the contribution to the strength of the friction in the material increases. Ultimately at high effective stress and low moisture content the matrix is fully ruptured and the tailings will behave in the same manner as the uncemented material.

The results above indicate that, since in cemented tailings the moisture content at a particular confining pressure depends on the initial moisture content when mixed, the undrained strength parameters of normally consolidated cemented tailings (with a consistent ratio of tailings to cement) are a function of voids ratio and not confining pressure. Whereas in naturally occurring normally consolidated soils since voids ratio is a function of confining pressure the undrained strength parameters are a function of either.

Results of Mixture Tests

Compaction of specimens: Compaction of the mixture of tailings and coarse material resulted in specimens with values of dry unit weight between 1.681 and 1.767 Mg/m^3. The moisture content, w, of the specimens following consolidation to the required confining pressure lay between 10.95 and 17.34%. These values are all within the range of typical values recorded on existing spoil heaps constructed from cement stabilised material [5].

Triaxial Tests - Compressibility: Prior to the shearing stage of the triaxial tests the specimens were consolidated to the required preconsolidation pressures. Values of the coefficient of volume change (for isotropic consolidation) for an initial increment of 100kPa were similar for all mixtures - the average being 0.11 m^2/MN - reflecting the similar consolidation characteristics of cemented tailings at varying initial moisture content, w_i, at low stress levels, c.f. Fig.1. When consolidated to higher levels of stress the mixtures containing the higher quantities of tailings displayed a greater change in volume. However, no substantial difference was observed in the compression behaviour of samples with tailings at varying initial moisture content, w_i. This was most probably because the maximum consolidation pressure achieved (400kPa) was insufficient to cause substantial rupture of the cemented matrix in the tailings. Hence, the compression characteristics of cement stabilised mixed spoil in the stress range investigated are dominated by the quantities of cemented tailings and not their initial moisture content.

Triaxial Tests - Shear strength: The effective stress parameters obtained from the results of the triaxial tests are summarised in Table 4. The tests on the mixtures with the larger proportion of coarse material, i.e. AD and AW, indicated very similar parameters. This indicates that at this mixture of 89% coarse material to 11% tailings the initial moisture content of the tailings has very little effect on the effective strength parameters of the material. It has been shown above, that the strength parameters of the cemented tailings are strongly effected by w_i of the tailings. Hence, it may be deduced that for the ratio of coarse material to tailings in the AD and AW mixtures that the strength parameters are dominated by the coarse component.

TABLE 4 -- Results of triaxial tests on mixture

Test I.D.	Friction angle, ϕ'	Apparent Cohesion, c'
AD	32.3	30kPa
AW	33.44	25kPa
BD	28.5	175kPa
BW	32.8	80kPa

Markedly different behaviour was observed in the tests on the BD and BW mixtures, where it can be seen from table 4 that the strength parameters were significantly different in the two series of tests. The values of the effective strength parameters obtained from the BW tests are between those of the BD tests and the A mixture tests (AD and AW). This displays that the strength parameters of the B mixtures are effected by the initial moisture content of the tailings. A series of consolidated drained triaxial tests conducted on samples of uncemented mixed colliery discard indicated that the effective stress strength parameters for this material were $\phi' = 28^{o}$ and $c' = 0$. Thus the influence of the cemented tailings is to provide higher strengths at lower confining pressures.

COLLIERY SPOIL AS A LANDFILL

It is desirable that the areas of landfill on which construction is planned to take place provides a material which will minimise the cost of providing foundations. Thus placement should result in a material with as low a compressibility and as high a bearing capacity as can be achieved. To ensure this it is necessary to compact the landfill during placement.

As has been mentioned above, the high moisture content of the mixed spoil renders it a material which is very difficult to handle. For the discard discussed herein, the limiting moisture conditions for the uncemented material were assessed by means of the "Moisture condition test" (MCV) [9]. The limiting moisture content for accessibility of compacting plant being 11.5%. Triaxial testing of compacted cemented mixed shale indicated typical moisture contents varying between 10.95% and 17.34% following consolidation. From which it should be noted that the sample with the lowest moisture content lies close to the limit of acceptability. To avoid mixing the two components, and thus permit limited compaction, a technique of layering the coarse material and tailings has been adopted in the formation of colliery spoil tips.

Only the capping layer of coarse material is compacted. However, the compressibility of the tailings layer can result in settlements unacceptable for commercial development of a site. Experience has shown [1] that when small quantities of cement are added to mixed colliery discard compacting equipment can operate on the fill. The results of the experiments described herein, indicate that cement addition enhances the effective stress parameters of mixed discard, hence improving its bearing capacity.

The consolidation tests on the mixed spoil indicated that the quantity of tailings present did not greatly affect the compressibility characteristics. However, it should be noted that the maximum consolidation pressure in the testing program was 400kPa. At higher levels of confining stress it would be expected that mixtures containing larger quantities of tailings at high moisture content (i.e. BW mixture) would display greater compressibility due to consolidation of the cemented tailings. Therefore, when subjected to high stresses, differential settlement might occur on a site where the proportions of coarse discard to tailings varied greatly during placement. At lower stress levels, due to the meta-stable state of the tailings, cemented mixtures containing high quantities of tailings would display less compressibility than uncemented materials.

CONCLUSIONS

Investigations into the geotechnical properties of typical colliery discard materials following washing have been conducted to investigate the suitability of this material for use in landfill. The following conclusions may be made from the investigations.

1. The process of coal washing results in the ratio of tailings to coarser reject being highly variable. The coarse material typically constitutes 67%-89% of the reject from the process whilst the tailings, which have a moisture content varying between 25% and 66.6%, constitutes 11%-33%. Hence the "as tipped" mixture of coarse reject and tailings can be highly variable.

2. Testing of the mixed colliery spoil indicated that the compressibility, effective stress strength parameter and stress-strain behaviour of the specimens for the mixtures containing the least tailings were dominated by the properties of the coarse material. Conversely the mixtures containing the highest amounts of tailings were greatly influenced by the properties of the tailings.

3. The high moisture content of the tailings results in a material which is highly compressible and not suitable for compaction. Cement addition results in an increase in the undrained shear strength and a meta-stable material which displays reduced compressibility at low stress levels. However, as effective stress is increased the voids ratio of the cemented material converges on that of the uncemented material. Ultimately, at high effective stress and low moisture content, evidence suggests that the

cemented tailings return to the properties of the uncemented material.

REFERENCES

[1] Llewellyn, L.I.D, "Cement stabilisation of tailings and the security of tips", Mine and Quarry, May 1984, pp 28-33.

[2] American Society for Testing and Materials, "Annual book of ASTM standards: Soil and Rock; Building Stones; Geotextiles", Vol. 04.08, ASTM, Philadelphia, PA, 1989.

[3] Williams, K.P., Johnston, A.G. and Davies, M.C.R., "Cement stabilization of tailings", Report to British Coal, 1986.

[4] British Standards Institution, "Methods of test for soils for civil engineering purposes", BS 1377 : April 1975.

[5] Sir William Halcrow and Partners, "Methods of disposal of artificially cemented tailings", Report, 1982.

[6] Vaughan, P.R., Maccarini, M. and Mokhtar, S.M., "Indexing the Engineering Properties of Residual Soil", Quarterly Journal of Engineering Geology, Vol.21, No.1, 1988, pp 69-84.

[7] Skempton, A.W., Discussion on "The planning and design of a new Hong Kong airport", Proceedings of the Institution of Civil Engineers, 7, 1957.

[8] Schofield, A.N. and Wroth, C.P., Critical State Soil Mechanics, McGraw Hill, London, 1968.

[9] Parson, A.W. and Boden, J.B., "The Moisture Condition Test and its Potential Applications in Earthworks", Supplementary Report 552, Transport and Road Research Laboratory, U.K., 1979.

Demetrious C. Koutsoftas, and Michael L. Kiefer

IMPROVEMENT OF MINE SPOILS IN SOUTHERN ILLINOIS

REFERENCE: KOUTSOFTAS, D.C., and KIEFER, M.L. "Improvement of Mine Spoils in Southern Illinois", Geotechnics of Waste Fills - Theory and Practice, ASTM STP 1070, Arvid Landva, G. David Knowles, editors, American Society for Testing and Materials, Philadelphia, 1990.

ABSTRACT: Field tests were conducted to evaluate the feasibility of preloading and dynamic consolidation to improve the compressibility characteristics of a mine waste spoil approximately 30 m deep. The mine spoils consisted of coarse granular materials ranging in size from sand to boulders for the lower 11 m and a heterogeneous mixture of materials ranging from clays to boulder size rock fragments for the upper 19 m.

A test fill approximately 7.6 m high was constructed in 10 days, and settlements were measured at various depths over a period of 200 days. Settlements as large as 45 cm were measured and most of the settlement occurred as soon as the test fill was completed.

Dynamic consolidation was performed in two test areas using different pounding grid configurations. The test areas were approximately 0.2 and 0.4 hectares, respectively. Dynamic consolidation was performed with a 16 ton pounder falling approximately 20 m. Geophysical tests as well as pressure meter tests indicate significant improvement of the spoils over depths of 9 m to 12 m.

KEYWORDS: preloading, dynamic consolidation, settlement, compressibility, mine spoils, geophysical tests.

INTRODUCTION

A proposed industrial complex was to be constructed at a site of approximately 50 hectares in Southern Illinois, underlain by the wastes of coal strip mining operations. Preliminary investigations revealed that the waste fills were approximately 30 m deep, extremely heterogeneous, and quite compressible. Deep foundations were considered impracticable and prohibitively expensive because of the presence of large boulders within the fill and because

Mr. Koutsoftas is an Associate at Dames & Moore, 221 Main Street, San Francisco, CA 94105; Mr. Kiefer is a Principal at Woodward-Clyde Consultants, 203 North Golden Circle Drive, Santa Ana, CA 92705.

of the potential for downdrag loads that could be caused by the fill due to settlements under the fill's own weight. The objective of the study was to develop cost-effective techniques to improve the compressibility characteristics of the waste fills so that the proposed structures could be supported on shallow foundations with tolerable post construction settlements.

There has been relatively little experience with the compressibility of waste fills created by open strip mining and even less experience on the actual performance of structures constructed on mine waste fills. In England [1, 2], experimental houses were constructed over strip mining waste fills to evaluate their settlement behavior. Dynamic consolidation and preloading were found effective in improving the compressibility of wet, cohesive, heterogeneous mine waste fills. Similar studies performed elsewhere in England [3] confirmed the potential for large and variable settlements under even light loads, placed on unimproved waste fills.

In view of the limited prior experience with the construction of heavy facilities over mine waste spoils and in view of the potential for large and erratic settlements, a comprehensive test program was undertaken to evaluate the settlement characteristics of the waste fills and the feasibility of improving the compressibility of the fills by dynamic consolidation and preloading.

The site development history, geotechnical investigations, the field tests and their results are presented with an assessment of the feasibility for improving the waste fills by preloading and dynamic consolidation.

SITE GEOLOGY AND MINING HISTORY

Geology

The site is located in Perry County in Southern Illinois. Prior to mining, it was a moderately dissected till plain with an average surface elevation of approximately +143 m (above Mean Sea Level). The thickness of overburden soils, consisting of loess and glacial till was approximately 6 m. The overburden soils were deposited unconformably over the Pennsylvanian Carbondale formation, which formed the premining bedrock surface in the area. The Carbondale Formation consists of alternating layers of sandstone, shale, siltstone, limestone and coal (Figure 1). Within the site areas, two layers of coal are present within depths that can be commercially mined. The first coal layer known locally as the Herrin coal, was present at an average depth of about 19 m below ground surface with an average thickness of 1.7 m. The second coal layer known as the Harrisburg Coal was present approximately 6 m below the Herrin coal, with an average thickness of 1.2 m.

Mining History

Mining started in late 1964 or early 1965 and was completed by the end of 1968. The mining operations took place along north-south

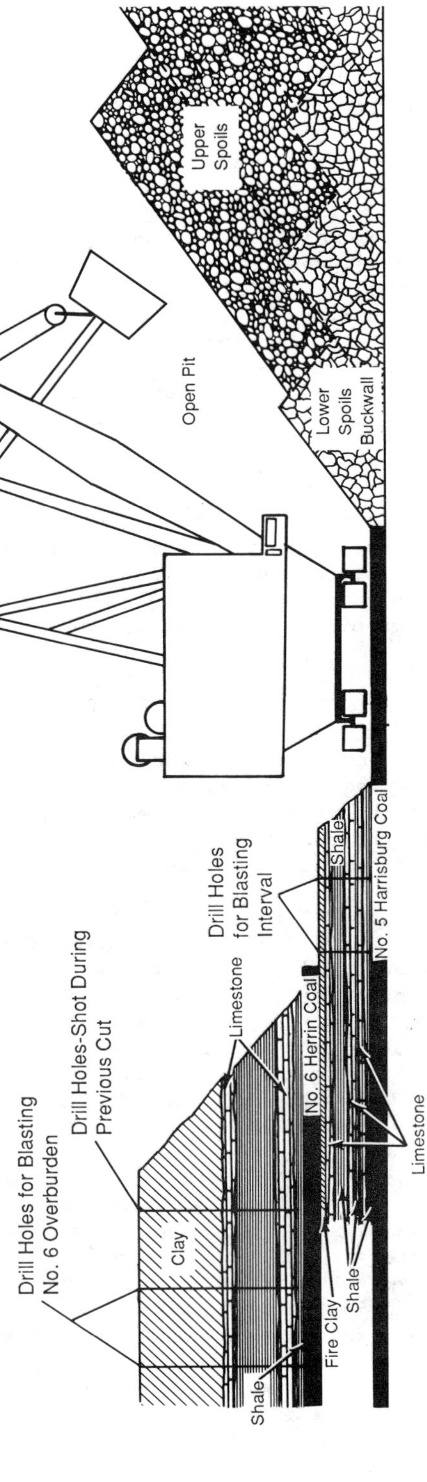

FIG. 1 Initial stratigraphy and mining sequence.

trending cuts (face of mine) and proceeded from east to west. Each separate north-south cut averaged about 20 m in width, and both coal seams were recovered in a single pass. The upper coal seam was mined first, after which the parting material between the two coal seams was blasted and removed to expose the lower coal seam. The parting material excavated between the two coal seams was placed behind (east of) the area being mined and stacked up to form a buckwall approximately 11 m high (Figure 1). During the next pass the material removed above the upper coal seam was placed behind and over the buckwall, forming a new lift of fill approximately 19 m high. All material above the top coal seam, including the topsoil, loess, till and Pennsylvanian rocks, was removed in a single pass, and thus the material above the buckwall consists of a heterogeneous mixture of materials ranging from clays to boulder size rock fragments. The lower portion of the fill, forming the buckwall consists primarily of rock fragments that are hard and durable with little tendency to break or disintegrate under the weight of the overburden or from exposure to air and groundwater. The buckwall material could be classified as rockfill, with weathered materials partially filling the voids.

The process of mining and disposing the spoils created a surficial topography consisting of a series of north-south trending ridges (windrows) separated by dips (gullies). On the average, the difference in levels between the ridges and the dips is on the order of 4.5 m, although greater variations in elevation were evident.

GEOTECHNICAL CHARACTERISTICS OF WASTE FILLS

The geotechnical investigation included boreholes drilled with conventional drilling equipment as well as boreholes drilled with a Becker Model 180 hammer drill. Because of the heterogeneous nature of the fill and the frequent presence of gravel and rock fragments, only drive type samplers could be used to sample materials from the boreholes drilled with conventional drilling equipment. A standard split spoon sampler and a larger drive sampler capable of retrieving samples 64 mm in diameter were used. Penetration resistances were recorded but it is quite likely that the measured blow counts (Figure 2) have little numerical significance other than serving as a comparative index of the range of materials encountered at any one level and of the change in penetration resistance with depth. It is evident from the measurements (Figure 2) that there is a gradual increase in resistance with depth, suggesting a possible increase in strength of the materials with depth.

Drilling with the Becker drill consisted of driving a double wall drill pipe with a double acting diesel hammer with a maximum rated energy of 10,850 joules. The pipe had an outside diameter of 168 mm and an inside diameter of 113 mm. The inside diameter of the bit was constricted to 100 mm so that the cuttings would be small enough to avoid becoming lodged in the drive pipe. Drill cuttings were brought to the surface by air forced from the surface down through the annular space between the interior and exterior pipes and up through the interior pipe. The cuttings were discharged to a cyclone for recovery. The pipe can be driven open-ended or closed-ended. The majority of the tests were performed with open-ended pipe. A small

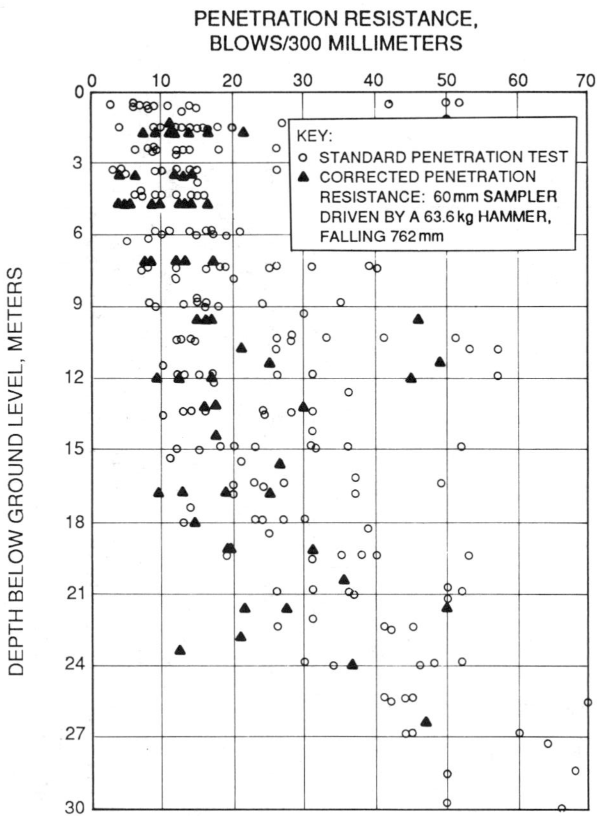

FIG. 2 Summary of penetration resistances from conventional boreholes.

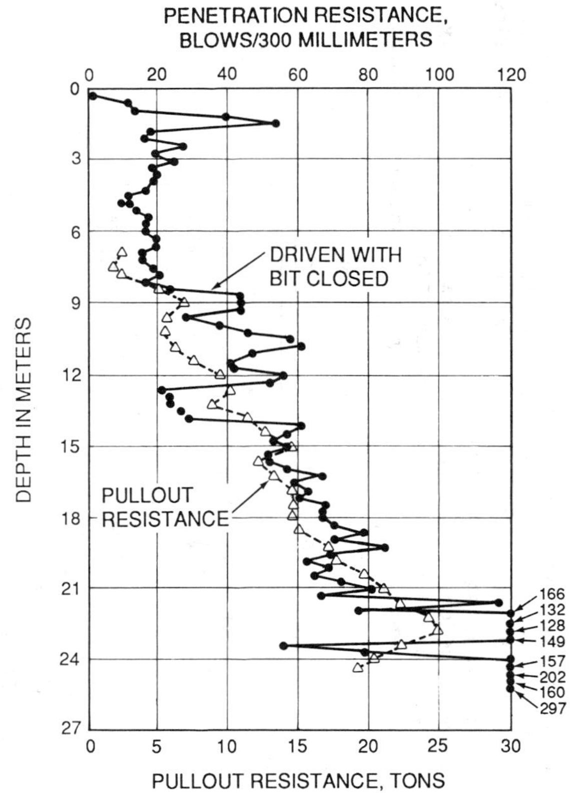

FIG. 3 Penetration and pull-out resistance; typical record of borehole drilled with the Becker drill.

number of tests were also performed with the tip closed. Driving a closed-ended pipe simulates pile driving and provides useful information regarding potential difficulties during installation of pile foundations. At the end of the drilling operation, the pipe was extracted by pulling with hydraulic jacks and the pullout force was recorded. Typical records of penetration resistances and pullout forces are presented (Figure 3). The pullout resistances were less variable because they were not affected by obstructions from rock fragments as was the case for the driving resistances. A summary of the pullout resistances recorded in 6 tests (Figure 4) shows a gradual and consistent increase in pullout resistance with depth.

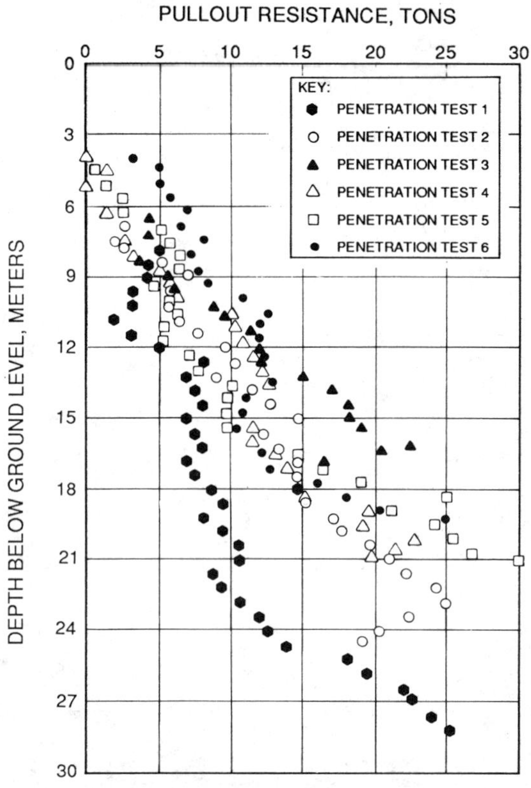

FIG. 4 Summary of pullout resistances from boreholes drilled with Becker drill.

Most of the samples recovered from the boreholes were essentially untestable either due to disturbance or because of the presence of gravel which would render the results of tests on small diameter samples virtually meaningless. Therefore, laboratory testing was limited to a small number of index tests with essentially no engineering property testing being possible.

Examination of soil samples recovered from conventional borings and the cuttings from the Becker drillholes confirmed that the waste fills were placed in a manner that essentially created two significantly different types of fill. The upper 19 m of the spoil was formed by completely mixing materials consisting of clays, silts, sands, gravels and rock fragments of varying sizes and hardness. The fill in this upper layer appears to consist of rock fragments within a silty or clayey soil matrix. It is similar in general appearance to a boulder-clay till. Groundwater observations revealed only limited free water in the boreholes, suggesting a deep water table consistent with the general hydrogeology of the area and the low permeability of the upper layer of spoils which prevents significant infiltration of water from surface runoff to form a perched groundwater table within the fill.

The lower 11 m of the waste fill consist primarily of a random mixture of fragments of shale, mudstone, limestone, highly weathered mudstone and occasional very stiff to hard clay. The lower 11 m contain significantly more rock fragments than the upper spoil, and have the general appearance of rockfill or rock fragments separated by weathered material or hard clay. The lower spoil is significantly more permeable as evidenced by the ability to pump water from the lower spoil with rates of discharge as high as 750 liters per minute.

Groundwater levels were generally encountered at depths ranging from 12 m to 21 m below existing ground level.

FIELD TESTING PROGRAM

Test Fill

The compressibility characteristics of the waste fills were evaluated by constructing a test fill approximately 24 m square at the top and about 61 m square at the base with average fill height approximately 8 m. There was considerable concern and deliberation during the planning of the test fill regarding minimum plan dimensions, test locations and whether the fill surface should be leveled (to fill the gullies) before the test fill was constructed. Cost and schedule considerations dictated the plan dimension and height of the test fill. One-dimensional conditions were desirable for ease of modeling of the test fill and for data interpretation. Furthermore, one-dimensional conditions simulate the condition of many of the structures such as slabs on ground, large diameter tanks and mat foundations. Recognizing that the seat of the settlement was most likely to be within the top 19 m of the spoil, the 24 m wide dimension at the top of the test fill is considered adequate in providing reasonably one-dimensional conditions under the center of the test fill. By appropriate instrumentation and data interpretation it was judged unnecessary to level the test area before filling began.

Instrumentation consisted primarily of deep settlement points (Fig. 5) as well as settlement measurements within the test fill. Optical surveying techniques were utilized to monitor settlements, always starting and closing back to a deep benchmark anchored in the natural rock formation near the test fill site.

Installation of pore pressure measuring devices was considered but did not seem practicable in view of the fact that the upper 19 m of

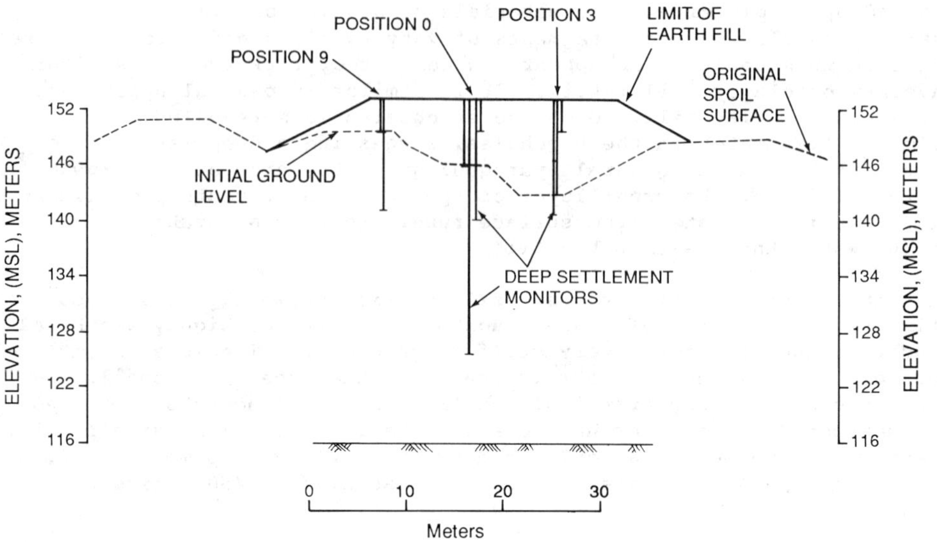

FIG. 5 Typical section through the test fill and settlement monitoring devices.

fill was probably partially saturated and the lower 11 m of fill was too porous and therefore pore pressure response would not be a significant indicator of the compression characteristics of this layer.

Construction of the test fill was completed in 10 days and involved placement of approximately 10,000 cubic meters of fill.

Dynamic Consolidation

Dynamic consolidation was performed on two test pads approximately 0.2 and 0.4 hectares in size respectively. The tests were performed by Menard Inc. and involved pounding the soil surface along predetermined grid patterns (Fig. 6). The pounder weighed approximately 16 tons and was dropped repeatedly at designated grid points from a height of 20 m.

The pounding operations were carried out in several phases and after each phase of pounding crushed stone was used to fill the craters formed by pounding. In the first area, labeled test area α (Fig. 6), the first two phases consisted of pounding over the primary grid points with phases 3 and 4 involving pounding over the secondary grid points. The last phase of pounding, called the ironing phase, involved pounding by a single drop per grid point along a 3 m x 3 m grid. Primary and secondary grid points were spaced on 7.6 m centers.

In the second area, termed area β, the first phase involved pounding over the primary grid points and the second phase involved pounding over the secondary grid points. The third phase involved pounding on a 6 m x 6 m grid within an area approximately 18 m x 18 m at the center of area β. The final phase involved ironing as

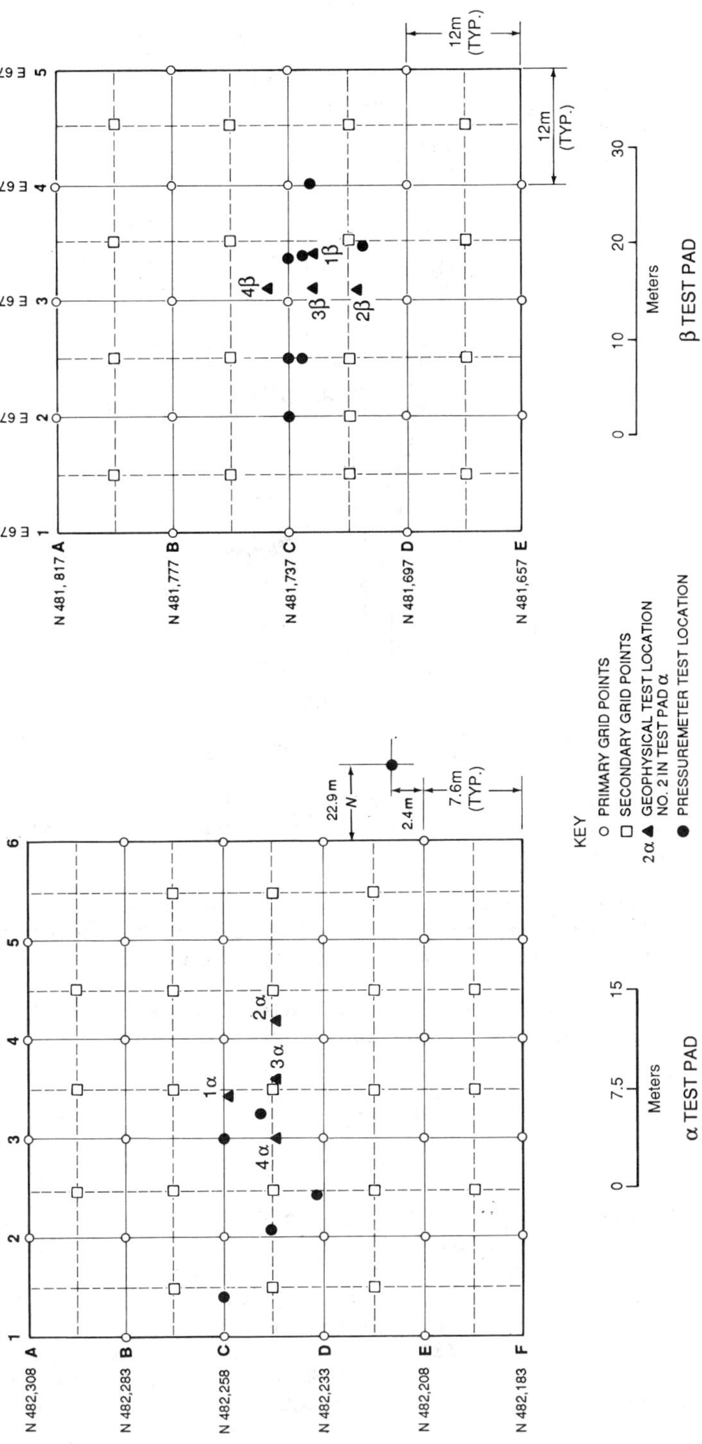

FIG. 6 Grid pattern and test locations of areas treated by dynamic consolidation.

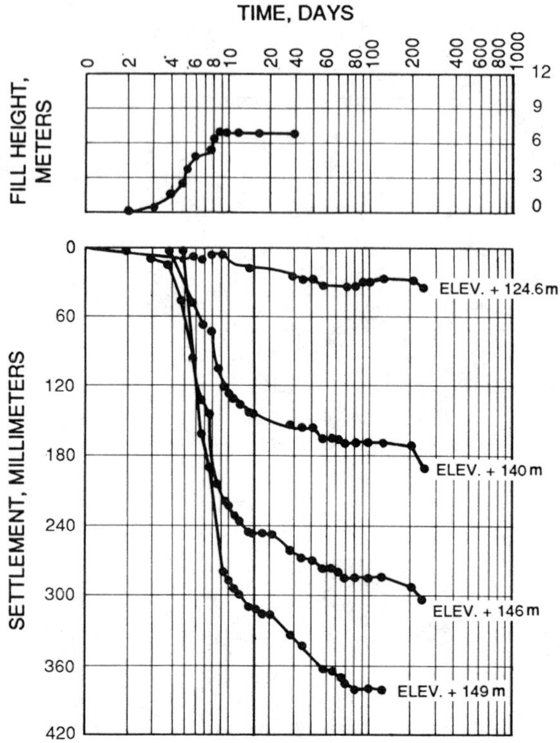

FIG. 7 Typical settlement records.

described for area α within the central portion where the third phase pounding took place. Primary and secondary grid points were spaced on 12 m centers.

Pressuremeter tests were performed at each of the two test pads before dynamic consolidation began and after each phase of pounding to monitor the progress of improvement of the spoil.

Before and after the test, cross hole geophysical tests were performed to verify the degree and the depth of improvement.

RESULTS OF THE SOIL IMPROVEMENT TESTS

Test fill

Typical records of fill placement and the development of settlements with time at the center of the test fill are presented (Fig. 7). The measurements show a very rapid development of settlements with consolidation completed essentially in 200 days. Settlements measured at various levels at the center and at quarter points along the perimeter of the test fill (at the top), were correlated with the height of fill (Figs. 8 and 9). It is evident by comparison of the data in Figures 8 and 9 that approximately one-half of the measured settlements occurred as a result of compression of the fill between elevations +149 m and +140 m (MSL).

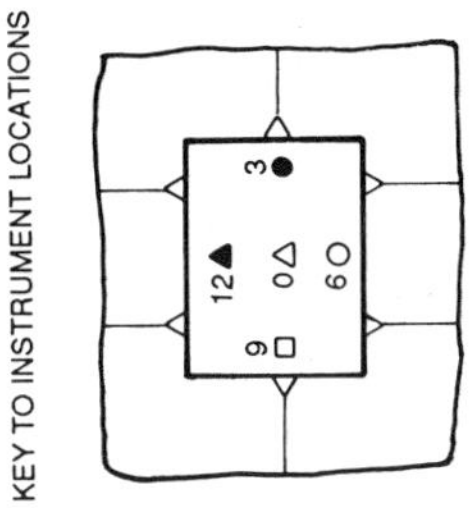

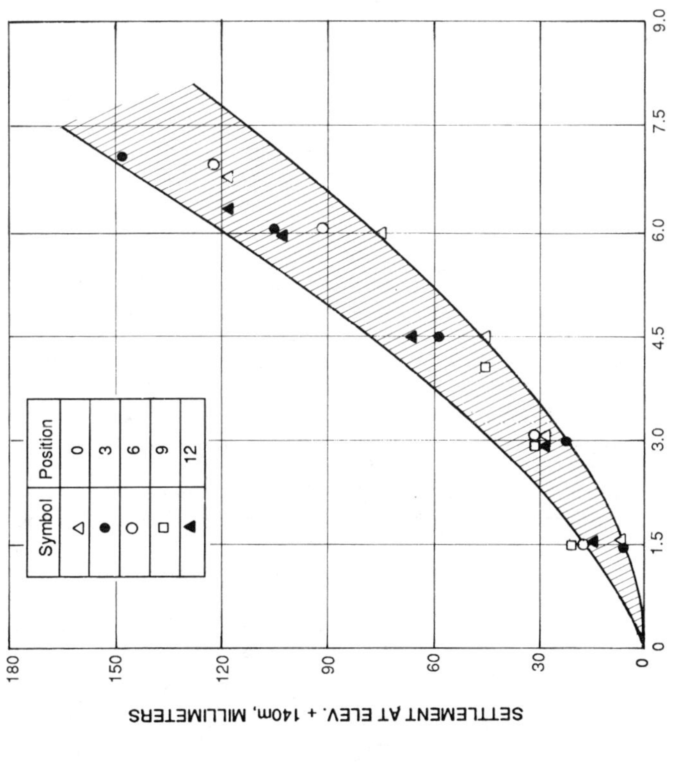

FIG. 8 Variation of settlements measured at Elevation 140 m.

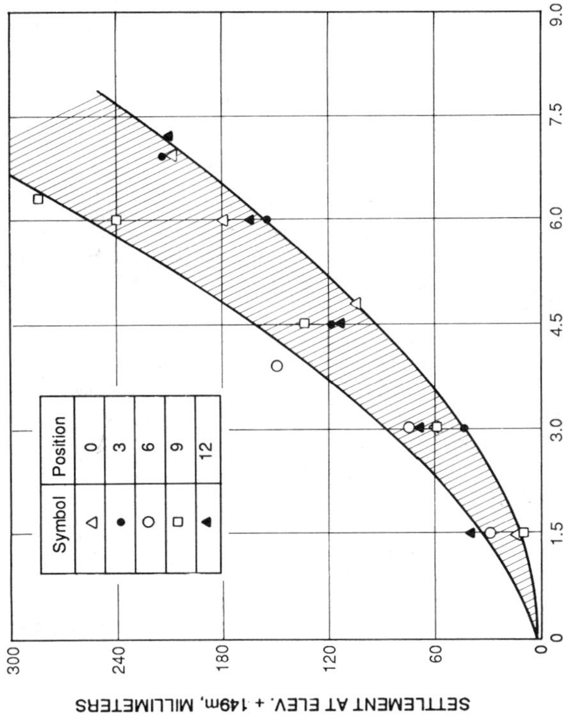

FIG. 9 Variation of settlement measured at Elevation 149 m.

The distribution of settlements with depth was further evaluated from plots of settlements measured at various levels below the test fill (Figs. 10 and 11). It is evident that 70% to 80% of the total settlements occurred as soon as construction was completed. The rapid development of the settlements is consistent with the partially saturated conditions of the upper fills and the high permeability of the lower 11 m of waste fill.

The settlement measurements were analyzed using Janbu's method [4], assuming that the settlements were entirely due to one-dimensional consolidation. It is recognized that some lateral deformations must have developed under the test fill as a result of instantaneous lateral spreading of the foundation soils. This immediate lateral spreading would have caused some settlement which would be in addition to settlements due to one-dimensional consolidation. Therefore, one dimensional conditions were not strictly applicable to this test fill. However, the assumption of one-dimensional consolidation is quite instructive in providing an approximate assessment of the compressibility characteristics of the waste fills.

The results of the analyses indicate a range of constraint modulus for effective stress levels up to 100 KP_a, ranging from 2.4 MPa to 7.2 MPa with an average modulus value of 4.8 MPa. At higher stress levels the constraint modulus increases essentially linearly with consolidation stress level in accordance with Eq. 1.

$$M = m\sigma_v' \tag{1}$$

where:

M is the constraint modulus
m is a constant, and
σ_v' is the vertical effective consolidation stress

The results of the analyses indicate m values ranging from 40 to 55.

Dynamic Consolidation

The effectiveness of the dynamic consolidation treatment method was evaluated from the results of pressuremeter and geophysical tests performed before and after treatment of the two test areas.

The results of the geophysical tests (Fig. 12) indicate substantial improvement within depths of 6 m for area α and 8 m for area β. The results of pressuremeter tests were in good agreement with the results of the geophysical tests, regarding the depth of improvement accomplished by dynamic consolidation. These depths are generally consistent with published correlations [5] between depth of improvement and energy per drop applied to the soil.

The total energy input was approximately 300 ton-meters per square meter of area treated, for both area α and β. The significant differences in improvement demonstrate the important effect of details in procedure in terms of energy application in determining the effectiveness of the method.

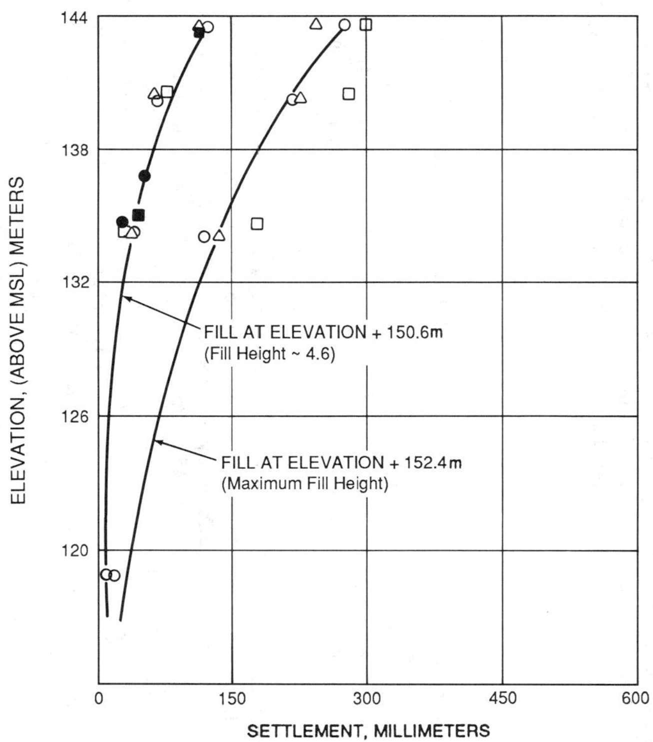

FIG. 10 Immediate settlements during construction of the test fill.

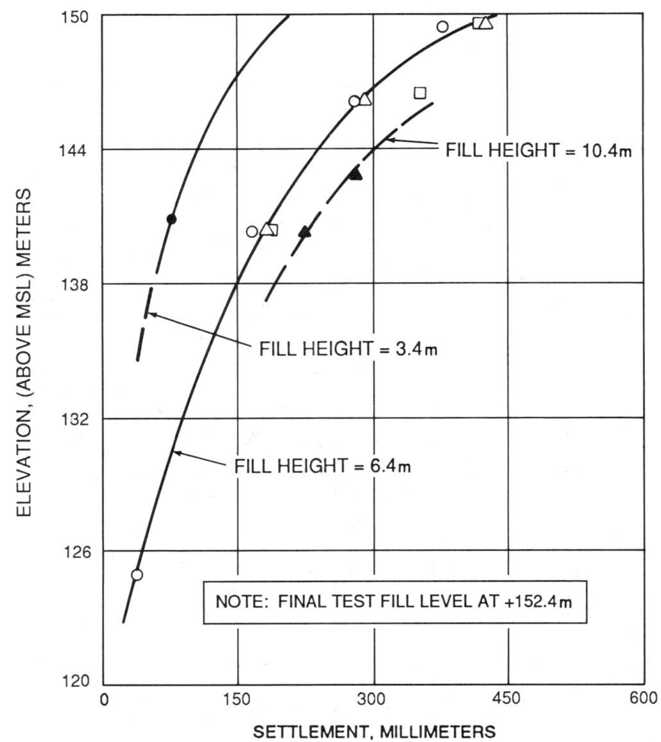

FIG. 11 Measured total settlements at 100 days, after test fill construction.

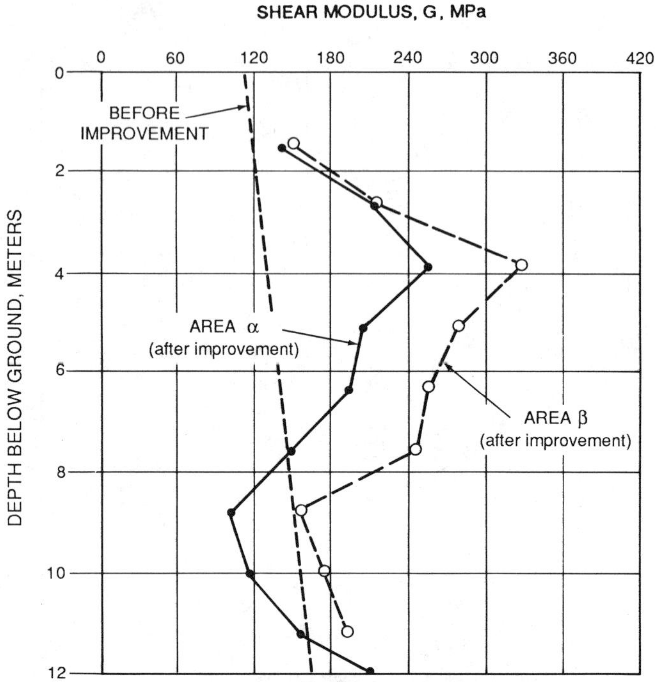

FIG. 12 Improvement in shear modulus, after dynamic consolidation.

The results of pressuremeter tests showed only marginal improvement during the initial phase of the treatment with most of the improvement occurring during the later phases of the treatment (prior to the ironing phase).

CONCLUSIONS

From the results of the field tests described earlier, the following conclusions can be drawn:

- Preloading is a cost-effective and rapid method of improving the waste fills encountered at the southern Illinois mining site.

- Dynamic consolidation was effective in treating the mine spoils; however, the depth of improvement is limited and can be seriously affected by procedural details of the treatment method. The need to fill the craters formed during pounding with crushed stone, (for the conditions encountered at this site) to effectively transfer the energy to the treated soil, is a drawback and may have a significant cost impact depending on availability of crushed stone in the vicinity of the site.

- A combination of dynamic consolidation followed by preloading may provide excellent ground conditions for the support of even heavy loads using shallow foundations.

ACKNOWLEDGMENTS

The authors wish to acknowledge the contributions of colleagues William Babcock, Roy Bell, and John Trapp who were responsible for much of the planning and execution of the work.

REFERENCES

[1] Charles, J. A., Farle, E. W., and Burford, D., "Treatment and Subsequent Performance of Cohesive Fill Left by Opencast Ironstone Mining at Snatchhill Experimental Housing Site, Corby," in Clay Fills Proceedings of the Conference held at the Institute of civil Engineers, London, 1978, pp. 63-78.

[2] Penman, A. D., and Godwin, E. W., "Settlement of Experimental Houses on Land Left by Opencast Mining at Corby," in Settlement of Structures, Proc. of the British Geotechnical Conference, Cambridge, 1974, pp. 53-61.

[3] Charles, J. A., Naismith, W. A., and Burford D., "Settlement of Backfill at Horsely Restored Opencast Coal Mining Site," Building Research Establishment Paper 46/77, 1977.

[4] Schmertmann, J. H., "Dilatometer to Compute Foundation Settlement," in Proceedings of the Conference on Use of In Situ Tests in Geotechnical Engineering, ASCE, Geotechnical Special Publication No. 6, 1986, pp. 303-321.

[5] Leonards, G. A., Cutter, W. A., and Holtz, R. D., "Dynamic Compaction of Granular Soils", Journal of the Geotechnical Engineering Divisions, ASCE, Vol. 106, No. GT1, pp 35-44.

NASSEF N. SOLIMAN

LABORATORY TESTING OF LIME FIXED FLYASH AND FGD SLUDGE

REFERENCE: Soliman, N.N., "Laboratory Testing of Lime Fixed Flyash and FGD Sludge," Geotechnics of Waste FIlls - Theory and Practice, ASTM STP 1070, A. Landva and G.D. Knowles, Eds., American Society for Testing and Materials, Philadelphia, 1990.

ABSTRACT: The operation of coal-fired power plants results in generating a considerable amount of flyash and flue gas desulfurization sludge. The properties of such material have not been widely reported and little is known about the field performance of this material when placed in landfills. This paper presents the results of extensive laboratory testing programs on six different mixes with variable flyash to sludge ratio. Lime was added as a fixing agent. Also, field samples from two operating power plants were tested in similar fashion, and results are presented and compared to laboratory testing results. In addition, a test pad was constructed from material produced by one of the operating power plants, and samples taken from the pad were tested.

Fixation of the flyash and FGD sludge by combining them with lime produces a mixture with improved properties over the flyash and the sludge. Generally there is significant improvement in structural characteristics, especially strength, and reduction in permeability by more than an order of magnitude. Therefore, the fixed material, with good quality control, could be handled and compacted easily in landfill.

Observation indicates that strength increases with time and with increase in density. The increase in strength when the mix was prepared utilizing brackish water is an important aspect observed during the study. Permeability results measured on samples prepared in the laboratory were found to be much lower than those obtained from field samples. Also, permeability was found to decrease with increasing density.

KEYWORDS: Fixed Flyash, Flue Gas Desulfurization Sludge, Strength, Permeability, Durability.

Nassef Soliman is a Senior Professional Associate at Parsons Brinckerhoff, Quade and Douglas, Inc., One Penn Plaza, New York, New York, 10119.

INTRODUCTION

Coal-fired power plants and the expansion in the use of coal will result in generating a considerable amount of flyash. In addition, the use of Flue Gas Desulfurization (FGD) units, required to minimize the sulfur content in the air, will result in additional amounts of sludge. Therefore, the disposal of flyash and sludge in an environmentally sound manner is becoming the major concern of utilities seeking to use coal as fuel.

There are several known processes for mixing flyash-sludge material and adding fixing agents such as lime. The result is a cementitious type of reaction in the sludge and flyash which binds the solids together to reduce the permeability and increase the shear strength, thereby enhancing physical stability and environmental acceptability.

The objective of combining FGD sludge with dry flyash and lime is to obtain a mixture suitable for landfill disposal or ocean dumping. A mixture at or near the optimum moisture content for compaction is the most desirable end product. When initially placed in landfill, the behavior of fixed ash-sludge is much the same as a silt material. Upon curing, the material gains strength, develops concrete-like properties, and does not show any tendency to reslurry when exposed to water.

The lime-flyash reaction (pozzolanic) is a slow process that takes place without any heat generation. However, external heat will accelerate curing. When water and flyash are mixed with lime, the silica and alumina will react with lime and the resulting product is tricalcium silicate and tricalcium aluminate. Further reaction takes place when the product is mixed with the scrubber sludge. The calcium sulfite and/or sulfate in the sludge will react with the tricalcium aluminate to form Tricalcium sulfate/sulfite Aluminate [1] [2].

This paper presents the results of field and laboratory testing that were made on six mixes of fixed ash sludge to determine their geotechnical properties, in connection with a new coal-fired unit located on Maryland's eastern shore.

LABORATORY TESTING

The laboratory testing program was designed to provide enough reliable data to assist in evaluating the properties of the fixed flyash/sludge material (fixed material). All testing was done according to ASTM specifications [3]. Six mixes labeled A to F of fixed material were prepared in the laboratory for testing. The special mixes were prepared using flyash from Delmarva Power & Light's Indian River plant and scrubber sludge from Gulf Power's Scholz plant. In addition, calcium oxide (quicklime) was added to Mixes A through E, and chlorides (in the form of calcium chloride, sodium chloride, and magnesium chloride) were added to Mixes A and B.

The fixed material of each batch was mixed thoroughly in electric mixers and stored in refrigerators during sample preparation as the pozzolanic reaction is very slow at low temperature.

In addition to the fresh samples prepared in the laboratory for testing, all samples obtained from field cured material were also tested, including:

- Hand-trimmed samples from block masses obtained from Plant X
- Cored samples from block masses obtained from Plant Y (six to eight months old)
- Cored samples from block masses obtained from the test pad (four years old)

TABLE 1 -- PHYSICAL PROPERTIES

Mix	Ratio of Fly Ash/ Sludge	Percentage of Lime Added	Sludge Chlorides ppm	Specific Gravity[3]	Maximum Dry Density (pcf)	Optimum Moisture Content (%)	Target Density (pcf)	Moisture Content as Received (%)	pH
A	0.7:1	2.5	14,600	2.31	95.9	16.0	88	19.9	12.1
B	0.7:1	2.5	44,000	2.32	97.0	15.5	88	20.6	12.0
C	0.7:1	2.5	50	2.34	92.3	16.4	81	25.4	12.0
D	1.34:1	2.5	50	2.34	96.0	15.5	88	15.6	12.1
E	0.54:1	2.5	50	2.33	98.3	13.0	91	18.0	12.1
F	0.7:1	-	50	2.33	96.0	15.5	88	18.1	8.0
Pad[1]	2.1:1	2.5	-	2.41	82	26	69	44.9	11.3
Pad[2]	1.3:1	2.5	-	2.4	-	-	64	52.4	-
	1.7:1	2.5	-	2.4	-	-	70	39.5	-
Sludge	-	-	-	-	-	-	-	15.7	7.5
Fly Ash	-	-	-	-	-	-	-	3	4.3

[1] Fresh material from Plant Y for lab testing.
[2] Fresh material from Plant Y used at the test pad.
1 pcf = 0.197 kn/m^3
[3] Specific gravity of Mix.

Classification Tests

Classification tests such as grain size distribution, specific gravity, and moisture-density compaction were performed on all samples (Table 1). For Mixes A to F, a target density was selected to correspond to a moisture content equal to what is expected during plant operation. The material was then compacted into tubes (2.4 inches by 6 inches, 2.875 inches by 6 inches and 2.4 inches by 1 inch) and stored in humidity room for curing. For fixed material obtained from the test pad, the target density was chosen to correspond to the moisture content equal to those measured in situ at the test pad during construction.

Strength Tests

To evaluate the strength of the fixed material, a comprehensive testing program was undertaken. Seventy-five unconfined compression tests and thirty unconsolidated undrained tests were performed on laboratory cured samples at 7, 14, 28, 56, and more than 56 days. Only maximum shear strength results are presented (Table 2).

TABLE 2 -- SHEAR STRENGTH OF FIXED FLYASH SLUDGE

	Maximum Sheer Strength ksf						Freeze/Thaw Test	
Mix	UC	Days Cured	UU	Days Cured	CD	Days Cured	UC	Days Cured
A	126.00	28	120.00	28	68.97	48	68.0 - 120.0	28-28
B	126.00	56	120.15	28	_	_	75.6 - 98.4	28-56
C	68.40	28	62.52	28	43.2	31	25.3 - 69.2	56-28
D	109.35	56	116.76	28	95.04	28	53.0 -102.0	28-56
E	81.36	56	89.04	28	67.00	33	31.2 -101.2	56-28
F(1)	-		-		-		-	
Test Pad								
Lab Cured	62.88	66	59.40	28	44.26	28		
Field Cured	58.60	1,400	_	_	_			
Plant X	45.99	180	45.12	180	24.93	180		
Plant Y	51.28	240	_	_	34.20	240		

1 ksf - 47.9 kN/m^2
UC: Unconfined compression test
UU: Unconsolidated undrained
CD: Consolidated drained
(1) Samples collapsed before testing.

The results indicate that the shear strength increased with curing time up to 40 days, then generally tended to level off. The shear strength varied from one mix to another. Mix A and B yielded higher shear strength, 126 ksf, compared to Mix C which yielded a maximum shear strength of 68.4 ksf. The three mixes (A, B, and C) have the same flyash/sludge ratio, 0.7:1; however, Mixes A and B were prepared using brackish water. Mixes D and E yielded 109.35 ksf and 81.36 ksf respectively. It appears that a flyash/sludge ratio of 1.34:1 is the best (in the C, D and E group) because of the high shear strength obtained by this mix as long as all other factors are the same.

Fresh materials from the test pad were tested at 7, 14, 28 and 66 days. Results indicate that the maximum shear strength is 62.88 ksf, which is lower than Mix D that has more or less the same flyash/sludge ratio. This is attributed to the care in mixing and handling of the special mixes in the laboratory compared to quality control at operating plants that handle large quantities.

Samples obtained from the test pad (4 years old) were tested and results indicated that the field cured material has a shear strength from 38.88 ksf to 58.61 ksf. Similarly, samples obtained from the Plant X and Plant Y landfill operation were tested and results indicate a shear strength of 45.99 ksf for Plant X and 51.28 ksf for Plant Y. Results from Plant Y are comparable to those obtained from test pad. Some of the reasons for the difference in performance between the laboratory mixes and those of the field pad/plant x and plant y is the source of the flyash/sludge and the high moisture content of the material. The source of the fly ash resulting from coal burning is different in both cases. Any variation in the sulfur content has an effect on the chemical reaction. The high moisture content of the material obtained from operating plants resulted in lower density. These factors have a direct impact on the strength of the material.

Fourteen consolidated drained triaxial tests were performed to evaluate the behavior of the fixed material under drained condition (long term). The results were generally lower than those obtained from unconfined compression test. It is probable that saturations utilizing back pressure may have an adverse effect on the shear strength.

Freeze/Thaw Tests

To evaluate the durability of the fixed material, tests for wetting and drying and for freezing and thawing were performed on special Mixes A through E. Since fixed ash-sludge differs from soil cement mixtures designated in ASTM specifications (D-559, D-560) in its components, the following modifications were made to the specifications:

- The fixed ash-sludge was not mixed with cement; however, the lime added to each mix served as the cementing agent.
- Samples were heated at 120°F (49°C), and dried at 135°F (57°C), temperatures which are lower than the specified 221°F (105°C). The temperature was lowered to avoid dehydrating the lime.
- Samples were compacted according to ASTM D1557, Method D, consistent with the preparation of all the samples.

Four types of freeze/thaw tests were designed to evaluate the behavior of the fixed ash-sludge in response to variable weather conditions. Eight samples from each mix were subjected to different cycles of freezing, thawing, and heating from 10°F (-12°C) to 120°F (99°C). All samples were checked for visible sample deterioration and change in dimension (shrinking or swelling). At the end of the cycling process for each test, samples were tested for unconfined shear strength. For all the tests, cycle 1 started immediately after compaction. Samples were frozen, thawed in the humidity room, moved into the oven at 120°F (49°C), then cooled at room temperature and placed back into the freezer for the next cycle, according to following plans:

- Two samples from each mix were subjected to three 48-hour freeze/7-day thaw cycles (28 days).
- Two samples from each mix were subjected to one 48-hour freeze/26-day thaw (28 days).
- Two samples from each mix were subjected to fourteen 48-hour freeze/48-hour thaw cycles (56 days).
- Two samples from each mix were subjected to four 7-day freeze/7-day thaw cycles (56 days).

Consolidation Tests

To evaluate the compressibility of the fixed material under loading, samples from laboratory cured fixed material of the special mixes and field cured material from Plants X and Y were tested. Consolidation tests indicated that the fixed material behaves like a very stiff soil. Mixes A through F had a void ratio of 0.63 to .73. Fresh material from Plant Y had a void ratio of 1.71 and field cured material from Plant Y had a void ratio of 1.67 and Plant X 2.47. Generally, material exhibited very little deformation up to 10 ksf.

Permeability Tests (cm/sec)

Permeability tests were performed to estimate the coefficient of permeability of the fixed material. Two types of tests were run: forty four falling head permeability in a permeameter, and forty eight constant head tests in the triaxial cell. Surcharge and head were varied for each test and from one test to another to determine whether these factors had an effect on the measured permeability (Table 3).

TABLE 3 -- PERMEABILITY RESULTS

Mix	% Lime	28 Days	Sealed 28 Days Only	More Than 56 Days	Sealed Samples More Than 56 Days
A	2.5	$1.76x10^{-6}$	$8.7x10^{-7}$		
B	2.5	$5.61\text{-}10^{-6}$	$1.93x10^{-6}$		
C	2.5	$5.53x10^{-6}$		$1.67\text{-}10^{-6}$	
D	2.5	$2.89x10^{-6}$	$3.86x10^{-7}$	$2.34x10^{-6}$	$2.34x10^{-6}$
E	2.5	$1.76x10^{-5}$	$3.2x10^{-5}$		
F	-	$4.91x10^{-5}$			
C	4	$1.32x10^{-6}$	$3.05x10^{-7}$	$4.85x10^{-6}$	
D	4	$1.32x10^{-6}$			
Flyash		$1.8x10^{-5}$			
Sludge		$8.7x10^{-5}$			
Plant X	2.5	—	$2.98x10^{-6}$		
Plant Y	2.5	—	$4.44x10^{-6}$		
Test Pad					
Lab Cured	2.5	$1.71x10^{-5}$	$1.43x10^{-6}$		
Field Cured	2.5	—	—	$3.x10^{-7}$	

Test for Interaction Between Fresh and Cured Material

The fixed material could be placed in lifts and stages during actual landfill operations. To simulate the gain in shear strength at the interface between freshly placed material on top of cured material, a special direct shear test was performed. Four samples, 2.4 inches in diameter and 1 inch in height, were prepared in the following manner.

Material was compacted into a ring and allowed to cure in the humidity room for seven days. After seven days, one-half of this material was extruded, leaving 1/2 inch empty in the ring. Fresh material was then compacted in a new ring, and the fresh material was pushed into the first ring until contact was made with the cured material. The remaining fresh material was then trimmed and a ring containing one-half new material and one-half cured material was obtained. The rings were allowed to cure together for various time periods and were sheared with various normal pressure. The results represent the shear strength at the interface between cured and fresh mixes (Table 4).

Special Testing

To evaluate the effect of curing under different conditions on the shear strength of the fixed ash-sludge, several samples were subjected to the following tests and then tested for unconfined shear strength (Table 5).

- Samples were cured for one week, then submerged in water for 28 days.
- Samples were placed in a heater at 120°F (49°C) for 28 days.
- Two samples from Mix D were refrigerated (40°F) (4°C) for 28 days.
- Samples were cured for one week, then sealed in plastic for 28 days.

DISCUSSION

The fixed ash-sludge material consists mainly of scrubber sludge, flyash and fixing material. Lime was used as fixing (cementing) agent for all the mixes and field cured material tested for this study. The properties of any fixed or stabilized material depend mainly on the following factors:

- The physical behavior of the individual components of the material.
- The ash-to-sludge ratio.
- The sulfate-to-sulfite ratio.
- Curing time and condition.
- Percentage of the fixing agent (cementing).
- Degree of compaction, i.e., density.
- Grain size distribution of each individual component.
- Degree of mixing and how thoroughly the fixing agent is mixed.

From these factors and the test results, it can be seen that the properties could vary even within one mix due to slight variation in lime content density or degree of mixing.

Moisture Content and Dry Density

Fresh fixed material could be compacted in landfill operation. Compaction tests have indicated that the material behaves like silt soils in general. To obtain the best performance from any material, it should be compacted to the maximum dry density, which can be achieved only at the corresponding optimum moisture content. For the fixed material that would be uneconomic and impractical because it would require drying to bring the moisture content to the optimum value. It is more practical to compact the material to the anticipated moisture content during plant operation. During the placement of the waste material, compaction to the target density could be achieved by placing the material in 18- to 24-inch lifts and compacting it with a few passes by a bulldozer or a sheepsfoot roller. In addition, the use of sheepsfoot roller will allow better interlocking between the consecutive layers, thus creating better healing and increased bond. Gradation curves are presented in Figure 1, and relationships between dry density and moisture content are presented in Figure 2.

Strength

The strength of any material is an indication of its stability and its ability to support loads and withstand severe weather conditions. The stronger the material, the better it will perform under weather changes.

The strength of the fixed material is a function of the curing time. Also, temperatures higher than 40°F (4°C) accelerate the curing process.

TABLE 4 -- RESULTS OF DIRECT SHEAR TESTS
Interface Strength Measurement

Mix	Days Cured(1)	Normal Pressure (ksf)	Maximum Shear Strength (ksf)
C	7;7	1.0	1.05
C	7;7	3.0	3.05
A	8;17	1.0	0.75
E	7;29	1.0	1.05

In all cases, the samples were sheared along the interface of the two lifts.
ksf = 47.9 kn/m^2

TABLE 5 -- SPECIAL TESTING

	Submerged 28 Days			Sealed 28 Days		
Mix	Unconfined Compressive Strength (ksf)	Dry Density (pcf)	M.C. (%)	Unconfined Compressive Strength (ksf)	Dry Density (pcf)	M.C. (%)
A	77.23	88.8	16.1	62.07	89.5	12.3
B	73.33	90.2	16.0	83.16	90.9	14.6
C	29.63	86.1	18.7	32.18	80.5	19.1
D				109.80	88.7	6.6
E	75.84	93.3	13.2	150.00	93.4	8.0

	Heated 28 Days			28 Days Refrigerator		
Mix	Unconfined Compressive Strength (ksf)	Dry Density (pcf)	M.C. (%)	Unconfined Compressive Strength (ksf)	Dry Density (pcf)	M.C. (%)
A	92.76	89.7	3.2			
B	120.00	89.3	5.1			
C	65.88	82.7	5.4			
D	92.16	89.4	0.7	48.20	89.3	2.5
E	101.04	88.6	2.4			

ksf = 47.9 kn/m^2

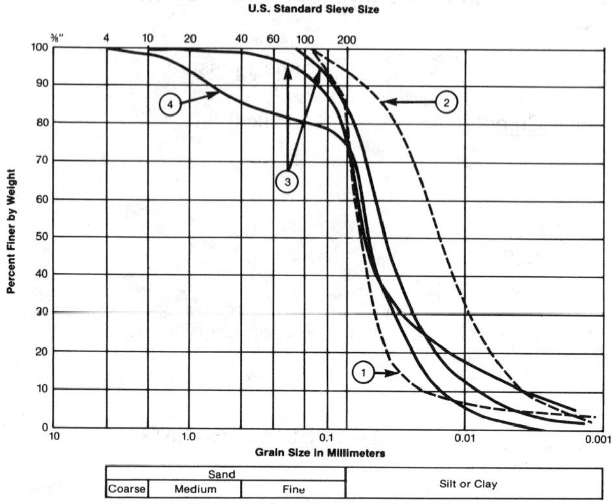

Gradation Curve

1—Sludge
2—Fly Ash
3—Range Mix A to E
4—Plant Y

Figure 1

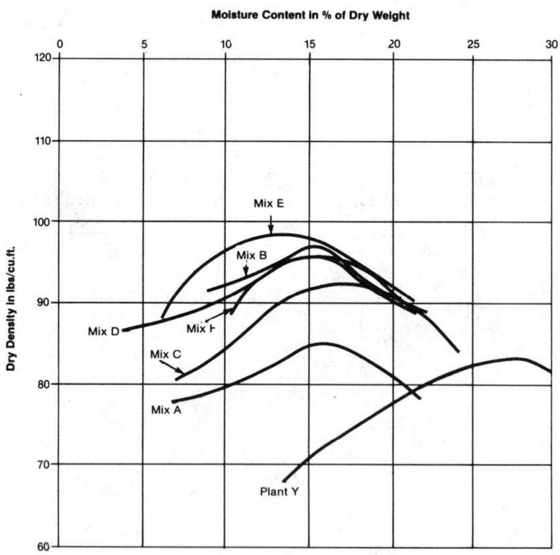

Relation Between the Dry Density and Moisture Content

pcf = 0.16 kN/m³

Figure 2

During testing, it was observed that the majority of the samples failed along a vertical plane, usually near the center of the sample which indicates that failure is structural and not a result of friction [4]. This is consistent with the fact that the strength of this material is gained through cementation. Within the anticipated height of a waste disposal pile and the corresponding stresses, no relation between the confining pressure and shear strength was found. Reported values of angle of internal friction for the fixed material could not be confirmed during this testing program [5] [6].

Strength test results for the special mixes seem to be generally within the published ranges [7] [8] [9] [10] [11] for the fixed ash-sludge, with the exception of Mixes A, B, and F. Mixes A and B are generally higher which is attributable to the calcium chloride that had been added to simulate the effect of the use of brackish water as the scrubber makeup. Yoder and Witczak [12] reported that calcium chloride can be added to lime-soil mixtures to improve the physical characteristics of the mix and to speed up the reaction. Calcium chloride benefits the reaction by holding the compacting moisture in the soil and by providing additional calcium ions to the mix. Mix F exhibited very low strength characteristics, obviously due to the absence of the fixing agent (lime). The majority of the samples from this mix collapsed during handling and those which came into contact with water disintegrated.

In order to evaluate the long-term shear strength, results are plotted versus curing time (see Figure 3) for Mix D as a sample. Because of the wide scatter in data, a least squares curve fitting technique was employed to obtain the strength- versus-time curves. Data from unconfined compression, unconsolidated undrained, and consolidated drained tests were used in a computer program which provided the "best fit" for log log, semi log and 1/x curves. The consolidated drained tests were given more weight than the others, because this test is representative of the long-range strength of the material.

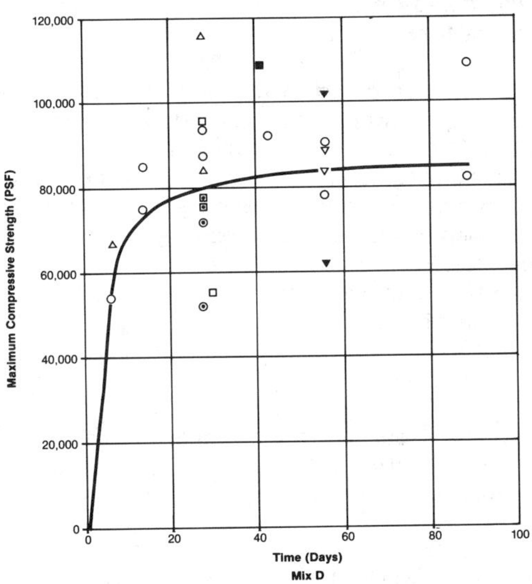

Strength vs Time

psf = 0.0479 kN/m²

Key
○ Unconfined Compression
△ Unconsolidated Undrained
□ Consolidated Drained
⊙ Freeze Thaw, Test 1
⊞ Freeze Thaw, Test 2
▽ Freeze Thaw, Test 3
▼ Freeze Thaw, Test 4
■ Sealed Samples

Figure 3

Examination of the data showed that the 1/x equation was most representative of the data because the curve leveled off with time which is expected for the strength. The general equation used was:

$$y = A + B/x$$

where A and B are constants. The curves present the recommended values with time for the different mixes (see Figure 4). Based on this data, well-prepared fixed material similar to Mixes A, B and D could have a shear strength of 80 ksf. For field operation with less quality control, a shear strength of 50 ksf is more practical.

To evaluate the effect of density on strength, several samples from Mixes A, D and E were prepared to densities ranging from 60 pcf to 90 pcf. The results indicate increase in strength occurs at approximately 70 pcf (see Figure 5).

The behavior of the special mixes under extreme weather conditions, was evaluated by subjecting samples to variable cycles of freeze and thaw. Evaluation of the material behavior indicated that no cracking or deterioration were apparent. All samples were tested for unconfined compression strength and are plotted in Figure 3 for comparison with other results. Generally, the strength values for these tests were less than those obtained from normal curing. However, durability tests performed on special mixes indicated that the fixed material will deteriorate slightly during a freeze-thaw-freeze cycle if water is made available to the material during the thawing period. It should be anticipated that during winter the upper few inches of the surface of the compacted material will deteriorate when subjected to a thawing period with rain followed by a freezing period.

To further evaluate the strength of the fixed material if subjected to variable conditions, a sample of each mix was cured for 28 days under the following conditions: submerged, sealed, heated, and kept at 40°F.

Compared to unsealed samples cured in the humidity room for the same period, the submerged samples yielded a lower compressive strength. The sealed samples yielded lower strengths except for Mixes D and E, which yielded higher strengths. Heated samples yielded generally the same strengths with the exception of Mixes B and E, which gave lower values. Apparently the number of samples tested (one each) was not enough to give a representative value for each case. Two samples prepared from Mix D were kept at 40°F (4°C) for 28 days, then tested for unconfined compression strength. Strength of these samples was about half of those for samples cured at 70°F (21°C) for 28 days. This shows that the curing rate is affected by temperature (Table 5).

Permeability

During the testing program, it became obvious that the permeability determined in the laboratory seemed to be generally lower than some described in published data [8] [9] [10]. Accordingly, several efforts were undertaken to evaluate the results and the methods of testing used.

Samples for special mixes were cured in the humidity room but they were not sealed in plastic bags to contain the moisture content of the samples. During testing it was noticed that moisture content dropped dramatically with time. Additional samples (C and D) were prepared. Most of the samples were sealed and cured in humidity room, while some were kept unsealed. Permeability tests were performed at 7, 14 and 28 days. No obvious difference was found between results obtained from tests performed on sealed samples and unsealed samples for curing time up to 28 days. However, samples left to cure more than 28 days yielded lower permeabilities than those obtained from previous mixes

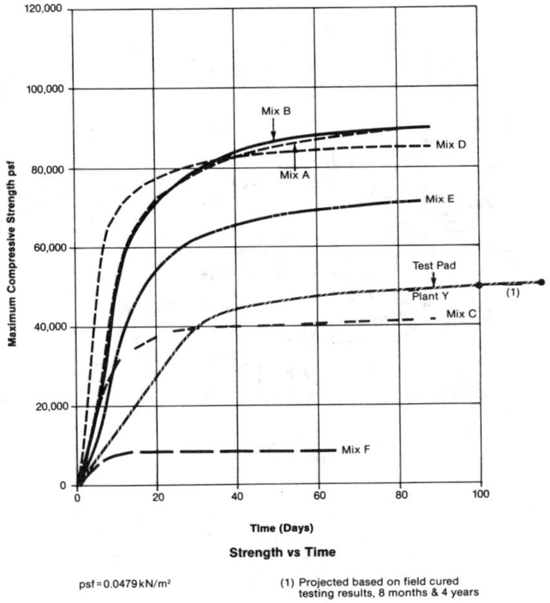

Figure 4

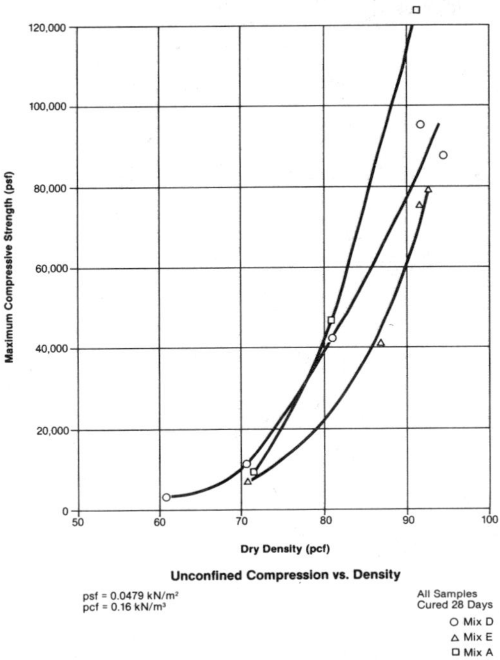

Figure 5

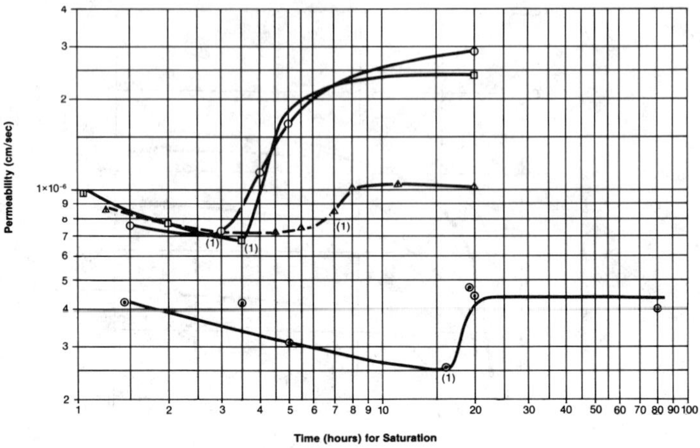

Effect of Saturation on Coefficient of Permeability

Key
⊙ Mix D; 64 Days, Sealed
▲ Mix D; 14 Days, Sealed Soaked 5 Minutes
◯ Mix C; 28 Days, Sealed Soaked 10 Minutes
◫ Mix D; 28 Days, Unsealed Not Soaked

Notes: (1) Tests were stoped at these points, and samples were saturated in triaxial cells, then tests continued.

Figure 8

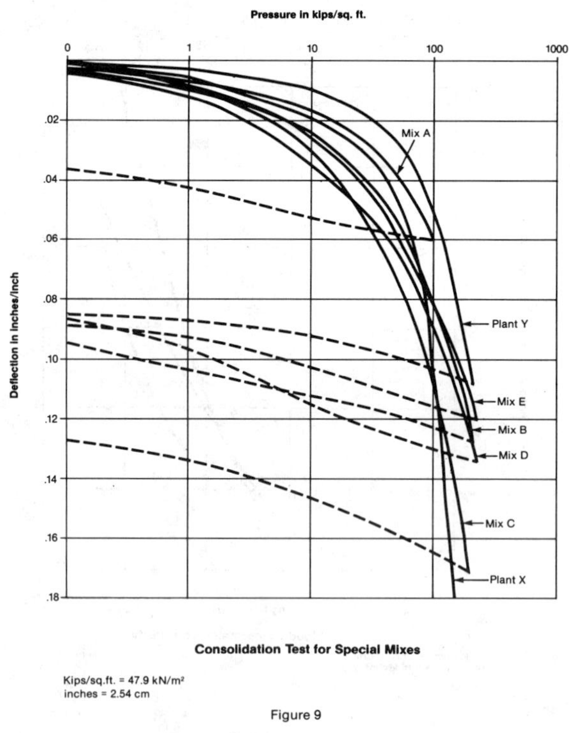

Consolidation Test for Special Mixes

Kips/sq.ft. = 47.9 kN/m²
inches = 2.54 cm

Figure 9

Compressibility

Several consolidation tests were performed on samples from Plant X, samples prepared from fresh material from Plant Y, and samples prepared from special mixes. All samples exhibited very small elastic deformation until 10 to 12 kips per square foot (ksf); then the samples exhibited a plastic deformation. The compacted fixed ash-sludge is very incompressible within 10 to 12 ksf, which is a normal range of working stresses. However, beyond that range the material experienced a relatively higher deformation, probably due to structural failure (see Figure 9).

Interaction Between Fresh and Cured Material

During the landfill operation, fresh material will often be placed on cured material. In order to evaluate the gain in shear strength between freshly placed material on top of cured material, a special direct shear test was performed. The results indicate that friction resistance between cured and fresh material takes place. The gain in frictional resistance is a function of normal stress. There is no obvious increase in frictional resistance with time (Table 4).

Durability

The durability tests are considered to be severe; however, the majority of samples survived the 12 cycles and generally performed very well. As these tests are designed to evaluate the ability of the material to resist stress, mixes with high strength were expected to be more durable than those with low strength. It was found that during freeze-thaw test, material loss from Mix B (highest strength) was minimal, while material loss from Mix C was maximal. It was also observed that the material losses for wet-dry were generally less than those observed for freeze-thaw tests. This is attributed to the water which was made available between freeze and thaw cycles. After the water penetrated the sample by capillary action, it became frozen and increased in volume during the freezing cycle. The increase in frozen water volume resulted in internal stresses which affected the material performance. Based on this observation, it is anticipated that a freeze-thaw cycle will cause slight deterioration in the upper few inches of the surface of the waste material. Field observation of the test pad seems to agree with this conclusion.

Conclusion

Fixation of the flyash and FGD sludge by combining them with lime produces a mixture with improved properties over the flyash and the sludge. Generally there is significant improvement in structural characteristics, especially strength, and reduction in permeability by more than an order of magnitude. Therefore, the fixed material, with good quality control could be handled and compacted easily in landfill. This will result in less volume of material which require less use of land. The decrease in the coefficient permiability will reduce the water perculation into the landfill and the tendency of leaching. The increase in shear strength will permit land filling with stable steep slopes.

The study concluded:

- The strength of fixed material increased with curing time, with the increase in density, and when mixed with brackish water.
- The fixed material could be compacted into blocks and dumped in the ocean to create a reef, similar to a coral reef, as the ocean water will enhance the strength of the material. This could solve the problem of disposal of the fixed flyash sludge material generated by the power plants.

prepared with the same ratio of flyash to sludge. These lower values are obviously due to sealing effect (Table 3).

Several relationships were developed to better understand and delineate the relation between permeability and other physical properties. The strength of the material and the corresponding permeability were plotted on a semilog chart. It was noted that the permeability decreased with increase in strength, which means that curing time and percentage of fixing agent (lime) are factors that affect permeability results (see Figure 6). Also, the dry density versus the permeability was plotted on a semilog chart. It was observed that the permeability decreased slightly with the increase in dry density, until about 83 pcf, when the observed decrease in permeability became larger (see Figure 7). This relationship is limited, however, by the maximum practical bulk density that could be achieved in the field. All samples were compacted in the laboratory to the maximum anticipated dry density.

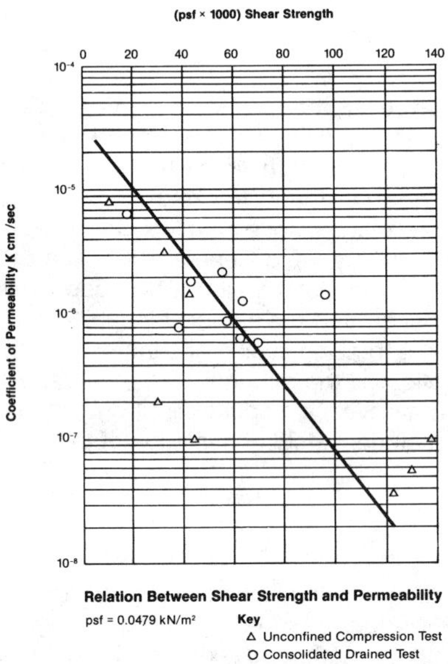

Figure 6

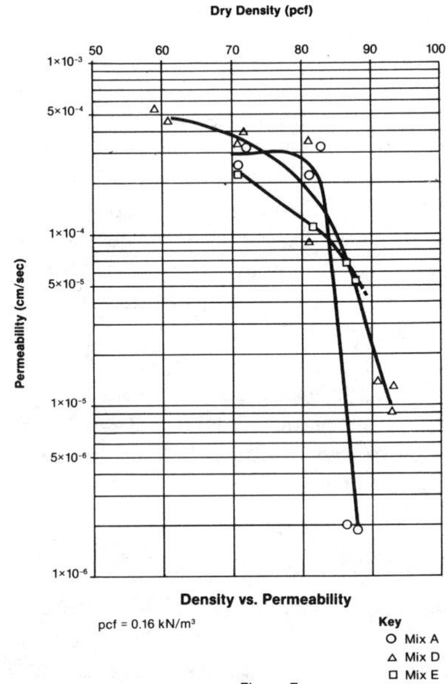

Figure 7

To evaluate the effect of saturation, three samples were tested in the triaxial equipment. Two of the samples were soaked in water for five and ten minutes, respectively; one was not soaked; and then permeability values were computed as a function of time. The results were plotted on a log chart (see Figure 8). From this chart, it can be seen that permeability decreased with time until the sample began to become saturated; then the coefficient started to increase with the increase of the degree of saturation. When the samples approached 100% saturation, the permeability coefficient exhibited very small changes.

- The permeability decreased with time and sealing of the samples which simulate field conditions. However, the study could not confirm published values for the fixed material of 10^{-8} to 10^{-10} cm/sec.

- A direct relationship exists between permeability and unconfined compressive strength of the material.

- The permeability coefficient decreases exponentially with linear increase in unconfined compressive strength. the permeability of the special mixes were lower by an order of magnitude than material from Plants X and Y.

ACKNOWLEDGMENTS

Laboratory testing for this study was done at the University of Maryland, College Park; Dames and Moore, Crawford, N.J.; and I.U. Conversion System, Inc., Philadelphia, Pa.

REFERENCES

[1] FGD Sludge Disposal Manual, 1979. Electric Power Research Institute, Final Report, Prepared by Michael Baker, Inc., January.

[2] GAI Consultants, Inc., "Ash Disposal Reference Manual, Draft Report." Prepared for Electric Power Research Institute.

[3] Annual Book of American Society for Testing and Materials Standards, 1979, Part 19.

[4] Rossoff, J. et al., of the Aerospace Corporation, 1977. "Disposal of By-Products from Non-regeneratable Flue Gas Desulfurization Systems, Second Progress Report." Prepared for EPA, May, PB-271 728, EPA-600/7-77-052.

[5] IU Conversion System, Inc. "Poz-O-Tec Process for Economical and Environmental Acceptable Stabilization of Scrubber Sludge and Ash." University City Science Center, Philadelphia, Pennsylvania 19104.

[6] IUCS: A&H Corporation, 1978. "Consultant's Report on Laboratory Testing and Analysis of Poz-O-Tec Material at Columbus and Southern Ohio Electric Company's Conesville Generating Station." Conesville, Ohio.

[7] San Juan Testing Laboratory, Inc. "Laboratory Tests Completed on Poz-O-Tec Stabilized Fly Ash at Four Corners Power Plant, Fruitland, New Mexico." For IU Conversion Systems, Inc., May 1975.

[8] Ullrich, C.R.; Hagerty, D.J.; and Van Ness, R.P. "Physical Properties of FGC Waste Deposits at the Cane Run Plant of Louisville Gas and Electric Company."

[9] IUCS: GAI Consultants, Inc. "Evaluation of Poz-O-Tec Material for Use as Landfill - Demonstration Project, Phillips Generating Station, South Height, Pa." Consultant's Report to IU Conversion Systems, Inc., Philadelphia, Pennsylvania.

[10] Hurt, P.R., et al., of Aerospace Corporation. "Flue Gas Desulfurization Sludge Disposal - The Shawnee Field Evaluation Project." Presented at U.S. Environmental Protection Agency Fifth Industry Briefing on Results of Lime/Limestone Wet Scrubbing Programs, Raleigh, North Carolina, December 1979.

[11] Rossoff, J., and Rossi, R.C. "Flue Gas Cleaning Waste Disposal - EPA Shawnee Field Evaluation." Prepared by Aerospace Corporation for presentation at EPA Flue Gas Desulfurization Symposium, New Orleans, La., March 8-11, 1976.

[12] Yoder, E.J. and Wirczak, M.W., 1975. Principles of Pavement Design, 2nd Edition.

[13] Soliman, N., Houlik, C., and Schneider, M., "Geotechnical Properties of Lime Fixed Flyash and FGD Sludge," International Conference on Environmental Geotechnology, Lehigh University, Pa., 1986

Joseph P. Martin, Francis J. Biehl,
John S. Browning III and Edward L. Van Keuren

CONSTITUTIVE BEHAVIOR OF CLAY AND POZZOLAN-STABILIZED HYDROCARBON REFINING WASTE

REFERENCE: Martin, J. P., Biehl, F. J., Browning, J. S., III, and Van Keuren, E. L., **"Constitutive Behavior of Clay and Pozzolan-Stabilized Hydrocarbon Refining Waste,"** Geotechnics of Waste Fills - Thoery and Practice, ASTM STP 1070, Arvid Landva and G. David Knowles, Eds., American Society for Testing and Materials, Philadelphia, 1990.

ABSTRACT: A project to stabilize acidic hydrocarbon sludge lagoons for landfill deposition is described. The sludge will be microencapsulated in a clay matrix, neutralized and cemented with a lime and fly ash pozzolanic mixture. The mechanical improvement and contaminant immobilization provide dimensional stability and low leachate production rates. Two types of clay were used as the aggregate in the stabilization mixture: a spent processing attapulgite and an onsite silty clay. Results of leaching, strength, and compression tests are presented for mixtures using each clay, as is preliminary analysis of the porous monolith response to in-situ stresses.

KEYWORDS: STABILIZATION; MICROENCAPSULATION; SLUDGE; IMMOBILIZATION; LANDFILL; COMPRESSIBILITY; POZZOLANS

INTRODUCTION

Modern landfills provide space for waste disposal while minimizing the potential for release of gases, liquids and particulates. Most landfills rely upon external containment (liners and caps) to isolate solid wastes from the local environment, and drains (leachate collectors and gas vents) to control contaminated fluids. Sludges present special problems. In the past, they were often placed in landfills

J. Martin is Assoc. Professor of Civil Engineering, Drexel University, Philadelphia, PA 19104. F. Biehl is an Environ. Engineer at DuPont, Newark, DE 19714. J. Browning is a Senior Engineer at NTH-Russell and E. Van Keuren is a Project Engineer at ERM Consultants, both Exton, PA 19341.

or stored indefinitely in lagoons, neither practice being currently acceptable. When destruction or reprocessing of historic sludge deposits are not feasible, modification is necessary for landfill deposition. The minimum acceptable treatment is dewatering to remove free liquids, but stabilization, i.e., improving mechanical behavior and decreasing contaminant mobility, is more desireable.

Sludge stabilization often employs an admixture of hydraulic cements, often including byproducts such as fly ash and kiln dust as pozzolans(1,2,3). This method of waste treatment would be expected to generate lower maintenance costs after closure than conventional cellular solid waste landfills (4,5,6). Solidification (the mechanical effect) to produce a dimensionally stable monolith provides a firm subgrade for the final cap. Contaminant immobilization reduces dependency on leachate and gas collectors. The immobilization may consist only of physical isolation in an impermeable matrix (microencapsulation), but chemical fixation mechanisms may also be active, such as neutralization, precipitation, partitioning and sorption.

This paper describes a project to stabilize a group of acidic hydrocarbon sludge lagoons. The waste is an asphaltic emulsion with a wide range of low volatility compounds, the byproducts of a discontinued petroleum refining process. Conventional solidification with portland and pozzolanic cements only gave poor results. However, sludge encapsulation in a clay matrix bound with a lime-fly ash admixture resulted in a material of medium stiff consistency, low permeability, and low carbon solubility. A conceptual model of the structure is shown on Figure 1.

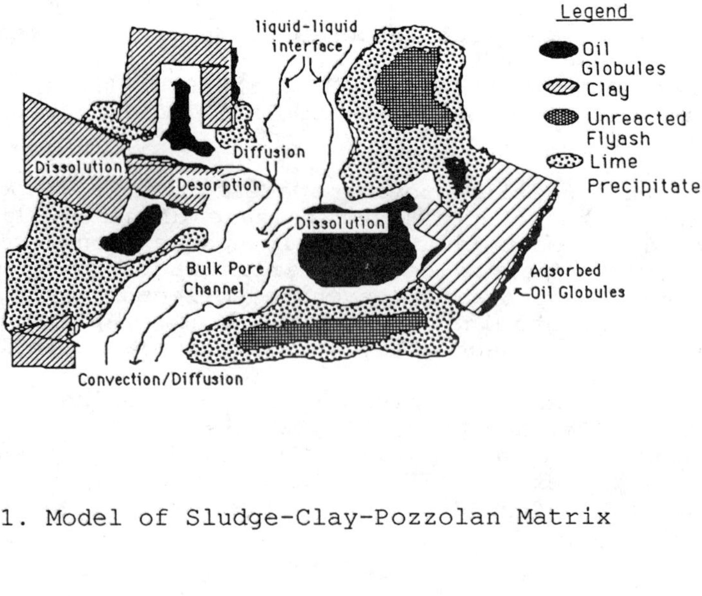

Figure 1. Model of Sludge-Clay-Pozzolan Matrix

The clays available are an attapulgite-based spent fuller's earth and the local silty clay, a residual soil. The properties of the sludge-clay-pozzolan mixtures depend upon component proportioning, moisture content, age, sludge consistency and compactive effort. These factors were varied in an empirical study until they yielded mixtures of a soil-like consistency that produced basic effluent of low carbon content when permeated with distilled water or dilute acidic solvents (8,9,10). A field study including construction of test cells demonstrated the practicality of the method (11). Development of the mixtures is described in detail elsewhere (10,12).

The concentration herein is on setting site standards and laboratory studies associated with developing a product to limit leachate generation and provide dimensional stability in the in-situ environment. The analysis employs adaptations of traditional soil mechanics methods.

CUSTOMARY STABILIZATION CHARACTERIZATION

Solidification with hydraulic cements was initially developed with inorganic wastes such as electroplating and flue gas desulfurization sludges (3). This practice uses established and economical construction technologies, and results in concrete-like materials. Contaminants are either incorporated in the hydration precipitates or isolated in the hardened paste (7). The more recent focus has been stabilization of sludges with high organic content, raising a concern with regards to interference with cementing reactions. The result is more likely to be encapsulation of organics in a porous matrix, and rely upon reversible fixation mechanisms such as neutralization (2,7).

Characterization of a stabilization technique or formulation is customarily done with material property tests, including unconfined compression (ASTM D-2166), EP Toxicity or TCLP procedures, and saturated permeability (ASTM D-2334 or falling-head tests). These tests indicate mechanical, chemical fixation, and transport behavior, respectively (13). Durability tests also indicate resistance to matrix deterioration in aggressive climates.

A landfill is subjected to a complex mechanical and biochemical stress environment, as illustrated on Figure 2. Understanding of the relationship between test results and the in-situ performance of a stabilized deposit is still under development. Stabilized landfills are similar to earth dams with respect to the need for self-support and restriction of fluid movement, but the fluids are retained within the embankment. Quantifying deformations, leachate and gas production rates and quality, and matrix longevity requires detailed engineering analysis. Such analysis would seem to be necessary to provide the basis for specifying appropriate material properties. Nevertheless, certain regulations and rules of thumb are also applied.

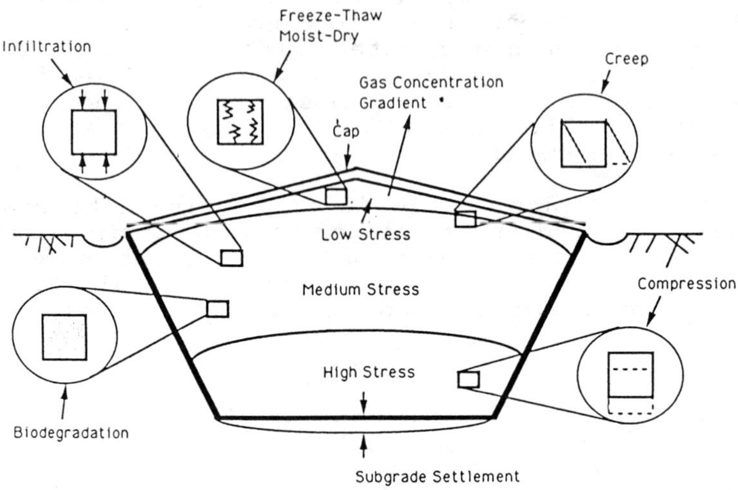

Figure 2. In-situ Environment of a Stabilized Landfill

The unconfined compression test is readily understood by non-technical persons, and is adaptable to empirical optimization of proportioning, mixing methods, mixing time effects, etc. In stabilization by microencapsulation, the strength indicates entrapping structure development, but does not in itself assure isolation has been achieved. This test may also be used as a design standard for mechanical stability, and an unconfined compressive strength (q_u) of 340 kN/M^2 (50 psi) is often specified for stabilization of hazardous wastes. This standard is readily met with many inorganic wastes, but can be unattainable with organics. In any case, the scenario of Figure 2 doesn't require such strength for physical stability.

The key mechanical concern is dimensional stability, indicated by the stiffness. If unconfined compression results show very high stiffness, low monolith deformations are probable. For porous monoliths, however, deformation behavior is better predicted with results of the one-dimensional compressibility or consolidation test (ASTM 2435). This allows calculation of the total and differential settlements due to gravity (self-weight) and cap loads. Such data is important in predicting cap alignment stability and rainfall runoff efficiency. High settlements also indicate fluid (gas and leachate) expulsion during consolidation. However, consolidation tests are time consuming and mask the effects of ongoing hydration. Consequently, they are best run on optimized mixtures after initial screening with the simpler unconfined compression test.

Characterization of immobilization is generally done with leaching and permeability tests. The former is often a batch equilibrium test that subjects pulverized samples to agitation in an aggressive acidic solvent, i.e., EP Toxicity and TCLP tests. Results are used in a pass-fail mode to determine if a waste or treated waste is hazardous, and also to indicate the types and relative order of mobilized contaminants (14). However, these tests provide little data on mobilization rates.

Sequential batch tests (15) or column permeation (16) provide more insight into the probable contaminant leaching rates and changes in leachate quality (strength and constituents) with accumulated solvent exposure. One issue is whether a leaching test should be done on pulverized or intact stabilized material. The former discounts the contribution of physical microencapsulation, and is thus more conservative. However, in a large monolith, only the surficial portion is subjected to expansion/contraction cycles under climatic variations. Below the surface layer, not only is the waste material insulated, but it is also laterally confined. Permeability tests on intact specimens would thus be more appropriate for predicting leachate quality, and also indicate internal transport ease as well.

Some adjustments to conventional stabilization practice were appropriate at this site. The waste is not hazardous, but classified as an industrial residue, which will have practical effect of allowing design to be more performance than prescription based. Capacity restrictions limited final landfill volumetric "swell" to at most 100% (relative to the existing sludge volume). The construction plan at this site was to place and compact a fresh mixture of a soil-like consistency and allow further hydration in place. A high strength cement-pozzolan stabilizing mixture with 50% sludge-filled porosity was developed. However, immobilization was poor, as indicated by results with water or dilute acid permeation. With substitution of cemented clay as the basic matrix, standards and analysis more appropriate to soils were applied. An as-compacted compressive strength of 50 kN/M^2 and a doubled value when hardened was specified to assure equipment support and slope stability. Settlement from gravity (self-weight) compression was seen as unavoidable.

WASTE MATERIALS

Hydrocarbon Sludge

The sludge consists primarily of long-chain aliphatics and polycyclic aromatics, but also has considerable sulfur.

It passes both EP Toxicity and TCLP tests in terms of hazardous material concentrations. Constituents classified as hazardous volatiles (Boiling point < 100°C) were not detected above the micrograms/liter (ppb) range. Some hazardous listed materials listed as semivolatiles and nonvolatiles are present up to the range of several hundred ppm. Only small traces of chlorinated compounds or heavy metals were detected.

The sludge from the most liquid lagoon used in all tests reported herein has an ash content of 4.5% and an organic carbon content(TOC) of 35.3%. The consistency is similar to that of a slow-curing roadway asphalt.

Spent and Native Clays

The spent clay or fuller's earth was used for color and metal removal in lubricant production. It was originally deposited in piles at 10% to 20% oil content, but expulsion of oils by consolidation and clay hydration has occured. The native clay is actually on the silt-clay borderline. Table 1 is a summary of clay properties:

TABLE 1

Clay Properties

Property	Spent Clay	Native Clay
Coarse fraction minerals	quartz	quartz
Clay fraction minerals	attapulgite	kaolinite
Cation exchange capacity (EPA method 9081)	105 meq/gm	49 meq/gm
In-situ moisture content	50% to 100%	10% to 20%
In-situ oil & grease content	4% to 8%	0
Index properties		
specific gravity	1.98	2.72
liquid limit	140%	31%
plastic limit	76%	22%
shrinkage limit	30%	12%
finer than #200 mesh	78%	86%
finer than 0.002 mm	10%	6%
Compacted properties		
max. dry unit weight	8.2 kN/m^3	17.5 kN/m^3
optimum moist. content	30%	16%

The specific gravity (G_s) was measured on samples oven dried at 105°C for 24 hours. Moisture contents are also based on this dry weight. However, the "moisture" lost in the drying includes volatile organics as well as water. The oil & grease content of the spent clay was reported by the site owner, and is part of the measured moisture content.

The high in-situ moisture of the spent clay posed potential construction problems. Admixing 5% to 10% lime allowed air-drying to about 60% moisture in a few hours, but longer exposures are required for further moisture removal. Overdrying could have negative effects, such as inadequate moisture availability for pozzolanic cement hydration. Consolidation tests also show significant swelling for moisture contents below 70%.

Unconfined compression tests on the spent clay indicate strengths increasing from 65 kN/m^2 (10 psi) to about 100 kN/m^2 (15 psi) as it dries from 70% to about 60% moisture content. Samples compacted at moisture contents over the 60% to 70% range consistently displayed permeabilities of the 10^{-8} cm/sec magnitude.

The less plastic native clay exists in friable, loamy aggregations. Its readily dried, pulverized and compacted to a continuous and remolded structure. Permeability and compressibility studies of this material were thus done under two conditions, presumably representing the extremes of the structure in which the sludge could be encapsulated:

-Static compaction to about 80% Standard Proctor unit weight at optimum moisture content (OMC), modeling an aggregated result of low mixing and compaction effort

-Kneaded compaction at 4% above the OMC to model a heavily remolded (mixing and compaction) effort, to about 93% of Standard Proctor unit weight

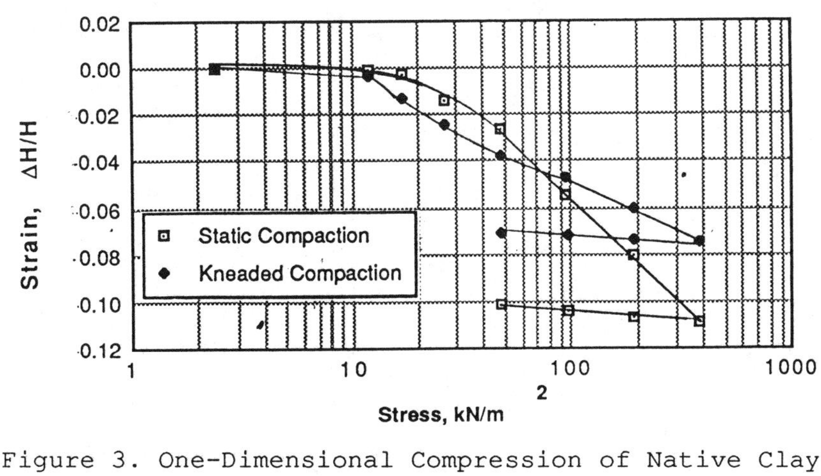

Figure 3. One-Dimensional Compression of Native Clay

Figure 3 shows results of one-dimensional compression tests on the native clay. The staticly compacted or aggregated sample was stiff at low pressures, but softened at the overburden stress range anticipated in the stabilized monolith, up to about 100-160 kN/m^2 for deposits 10-15m thick. The aggregated samples also displayed permeabilities around 5 x 10^{-5} cm/sec. Remolding and densification by kneading action improved stiffness and decreased permeability to about 2.5 x 10^{-8}cm/sec. With these results, it was determined that it was not advisable to preserve the clay structure in the final sludge mixture.

Unconfined compression tests also show significant strength gain with age. For the samples compacted with a Harvard Miniature apparatus around the OMC, as-compacted strengths were around 80 kN/m^2, increasing 50% to 75% with 30 days of "curing".

Additives

Lime is necessary for neutralization, reduction in clay plasticity, and participation in the cementing reaction. Hydrated dolomitic lime was used to minimize heat and volatilization.

Type F fly ash (not self-cementing) is available from nearby power plants. It serves as a moisture absorbent to improve sludge and clay blending, and then participates in the pozzolanic cementing. Fly ash also appears to serve as a continuing source of alkalinity during permeation with acidic solvents,thus improving the longevity of the stabilization process. The fly ash used has a specific gravity of 2.49, a median grain size (d_{50}) of 0.03mm, and 8% to 12% unburnt carbon content (Loss on Ignition test).

Pozzolan hydration is relatively slow, and it would be further retarded by the presence of organic matter. To establish the best-case scenario, lime-fly ash mixtures in the ratios 1:3 to 1:4, and at a range of water contents from 10% to 50% (based on total dry weight) were compacted in 2.5 cm diameter teflon molds. Specimens were then extruded and cured in sealed bags. Unconfined compressive strengths of 500 kN/m^2 to 700 kN/m^2 at moisture contents of 25% to 35% were obtained in 30 days. Water contents beyond this range gave poor results. A rich lime-fly ash additive blend in the ratio of 1:3 was chosen for the sludge stabilization to allow uptake of calcium by both the clay and the pozzolanic reaction.

Addition of small amounts of portland cement to the pozzolan blend substantially increased the strength. Portland cement will be used to improve or accelerate strength gain only if needed.

MIXTURES USED

In the empirical studies, mixtures were described by relative weight proportions in the order:

acidic sludge/clay/fly ash/hydrated lime

The clay proportion is on a dry weight basis. Other key variables are the consistency of the raw sludge, the water content of the clay used, and the moisture content of the complete mixture measured with 105°C oven drying. As noted above, the latter measure includes volatile organic material as well as water. It would be expected that the moisture content would decrease with time, indicating pozzolan hydration and time-dependent volatile adsorption.

The sludge fraction is the reference "1.0 parts", with the other components expressed as ratios to the sludge content. The mixture that appeared to be optimum with the spent clay is described:

1.0 sludge/1.0 spent clay/0.75 fly ash/0.25 lime

With the less plastic native clay, a smaller amount of that material was required to give the desired consistency, immediate strength and immobilization results:

1.0 sludge/0.75 native clay/0.75 fly ash/0.25 lime

Spent clay moisture affects both construction ease and mixture performance. One test series, called the "moist clay" mix, was made with the source clay at 60% water content, producing a net moisture content for the mixture of 36.2%. A second series employed clay at 50% moisture, resulting in a fresh mixture moisture content of 34.2%. While this is not radically drier, the texture was quite different. This set can be referred to as the "friable" or drier clay mix. Both the source clay and the mixture moisture content are above the OMC for the clay (Table 1).

The native clay water content is readily controlled, allowing a single clay moisture content to be used (17.5%), just above the optimum moisture content. The resulting mixture was 11.0% moisture content.

Each mixture was compacted in a Proctor mold at both full and 50% Standard Proctor effort as shown on Table 2. The specific gravity (G_s) of an organic mixture depends upon the definition of the solids. To compute the volumetric efficiency of encapsulating the sludge, a specific gravity can be calculated with the unreacted pozzolan and clay proportions. This yields $G_s = 2.24$ for the spent clay mix, and $G_s = 2.54$ for the native clay mix. On this basis, the computed expansion of each mixture meets the site capacity standard (< 100% expansion).

TABLE 2

Compacted Mix Volumetrics

Property	Spent Clay Mix	Native Clay Mix
Clay moisture (105°C)	50%	17.5%
Mix moisture (105°C)	34.2	11.1
Full effort unit weight	13.5 kN/M^2	15.4 kN/M^2
50% effort unit weight	13.3 kN/M^2	14.8 kN/M^2
Mixture specific gravity	1.95[1]/2.24[2]	2.08[1]/2.54[2]
50% effort porosity	48%	35%
50% effort saturation	72%	43%

Notes: 1. Measured G_s on complete mixture residue at 105°C
2. Calculated G_s based on original solids

In computations of seepage and compression, it was noted that the nonvolatile sludge fraction could be treated as part of the solids. The saturation, moisture content and porosity values shown on Table 2 are based upon specific gravities measured with 105°C oven drying residue, 12% to 20% below G_s values computed from the added solids.

EMPIRICAL STANDARDS

The proportionings cited above were clearly selected on the basis of the best that could be done with the given components while meeting the requirement for high porosity. Much of the rest of this paper describes the test results, but it is first necessary to establish acceptable values of permeability, strength, etc. One approach is to treat the original mechanical and hydraulic properties of the clays as thestandards. The basic premise is that the clay skeleton is weakened and expanded to accomodate the sludge volume. The pozzolanic cementing should thus at least restore the original clay matrix properties. The basis for this approach is twofold: steep clay slopes are stable at the site, and, despite the presence of the unstabilized sludge lagoons for up to one-half century, the permeability of the soils is sufficiently low that only minimal groundwater impact has been detected.

A second approach is to work from the intent of the project as a whole, reducing the risk of contaminant release. Analysis of each potential release mechanism and the in-situ environment would indicate the required characteristics of the stabilized mixture. Wastes pose a threat to the environment if a contaminant is in a mobile form, a transport pathway is available and a gradient exists to induce and sustain movement. Design thus centers on restricting one or more of these factors to restrict release of contaminated particulates, gases or liquids.

The first two potentially mobile contaminant forms, particulates and gas, are readily addressed at this site. Solidification will limit particulate erosion during construction, the cap will seal the waste after closure, and the monolithic subgrade itself helps to maintain the cap integrity. With a highly organic waste, vapors from direct volatilization or biodegradation would appear to be a serious concern. However, the sludge has been exposed for years, such that volatiles initially present near the surface are now gone, as confirmed by low vapor concentrations during the field study. The saturated hydrocarbons resist anaerobic degradation, and the rate of aerobic decomposition in the monolith will be restricted by oxygen diffusion rates. Consequently, post-closure gas generation rates will probably be very low.

Leachate is the major long-term concern at this site. Mechanisms that might generate leachate from landfill deposits include: a) free liquid drainage b) consolidation expulsion of pore water, c) surface or ground water infiltration and seepage, and d) molecular diffusion. A very basic goal of sludge solidification is to remove or absorb free liquid, and this was successfully done with the soil-like consistencies of the mixtures described above. Measurements of the lumped organic diffusion coefficient for the stabilized mixture yielded values in the range of 10^{-10} m^2/sec (17). Consequently, neither mechanisms "a" nor "d", above, are seen as serious concerns.

The stabilized but unhardened mixtures will be placed and compacted to about the degree of saturation noted in Table 2. Consolidation of hardening layers under subsequent overburden lifts will tend to expel pore gases in the clayey mixture rather than the semi-liquid entrapped sludge and retained water (18). Thus, as porosity is reduced, the degree of saturation will increase. Consequently, expulsion of leachate will not occur at a substantial rate until accumulated compression causes a layer to approach 100% saturation. While consolidation expulsion is a construction and short-term post-closure problem, it is likely to produce a highly concentrated leachate.

Preventing leachate generation by this mechanism requires limiting compression of the lowest layers. If the as-placed porosities and saturations are as indicated on Table 2, then the strain required to expell all gases and initiate liquid expulsion is about 12%. Depending upon the overburden stress on the lowest layer, this strain limit sets the minimum acceptable stiffness, or inversely, maximum compressibility.

The long term concern is leachate generation by infiltration. In conventional landfills, the waste deposit itself generally offers no resistance to infiltration or seepage. During filling, all incident rainfall produces leachate as the final cap has not yet been installed. Design of the leachate collection concentrates on limiting the depth of ponding on the liner,to suppres the gradient across a barrier that is relatively thin compared to the potential depth of liquid surcharge on it (19). After closure, limiting leachate production rates depends upon the cap to shed rainfall and restrict infiltration, and on the liner system (if any) to resist groundwater intrusion.

In contrast, both liquid entry into and passage through stabilized waste deposits is restricted by the permeability or hydraulic conductivity. During filling, most rainfall can run off with minimal contact with the waste mass. For the post-closure term, a worst-case condition can be envisioned. In this scenario, a film of water lies atop the deposit between a flat, leaking cap and the top of the monolith. In this case, assumed to be steady, the hydraulic gradient is unity due to gravity, and the velocity as described by Darcy's law is numerically equal to the permeability ($V = K$). Rainfall rates in excess of this amount runs off even if the cap is cracked.

With the equivalence of the vertical seepage rate and the permeability, an acceptable value of the latter can be set directly from the maximum allowable leachate production rate, often set by regulation. One aspect of the clay-based stabilization is that even if the cement deteriorates, the maximum seepage rate is still restricted by the loosened clay matrix permeability, a "self-lining" feature.

Another concern is leachate generation from innundation of lower layers by a rising water table in the distant future when leachate collection may be discontinued. If the monolith permeability is substantially below that of the local soil, the landfill is a barrier to regional ground water flow (9). Undisturbed samples of the native soil display a horizontal hydraulic conductivity of 3×10^{-5} cm/sec. A standard that might be applied is a requirement that the hardened sludge mixture be below 3×10^{-7} cm/sec, two orders of magnitude below the surrounding soil.

IMMOBILIZATION RESULTS

Permeability/Hydraulic Conductivity

Stabilized mixture permeability is used as a measure of its capability to physically isolate contaminants as well as an indicator of restricted internal fluid movement in general. About 200 g of each fresh mixture was compacted with either 50% or 100% of Standard Proctor effort in 6.35 cm diameter fixed wall plexiglass permeameters. Curing and saturation for 14 days was allowed under a reservoir of distilled water (pH = 6) 3cm deep, followed by falling-head permeability tests. The range of hydraulic gradient changes during permeation was restricted to maintain an average gradient of about 80 cm/cm. Some tests were run at lower gradients to investigate the relationship between gradient and permeability, but the results were inconclusive.

There were no measurable differences between the spent clay mixture samples compacted at each level of effort, but the moisture content of the mixture was influential. The moist clay mixture samples displayed permeabilities in the 10^{-8} cm/sec range, while the friable or drier clay samples showed permeabilities about one order of magnitude higher. Either result would be regarded as very low permeability and met the requirement of being two orders of magnitude below the surrounding soil.

Compaction effort did have a major influence on the permeability of the mixtures including the native silty clay as the matrix. Heavy compaction is apparently necessary to disrupt and remold the clay aggregates, as was described earlier for the clay by itself. Light (50% Proctor) compaction produced values of about 2 x 10^{-5} cm/sec. Heavy (100% Proctor) compaction of the same mixture produced similar unit weights (Table 2), but the resulting permeabilities were in the range of 3.5 x 10^{-7} cm/sec.

Microencapsulation with Permeation

In considering the overall goal of contaminant immobilization, the quality of the leachate is as important as the volumetric production rate. With the wide array of hydrocarbon fractions, the total organic content (TOC) of the permeameter effluent was used as the index of sludge microencapsulation. Alkalinity and pH were also monitored. Two permeants were used: distilled, de-aired water and the same liquid with 0.05N sulfuric acid (pH=2). However, the stabilized mixture alkalinity was so high that there was no noticeable difference in permeability or effluent quality results with either permeant, even after passage of eight pore volumes of solution. With lower plasticity and clay fraction, the native clay mixture had a richer lime content. This is reflected in the effluent pH. For the spent clay mixtures, the effluent pH varied from 6 to 8, while the native clay samples produced effluent pH's in the range of 10 to 11.

Figure 4 shows the results of extended permeation with distilled water of the drier spent clay mixture and the native clay mixtures at both compaction efforts. After the "first flush", or displacement of the pore liquid, the two clay mixtures were found to be of similar effectiveness in retaining hydrocarbons. The poorer initial quality of the spent clay mixture indicates that lower permeability does not compensate for the more open structure, i.e. higher porosity. The TOC of the first few samples may also reflect mobilization of the oil initially in the spent clay. Separate-phase liquid was not displaced in either case.

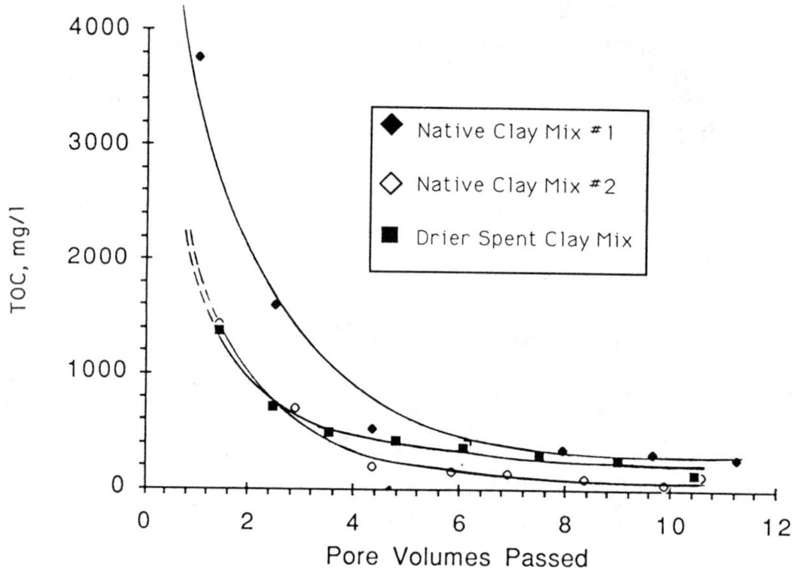

Figure 4. Results of Stabilized Mixture Permeation

It is useful to study leaching results in terms of mass balances. After 10 pore volumes of throughput, less than 3% of the hydrocarbon content of any specimen was mobilized and displaced. It is estimated that the liquid in the pores would have a TOC of about 20,000 mg/l if there were no immobilization.

MECHANICAL PROPERTIES OF STABILIZED MIXTURES

Unconfined Compressive Strength

Samples were compacted at 100% Proctor effort in teflon molds, extruded, and cured in sealed containers. Compaction at lower efforts did not yield consistent results. Some of the native clay mix samples were also confined in the molds for 60 days of curing before extraction and testing.

Stress-strain curves for the drier spent clay mixture

are shown in Figure 5, and the strength gains with time for the two moisture contents are illustrated in Figure 6. The curves shown in Figure 5 are typical for remolded clays, as expected, as the moisture content is above the OMC for the clay. Consistent increase in strength and stiffness with time is attributed to the pozzolanic cementing. This did not cause increased brittleness, as failure typically occured at at about 10% strain. The goal of meeting the source clay strength (65-100 kN/m^2) was met, but the hydrocarbons do seriously affect the cementing. A major strength increase would otherwise be expected according to soil stabilization practice with pozzolans (20).

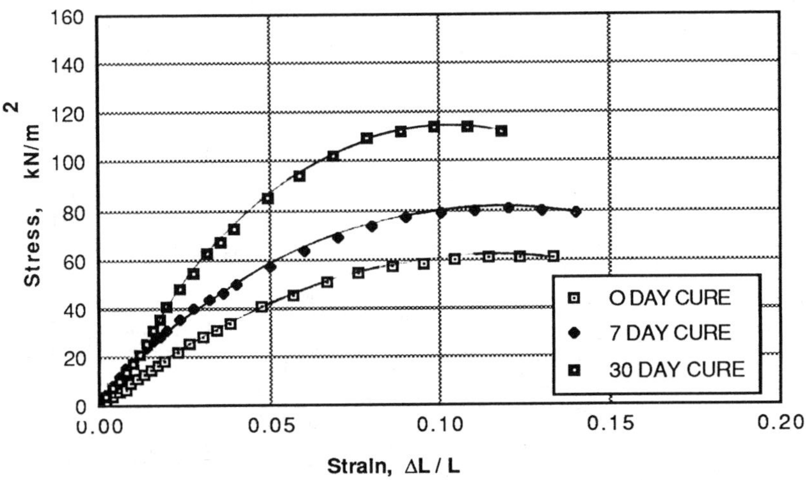

Figure 5. Stress-Strain Behavior of Spent Clay Mixes

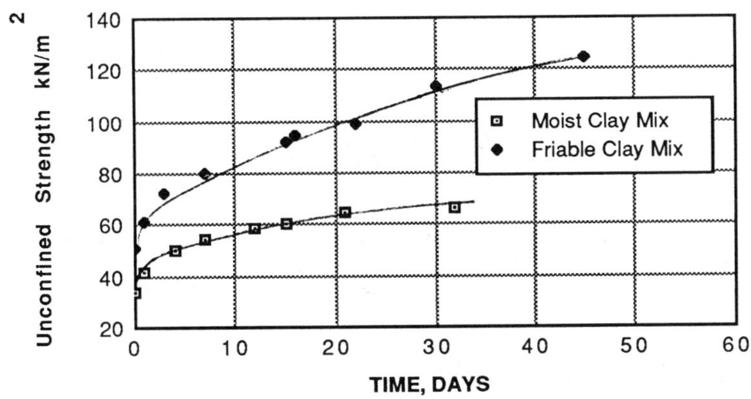

Figure 6. Strength Gain Rates of Spent Clay Mixes

Despite a concern for moisture availability for pozzolan hydration, the moist clay mix showed lower strength gain. Apparently, the extra moisture had a negative effect in reducing clay skeletal strength, masking any improvement in pozzolanic hydration. If the use of wetter clay were unavoidable, it would probably be necessary to add portland cement to bring strength up to standard.

Stress-strain results for the native clay mixture are shown in Figure 7. This mixture was drier and had a lower clay skeleton proportion than the spent clay mixtures (0.75 part vs. 1.0 part). The native clay mixture was stiffer in unconfined compression, with peak strengths obtained at 3% to 4% strain. Effects of lateral confinement during curing were also studied, with a 25% increase in strength and doubled stiffness after 60 days curing. Compaction will produce residual lateral stresses, so it is expected that field samples will be as strong as laboratory specimens.

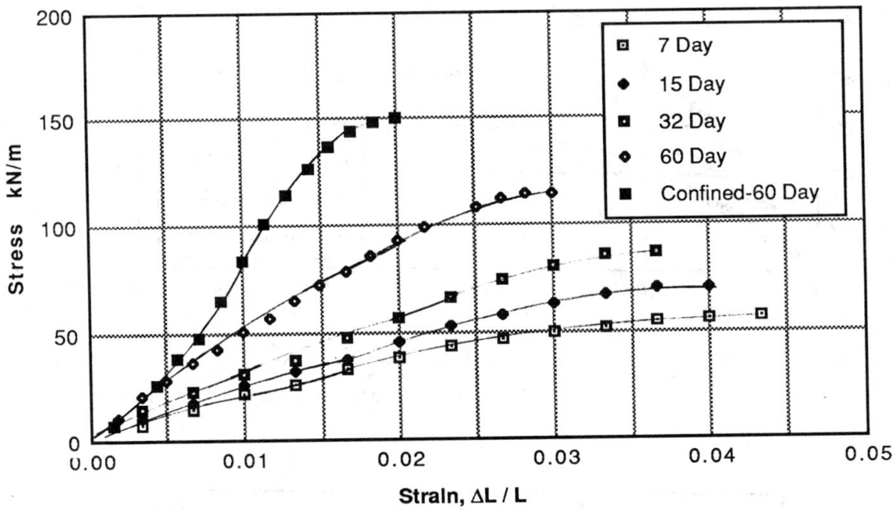

Figure 7. Stress-Strain Behavior of Native Clay Mixes

Replacing 25% of the lime in the native clay mixture increased strength to about 175 kN/m^2 with 60 days curing, but further cement addition at the expense of the lime reduced hydrocarbon fixation. Achieving the stabilization goal of 340 kN/m^2 (50 psi) strength cited earlier appears to be impossible with this set of constituents at the required volumetric efficiency for sludge encapsulation.

One-Dimensional Compression

The behavior of the stabilized sludge in laterally confined compression affects cap alignment, permeability and leachate generation. Test samples were obtained by pressing cutter rings into the mixtures compacted into the molds at 50% and 100% Proctor effort. Rings were then cured in sealed containers for one week, and then loaded as per customary procedures (doubled load at each increment), except as noted below. Each load was retained for three days to complete primary (hydrodynamic) consolidation. All data reported herein are for the 100% compaction specimens. Results for 50% effort samples were marginally different.

Figure 8 shows one-dimensional compression versus log stress results for the spent clay mixtures, including a replicate sample. While it is customary and conservative to run tests on saturated soil, the likely field condition is that most of the deposit will remain unsaturated. It was also desired to investigate swelling behavior as this was observed with the spent clay itself. As-compacted samples were loaded to an overburden pressure of 9.5 kN/m^2 before the consolidometers were flooded with water. Selection of an overburden equivalent to a depth of about 0.7 meters before flooding was arbitrary. Neither swelling nor collapse was observed, indicating that the clay skeleton was stabilized with respect to weather sensitivity.

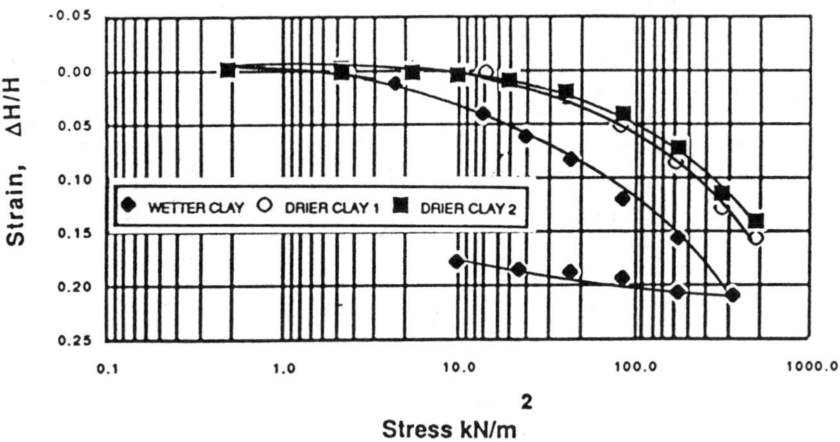

Figure 8. One-Dimensional Compression of Spent Clay Mixtures

The moist clay is much more compressible, as would be expected from the unconfined compression test results cited earlier. At an overburden stress of over 100 kN/m^2 (about 8 m depth), the predicted strain would exceed 12% and tend to cause leachate movement. The initial saturation of this mixture is about 75% at a porosity of 0.49. A 12% strain would reduce the porosity to about 0.4, bringing the saturation to the 95%-100% range if there is some rainfall infiltration during construction. In contrast, while the drier clay is only slightly less porous and saturated (Table 2), it would only undergo about 6% strain and thus, probably still remain unsaturated.

Figure 9 shows one-dimensional compression results for native clay mixtures, with samples extracted from the 100% Proctor effort molds in the same maner as described above. In contrast to the dramatic effect of compaction effort on permeability, the compression results were similar for both compactions. The variation shown on Figure 9 is the loading increment, with one curve representing the customary load doubling at each stage, and the other showing the effect of a constant or single load increment at high stress levels. This would more accurately represent the field condition.

The difference between the two loadings is insignificant, especially if cement hydration during the test is assumed to continue. The same time interval was used (3 days/point) in each test, and thus, at a given stress level, say 50 kN/m^2, the constant-increment sample was more fully cured, and presumably, stiffer.

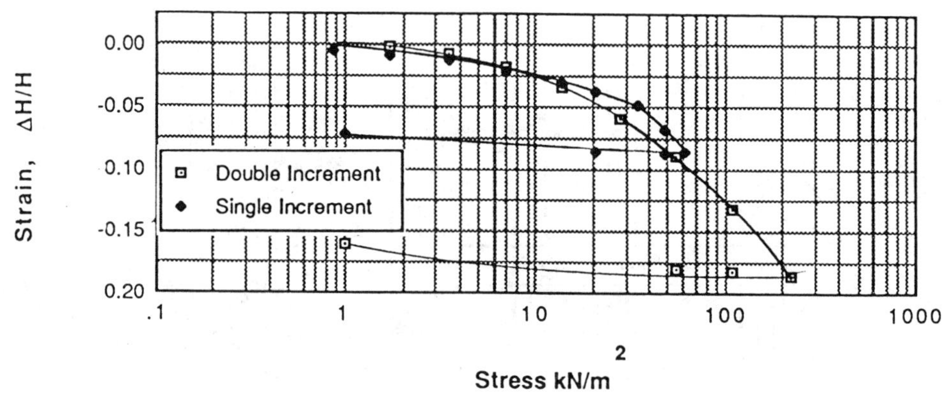

Figure 9. One-Dimensional Compression of Native Clay Mixtures

While the native clay mixture is much stiffer in unconfined compression, it is similar in stiffness to the moist spent clay for laterally confined conditions. A strain of about 12% is also predicted at a pressure of 100 kN/m^2. This would not, in itself, cause a leachate problem as the mixture is much drier.

SUMMARY

This project involved stabilization and redeposition of a non-hazardous hydrocarbon sludge. Volumetric constraints indicated that a very porous product was needed to encapsulate the sludge. Stabilization with a pozzolanic cemented admixture alone did not produce satisfactory results. An alternative procedure of encapsulation in a clay matrix was investigated, using a spent attapulgite and a native silty clay. Performance in mechanical stability, and contaminant immobilization by isolation and fixation. was considered. Empirical testing established optimum component proportionings, further studied to determine properties connected with prediction of monolith behavior.

Several sets of criteria were considered, including: a) customary standards applied to concrete-like stabilization, b) matching geotechnical properties of the source clays and c) performance criteria indicated by analysis of monolith behavior to limit pollutant release.

It was determined that the strength criteria of set "a" could not be met, but the consequences are debatable. Another standard criteria, passage of the EP Toxicity or TCLP tests, was already met by the source sludge. Without further leachate quality criteria, the fixation goal was simply to make leachate quality as good as possible.

The premise of set "b" was compensating for sludge inclusion in the clay with pozzolanic cementing. The original clay strength, compressibility and permeability were matched or exceeded by the stabilized mixtures. With regards to criteria set "c", it was determined that the erosion and vapor loss mechanisms of contaminant release were minor problems, but leachate generation by either infiltration or consolidation was important. Heavily compacted native clay and the drier spent clay mixtures showed good compressibility and permeability values.

CONCLUSIONS

A landfill is a form of artificial geology, and predicting the response to the internal and external environment is the logical basis for investigation and design. Stabilized waste monoliths can provide performance superior to conventional cellular solid waste landfills that rely on an external containment for isolation. The problem herein was formulation of porous mixtures to construct a dimensionally stable monolith that would limit release of encapsulated contaminants.

It was necessary to use local clays to form a porous skeleton, and to condition them structurally and chemically with a lime-fly ash pozzolanic admixture. The approach taken was to adapt traditional geotechnical methods of analysis, testing, etc. to the problem, essentially treating the waste deposit as an embankment In the empirical study, it was determined that the primary variable in the spent clay mixture was the moisture content, while performance of the native clay mixtures depended upon compactive effort and remolding of the natural aggregations. It was also determined that the additives would be best used in a 1:3 lime-fly ash ratio.

The result was that two sludge mixtures with reasonable material property values were obtained, suitable for a monoliths up to 12m thick in this hydrogeologic setting, i.e.: the cap would have a stiff and stable subgrade, deformations would not produce leachate, and there was little threat of infiltration and seepage. Approximate volumetric division of the stabilized mixture involved dividing the sludge into mobile and solid fractions.

The pozzolanic cemented clays encapsulate and also fixate the waste by complex and interrelated mechanisms that are not fully understood. It appears that fixation is more of a lime conditioning of clay that encourages organic partitioning in the pores rather than surface adsorption. However, empirical leachate quality result indicated that it was not as necessary to clarify fixation mechanisms as to be able to predict mechanical and seepage behavior.

ACKNOWLEDGEMENTS

The work was sponsored by Sun Refining & Marketing Co., Dr.Arthur J.Raymond, Project Director

REFERENCES

1.Smith, C.L. and D.J. Frost, "Secure Landfilling with Pozzolanic Cementing", Proc. 1st Ann Conf. on Hazardous Waste Mgmt., Phila, Pa., pp 153-160, 1983.

2. Morgan, D.S., Novoa, J.I. and A.H. Halff, "Oil Sludge Solidification using Cement Kiln Dust", J. Env. Eng. Div., ASCE 110 (EE5), pp 935-949, 1984.

3.U.S.EPA "Handbook for Stabilization/Solidification of Hazardous Wastes" EPA/540/2-86/001, 1981

4.Yen, B.C., and Scanlon, S., "Sanitary Landfill Settlement Rates", J. Geotech. Eng. Div., ASCE 101 (GT5), 1975.

5. Murphy, W.L. and P.A. Gilbert "Estimation of Maximum Cover Subsidence Expected in Hazardous Waste Landfills". Proc., 10th Ann. Research Symp., U. S. EPA, 0April 1984

6.Martin, J.P., Felser, A.J. and E.L. Van Keuren, "Hydrocarbon Waste Stabilization for Landfills", Proc. ASCE Specialty Conf. for Waste Disposal, Ann Arbor MI, pp June 1987.

7.Malone, P.G., and R.J. Larson, "Scientific Basis for Hazardous Waste Immobilization", Hazardous and Industrial Solid Waste Testing: Second Symp., ASTM STP 805, R.A.Conway and W.P.Gulledge, eds., 1983.

8.Martin, J.P., Robinson, W.T., and E.L. Van Keuren, "Large Volume Stabilization of Hydrocarbons with Spent Clay and Pozzolans", Proc. U.S.Dept. of Energy Oak Ridge Model Conf., Oak Ridge TN, Oct 1987

9. Martin, J.P., Biehl, F.L.,and W.T. Robinson, "Stabilized Petroleum Refining Waste Interaction with Silty Clay Subgrade" Proc. Conf. on Petroleum Spill Impact on Soil, Univ. of Mass., Amherst, Mass., Sept 1988.

10. Robinson, W.T. "Characterizing the Leaching Potential of Hydrocarbon Wastes from a Stabilized Mixture" M.S. In Environ. Engineering Thesis, Drexel Univ, Phila PA, 1987

11.Van Keuren, E.L., Martino, J., Martin. J. and A.Defalco,"Pilot Field Study of Refinery Waste Stabilization" Proc. 19th Mid-Atlantic Industrial Waste Conf., Bucknell Univ., Lewisburg, PA, June 1981.

12. Van Keuren, E.L., "Advanced Laboratory and Field Study of Hydrocarbon Waste Stabilization with Pozzolans", M.S. In Civil Engineering Thesis, Drexel Univ, Phila PA, 1987

13. Webster, W. C., "Role of Fixation Practices in the Disposal of Wastes" ASTM Standardization News, March 1984.

14. Rinaldo-Lee, M.B., Kunes, T.P., and D.G. Nicholas, "A Comparison of Groundwater Quality at Selected Landfills to Leaching Test Results", Hazardous & Solid Waste Testing, First Conf., ASTM STP 760, 1983

15. Bishop, P. L. "Prediction of Heavy Metal Leaching Rates from Stabilized/Solidified Hazardous Wastes" Toxic and Hazardous Wastes, Proc. 18th Mid-Atlantic Indust. Waste Conf., Virginia Polytechnic Inst., pp 236-252, 1986.

16. Maskarinec, M.P., and D.K. Brown,"Batch and Column Extraction Procedures for Organics in Solid Wastes" Hazard. & Solid Waste Testing, ASTM Second Conf., STP 805, 1985

17. Daniel, D.E. and C.Shackelford, "Engineered Barriers Designed to Release Contaminants at Disposal Sites Only by Molecular Diffusion". Proc., U. S. Dept. of Energy Oak Ridge Model Conference, Oak Ridge TN., Oct. 1987

18. Martin. J.P."Embankment Consolidation from Unsaturated to Saturated Conditions" PhD in Civil Engineering Dissertation. Colorado State Univ., Ft Collins, CO, 1983

19. Demetracopoulos, A.C., Korfiatis, G.P., Bourodimos, E.I., and E.G. Nawy " Modeling for Design of Landfill Bottom Liners" J.Env. Eng., ASCE Vol 110, (EE6), 1984

20. Winterkorn, H.T., "Soil Stabilization", Foundation Eng. Handbook, H.F. Winterkorn and H.Y. Fang, Eds., Van Nostrand Reinhold 1975.

Stability and Settlement Analyses

James K. Mitchell, Raymond B. Seed, and H. Bolton Seed

STABILITY CONSIDERATIONS IN THE DESIGN AND CONSTRUCTION OF LINED WASTE REPOSITORIES

REFERENCE: Mitchell, J. K., Seed, R. B., and Seed, H. B., "Stability Considerations in the Design and Construction of Lined Waste Depositories," Geotechnics of Waste Fills--Theory and Practice," ASTM STP 1070, Arvid Landva, G. David Knowles, Eds., American Society for Testing and Materials, Philadelphia, 1990.

ABSTRACT: Safe disposal and storage of solid hazardous waste now requires the construction and filling of repositories underlain by multi-layer liner systems. These lining systems typically contain a large number of material interfaces, many of which have low shear strengths. This introduces potential failure surfaces along the side slopes and base of the fill mass which may control the overall stability of the overlying waste fill during fill placement operations. This paper discusses lessons learned from the investigation of the Kettleman Hills repository stability failure of March 19, 1988, regarding evaluation of the shear resistances along the different liner interfaces, the factors that control overall stability of the waste fill mass, and application of these lessons to analysis and design of safe repository filling operations.

KEYWORDS: stability, waste repository, laboratory testing, analysis, interface shear strength

THE KETTLEMAN HILLS WASTE LANDFILL SLOPE FAILURE

Landfill Unit B-19, covering an area of about 120,000 m^2, forms part of a Class I hazardous waste treatment and storage facility at Kettleman Hills, California. Unit B-19 is a large, oval-shaped "bowl" with a nearly horizontal base, and side slopes of both 1 on 2 and 1 on 3. The lining of the northern end of the bowl, designated Phase I-A and covering about 50,000 m^2, was completed first. A plan view of the lined Phase I-A repository basin is shown in Fig. 1(a).

Placement of solid waste and soil cover into the Phase I-A portion of Unit B-19 began in early 1987, and proceeded at an essentially

The three authors are Professor, Associate Professor and former Cahill Professor of Civil Engineering, respectively, all at the University of California, Berkeley, CA 94720.

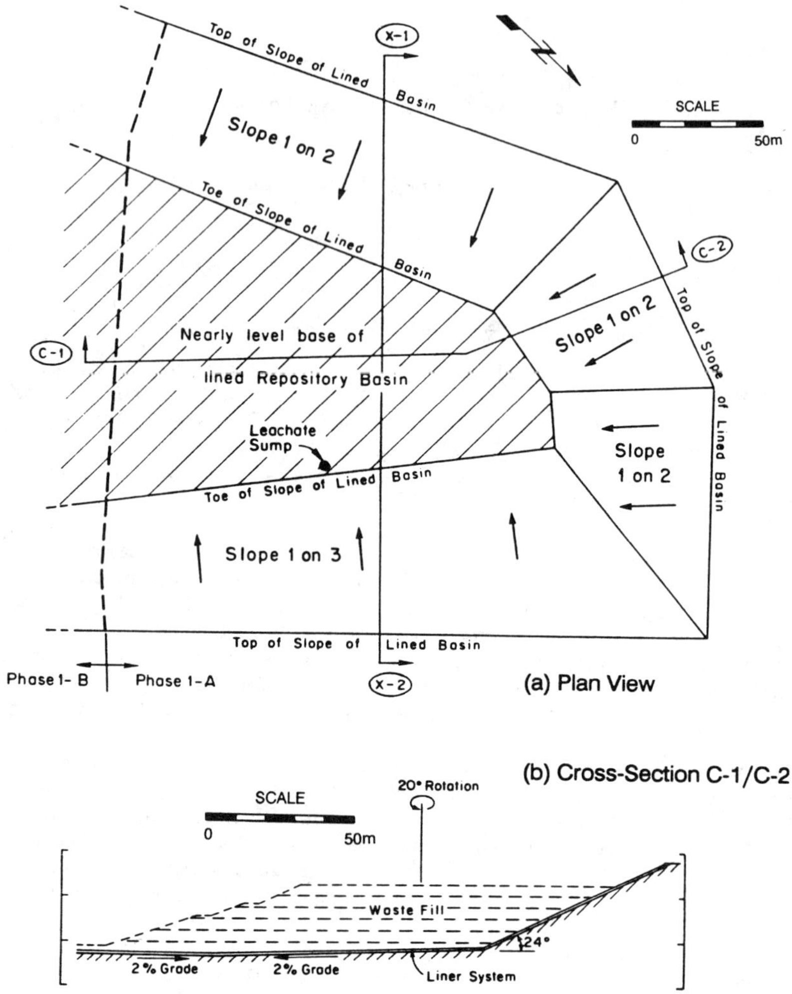

FIG. 1 -- Schematic illustration of lined repository basin and waste fill mass: Kettleman Hills Unit B-19; March 19, 1988.

constant rate until March 19, 1988. A cross section showing the elevation and profile of the fill at that time is shown in Fig. 1(b).

On March 19, 1988, a slope stability failure occurred which resulted in lateral displacements of the waste fill of up to 11 m., as shown by the displacement vectors in Fig. 2, and surface settlements of up to 4.3 m. Surface cracking was clearly visible, as also were tears and displacements on the exposed portions of the liner system. The failure developed over a period of a few hours, from early morning to early afternoon. No subsequent movements have been measured.

Based on field observations, photographic and survey records, and stability analyses, it was evident that the failure developed by sliding along interfaces within the multi-layer liner system, within

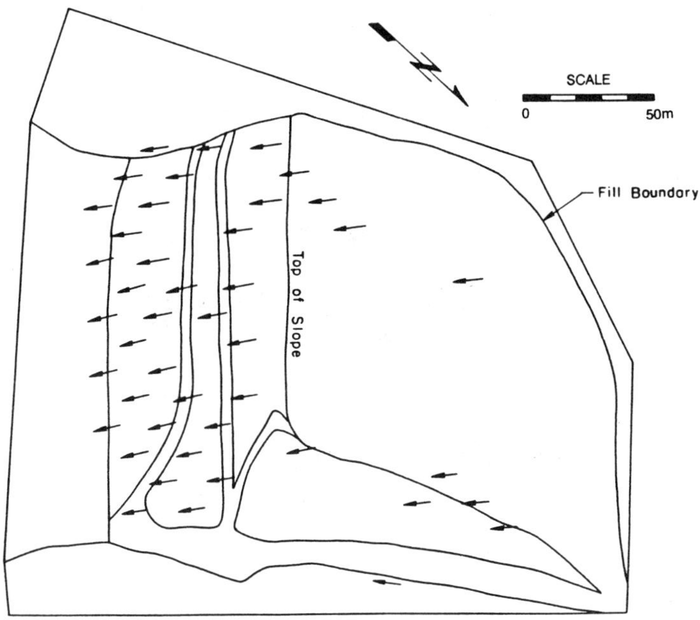

FIG. 2 -- Plan view showing displacement vectors representing measured lateral movements due to the slide of March 19, 1988

the compacted clay layers which form parts of the composite liner system, or along combinations of liner interfaces and through the clay.

A more comprehensive and detailed description of the failure, complete results of a testing program to evaluate system shear strength properties, and stability analyses to determine the cause of the failure are given by Seed et al. [1].

THE LINER SYSTEM AND FAILURE CONDITIONS

The composite flexible membrane liner (FML)-compacted clay double liner system used at Kettleman Hills is consistent with the type mandated by the 1984 Amendments to the Resource Conservation and Recovery Act. The base liner system, shown schematically in Fig. 3, provides two pervious layers for leachate collection and removal, two 60 mil thick Gundle high density polyethylene (HDFE) layers for leachate retention, two compacted clay layers for leachate retention, and geotextile layers (Trevira Spundbond No. 1145) for filtration. A layer of HDPE geonet (Polynet 300) for drainage purposes was included above the primary HDPE liner. The liner configuration used on the side slopes of the fill basin was similar to that used on the base.

A properly constructed composite double liner system of the type used at Kettleman Hills provides protection against transport of leachate out of the containment system. At the same time, however, the system contains a number of low strength interfaces which may act as potential surfaces of sliding whenever elevation differentials exist in

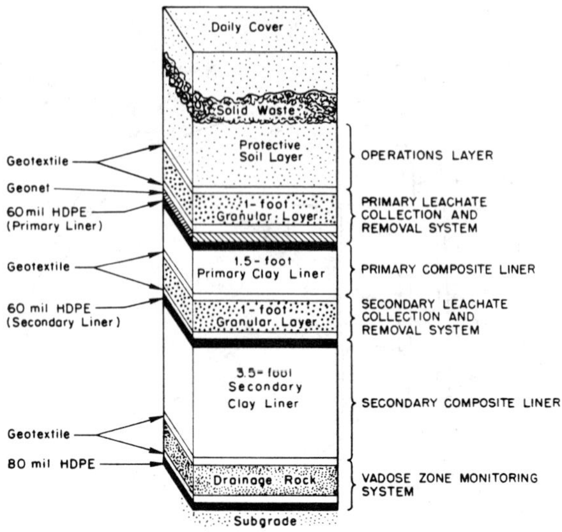

FIG. 3 -- Schematic illustration of the multi-layer liner system at the base of the Unit B-19, Phase I-A landfill

the waste fill such as was the case at the time of the landfill failure, Fig. 1(b).

From the results of a special testing program, which is described in the next section, the most critical interface strength values were determined to be those indicated in Table 1.

When the strength values indicated in Table 1 were used in two-dimensional stability analyses based on plane sections through the landfill, composite factors of safety somewhat greater than one were obtained. When three-dimensional analyses of the fill and liner system were made, however, factors of safety closer to unity were obtained for the geometry at the time of failure. The results of the stability analyses are listed in Table 2. The methodology is summarized in a later section of this paper and described in detail by Seed et al. (1988).

TABLE 1 -- Friction Angles or Shear Strengths on Critical Interfaces in the Kettleman Hills Landfill Liner System

Interface	Residual Friction Angle (ϕ_r) or Residual Undrained Shear Strength (τ_r) Along Saturated Base	Residual Friction Angle (ϕ_r) Along Dry Slopes
HDPE Liner/Geotextile	8° ± 1°	9° ± 1°
HDPE Liner/Geonet	8.5° ± 1°	8.5° ± 1°
HDPE Liner/Saturated Clay	45 ± 12 kPa	N/A

TABLE 2 -- Summary of the Results of Stability Analyses of the Unit B-19, Phase I-A Landfill at Failure

Base Wetting Conditions	Factor of Safety		
	2-D Analyses	3-D Analyses	Overall Best Estimate*
Wetting only in the Vicinity of the Leachate Collection Sump	1.2 to 1.25 (est.)	1.08	0.95-1.25
Full Saturation of Clay Along Repository Base	1.1 to 1.15 (est.)	1.01	0.85-1.15

*Authors' estimates taking into account uncertainties in liner system friction angles of ±10%, uncertainty in the 3-D analysis methods of ±15%, and uncertainty in the HDPE/compacted clay interface shear resistance of ±25%.

LINER INTERFACE STRENGTH EVALUATION

At the time of the Kettleman Hills Landfill failure few published values of the interface shear resistance between the geosynthetic components of liner systems or between geosynthetics and soils were available. Martin et al. [2] reported values for sand-geomembrane combinations, geomembrane-geotextile combinations, and sand-geotextile combinations. Only the HDPE geomembrane to geotextile combinations were representative of the potentially critical interfaces present in the Kettleman Hills liner system, and Martin et al. [2] reported a range of friction angles of 6 to 11 degrees for these.

Accordingly, it was necessary to peform a testing program to obtain the needed information. Both direct shear tests and pullout box tests were used. The direct shear tests offered the advantages of simplicity and the ability to test a number of interface combinations in a short time. They had the disadvantages, however, of limited possible shear displacement (0.75 cm. maximum) and rather small sample size (7.1 cm. by 7.1 cm.). The pullout box tests permitted shear displacements of more than 8 cm., with a larger initial interface contact area of 214 square centimeters.

Direct Shear Tests

A modified Karol-Warner direct shear testing apparatus was used to test interface combinations of HDPE liner/geotextile, HDPE liner/geonet, geotextile/geonet, HDPE liner/HDPE liner, HDPE liner/compacted clay, and geotextile/compacted clay. Interface specimens that did not involve compacted clay consisted of 7.1 cm. by 7.1 cm. square samples mounted with epoxy cement on 4 inch diameter platens as shown schematically in Fig. 4(a). Corrections were made for decreasing contact area between interfaces that developed with increasing shear

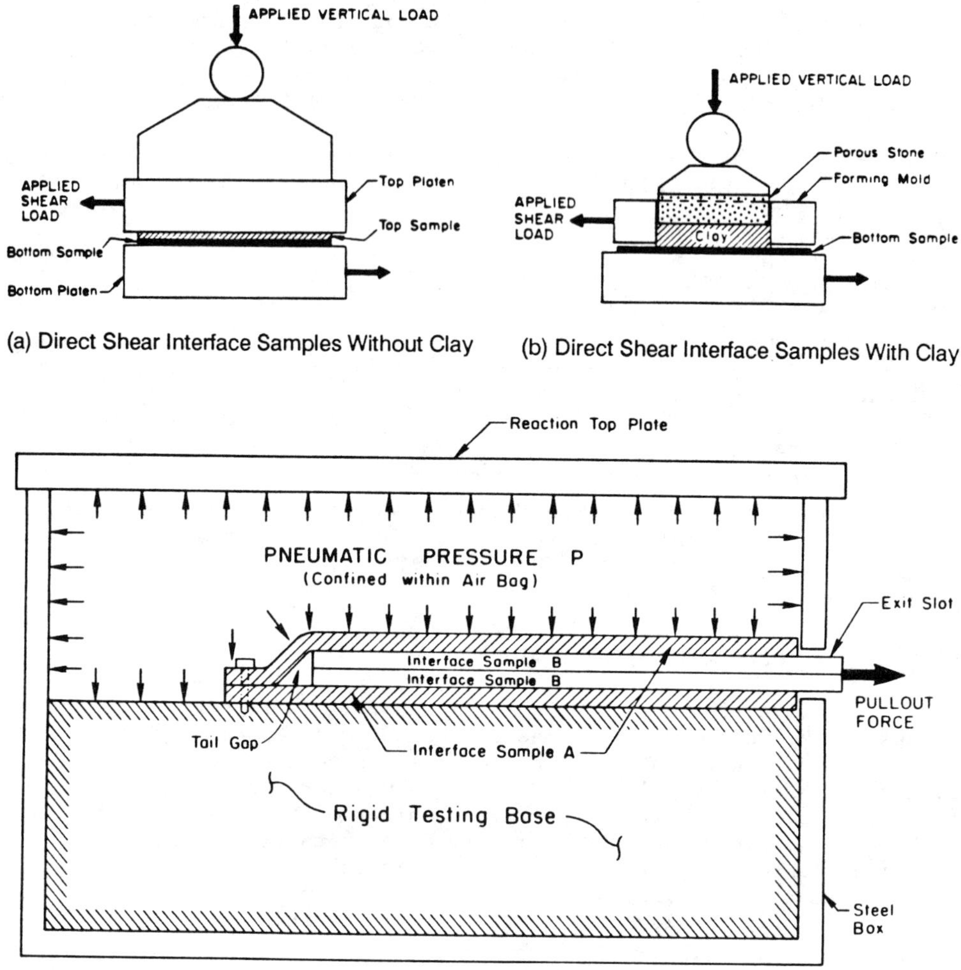

(a) Direct Shear Interface Samples Without Clay

(b) Direct Shear Interface Samples With Clay

(c) Schematic Illustration of Pullout Box Testing Apparatus

FIG. 4 -- Direct shear and pullout test configurations

displacement. Most interface combinations were tested both dry and submerged and over a range of normal stresses from 150 to 500 kPa. Shear displacement rates were 0.01 to 0.1 cm./min.

The sample configuration for interface combinations that included compacted clay is shown in Fig. 4(b). A 7.1 cm. by 7.1 cm. geosynthetic specimen was mounted on a 4 in. diameter round platen. The clay liner material was compacted above this within a 5 cm. by 5 cm. steel box to a thickness of 0.7 to 1.0 cm. using a Harvard Miniature Compaction Test pneumatic tamping piston. The clay liner material at the Kettleman Hills site was a mixture of on-site claystone, siltstone, and sandstone plus 5 percent bentonite. It was compacted to initial dry densities of 1450 kg/m^3 (± 3%) and initial water contents of 27 to 31 percent. These values represented an average relative compaction of 94 percent and a water content of 5 percentage points above optimum based on the Standard Proctor Compaction Test (ASTM D-698).

The results of the direct shear testing program are summarized in Table 3. The test results show that the residual interface shear resistance is mobilized at very small shear displacements. The results show also that wetting results in about a 1 degree reduction in the residual interface friction between polished geomembrane and geotextile. On the other hand, submergence resulted in a slight increase in the mean values of residual friction angle between the HDPE geomembrane and geonet and for HDPE geomembrane sliding on HDPE geomembrane. This behavior, as well as the somewhat surprising finding that the variability in residual friction angles for the HDPE liner to HDPE liner interface is greater than for any of the other combinations tested, suggests that small variations in geosynthetic material structure, surface texture, surface cleanliness, and sample orientation may all influence the interface shear resistance.

Residual friction angles for all interface combinations were generally from 0.5 degrees to 2 degrees less than the peak friction angles. Peak friction resistance was developed at shear displacements of 0.03 cm. to 0.65 cm., with a value of less than 0.12 cm. in most cases, and in most tests residual conditions were achieved at shear displacements of less than 0.25 cm.

HDPE liner/geonet interface strengths were found to be directionally dependent, so tests were done with different orientations of the geonet relative to the direction of shear. Visual differences were observed between the two sides of the geonet and of the HDPE geomembrane samples. The effects of these differences on the interface shear resistance were investigated and found to be small.

When HDPE liner/geotextile interfaces were tested, a tendency for the geotextile to "polish" the HDPE liner was observed. For this reason, samples were sheared repeatedly in order to evaluate interface friction under conditions ranging from "unpolished" (virgin samples) to "fully polished," a condition at which repeated interface shear caused no further reduction in shear strength.

The results in Table 3 show also that when shear between geonet and geomembrane is along the geonet rib direction, the residual friction angle is about 2 degrees less than for shear transverse to the ribs. The "transverse" shear values were developed for angles between the shear direction and the rib orientation of greater than about 15 degrees. Shear displacements in the Kettleman Hills failure were all in the "transverse" direction.

Pullout Box Tests

The pullout box tests were done using the apparatus shown schematically in Fig. 4(c). Interface Material A was fixed to the bottom part of the box. Two strips of interface Material B were cut to a width of 3.8 cm. and a length of 28 cm. and epoxied back-to-back with the surfaces to be tested facing outwards. This strip was placed on top of Material A. A second sheet of Material A was placed over the strip of Material B and fixed as shown in the figure. Spacers of Material B were used alongside the test strip to maintain the upper sheet of Material A at constant elevation when pneumatic pressure was applied using an air bag.

TABLE 3 -- Summary of interface direct shear test results - Kettleman Hills repository liner system components

Interface Components	Conditions	Number of Tests	Peak Friction* Angle (ϕ_p) or Strength (τ_r)	Displacement* at Peak, (cm.)	Residual* Friction Angle (ϕ_r) or Strength (τ_r)	
HDPE Liner/Geotextile	Dry, Unpolished	6	12.5 ± 0.7°	0.119 ± 0.025	10.6 ± 1.2°	
HDPE Liner/Geotextile	Dry, Partly Polished	13	10.6 ± 0.7°	0.107 ± 0.051	9.8 ± 0.7°	
HDPE Liner/Geotextile	Dry, Fully Polished	4	10.3 ± 0.9°	0.122 ± 0.025	9.6 ± 0.9°	
HDPE Liner/Geotextile	Submerged, Unpolished	4	10.4 ± 1.0°	0.056 ± 0.013	8.4 ± 1.2°	
HDPE Liner/Geotextile	Submerged, Polished	9	9.3 ± 1.0°	0.086 ± 0.033	8.4 ± 0.9°	Note 1
HDPE Liner/Clay	As Compacted	3	13.6 ± 2.4°	0.279 ± 0.025	12.4 ± 1.1°	
HDPE Liner/Clay	Saturated	5	50 ± 6 kPa	0.38 ± 0.23	45 ± 4 kPa	Note 2
HDPE Liner/Geonet	Dry, Transverse	3	9.0 ± 0.25°	0.076 ± 0.020	7.6 ± 0.3°	
HDPE Liner/Geonet	Submerged, Transverse	9	8.8 ± 1.2°	0.378 ± 0.287	8.3 ± 1.2°	Note 3
HDPE Liner/Geonet	Submerged Aligned	20	7.6 ± 1.3°	0.081 ± 0.032	6.3 ± 0.9°	Note 4
Geotextile/Clay	Saturated	3			>24°	Note 5
Geotextile/Geonet	Dry, Aligned	1	>20°			Note 6
Geotextile/Geonet	Submerged, Aligned	5			>10°	
HDPE Liner/HDPE Liner	Dry	9	9.9 ± 2.2°	0.048 ± 0.035	8.8 ± 2.4°	
HDPE Liner/HDPE Liner	Submerged	6	9.9 ± 1.8°	0.25 ± 0.20	9.2 ± 1.9°	

*Mean values ± standard deviation.

Note 1 A solitary value of $\phi_p = 12.3°$, $\phi_r = 10.0°$ not included.

Note 2 Samples were soaked overnight under a 3 kPa surcharge, a normal stress of 165, 330 or 500 kPa was applied, and the sample was rapidly sheared to failure.

Note 3 In three tests, not included in the listed values, the frictional resistance was still increasing after a shear displacement of 0.65 cm.

Note 4 In one test, not included in the listed values, the frictional resistance was still increasing after a shear displacement of 0.65 cm.

Note 5 Samples were soaked overnight under surcharges of 3 to 10 kPa. A larger normal stress was applied and the samples were rapidly sheared to failure.

Note 6 Test stopped to avoid damage to load cell.

The strip of Material B was pulled out at a constant rate of about 0.40 cm. per minute. The applied pullout force was measured using an load cell, and the pullout displacement was measured using a LVDT. The data were recorded and processed using an IBM-PC/AT microcomputer with A/D and D/A capacity.

Two "bridging" corrections were necessary to obtain the proper values of interface normal stress. One was to account for the slight gap along the sides between the edges of the Material B strip and the adjacent spacers. The other was to account for the tail gap (Fig. 4c) that formed as the test strip was pulled out. In each case the pneumatic vertical pressure applied to one-half of the bridging span was assigned to the interface contact pressures.

The use of double-sided pullout strips resulted in an initial interface contact area of 215 square cm., which was more than four times the surface contact area in the direct shear tests. Also, the pullout tests could be continued to shear displacements of more than 7.5 cm., as opposed to less than 0.75 cm. in the direct shear tests.

Pullout box tests were performed for HDPE liner/geotextile interfaces, HDPE liner/geonet interfaces, and HDPE liner/HDPE liner interfaces. Combinations including compacted clay could not be tested in the pullout box.

The pullout box test results are summarized in Table 4. Results of the direct shear tests are also shown for comparison. It may be seen that there is a very good agreement between the residual interface shear strengths determined using the two types of tests. This adds confidence to the use of the simple direct shear box tests for determination of interface strengths.

Finally, the results of the shear box and pullout box tests show that the lowest strength interface combinations for the materials in the Kettleman Hills liner system were HDPE liner/geotextile, HDPE liner/geonet, and HDPE liner/pre-soaked compacted clay interfaces.

TABLE 4 -- Comparison between pullout box and direct shear tests

Interface	Conditions	Number of Pullout Tests	Residual Friction Angle, ϕ_r	
			Pullout Tests	Direct Shear Tests
HDPE Liner/ Geotextile	Dry, unpolished	1	9.5°	9.5° to 12.5°
	Dry, polished	1	8.0°	8.5° to 10.5°
	Wet, unpolished	6	7.0° to 10.5°	8.0° to 10.0°
	Wet, polished	2	6.5° to 9.0°	7.0° to 9.5°
HDPE Liner/ Geonet	Submerged, Transverse	2	8.0° to 9.0°	7.0° to 10.0°
	Submerged, Aligned	6	6.0° to 8.0°	5.0° to 8.0°
HDPE Liner/ HDPE Liner	Submerged	6	7.0° to 13.5°	6.0° to 13.0°

STABILITY ANALYSIS CONSIDERATIONS

It was noted earlier in this paper that conventional two-dimensional stability analyses made using the measured interface strengths could not fully account for the Kettleman Hills landfill failure. From cross sections such as B-1 - B-2 shown in Fig. 5, plane section (2-D) factors of safety can be computed, giving the results shown in Fig. 6 (Seed et al., 1988).

Based on the interface shear strength testing program, the liner system interface strengths were taken as:

(1) ϕ_r = 8.5° for dry interface conditions representative of the sloping sides of the landfill basin, Fig. 1.

(2) ϕ_r = 8° for submerged or moist conditions assumed representative of the nearly level base whenever frictional resistance is less than the undrained clay/HDPE liner interface strength.

(3) τ_r = 45 kPa for submerged or wetted liner interface conditions whenever the height of fill times its unit weight (1750 kg/m^3) times tan 8° exceeds 45 kPa. This corresponds to fill heights of more than about 18 m.

Residual interface shear resistances were used because the peak resistances were exceeded at very small displacements, Table 3. Displacements of these magnitudes are likely to have occurred during liner construction or fill placement.

By weighting each plane section factor of safety in Fig. 6 in proportion to the mass of fill tributary to the plane section, overall factors of safety can be estimated to be about 1.15 to 1.25 for the probable minimum base wetting case and about 1.1 to 1.15 for the full base wetting case. The former assumes wetting only in the vicinity of the leachate sump, shown in Fig. 1(a). The latter assumes that all of the nearly level base in Fig. 1(a) is wetted. These factors of safety, while low, do not indicate sufficient instability to result in the observed slope displacements of up to 35 ft.

The fact that the side slopes of 1 on 2 or 26.6 degrees on the southwest and northwest sides of the basin and 1 on 3 or 18.4 degrees on the northeast side are considerably greater than the interface friction angle of 8.5 degrees is significant. It means that fill on

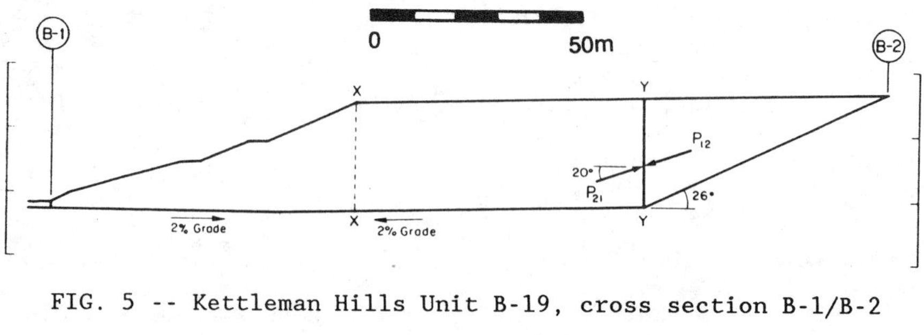

FIG. 5 -- Kettleman Hills Unit B-19, cross section B-1/B-2

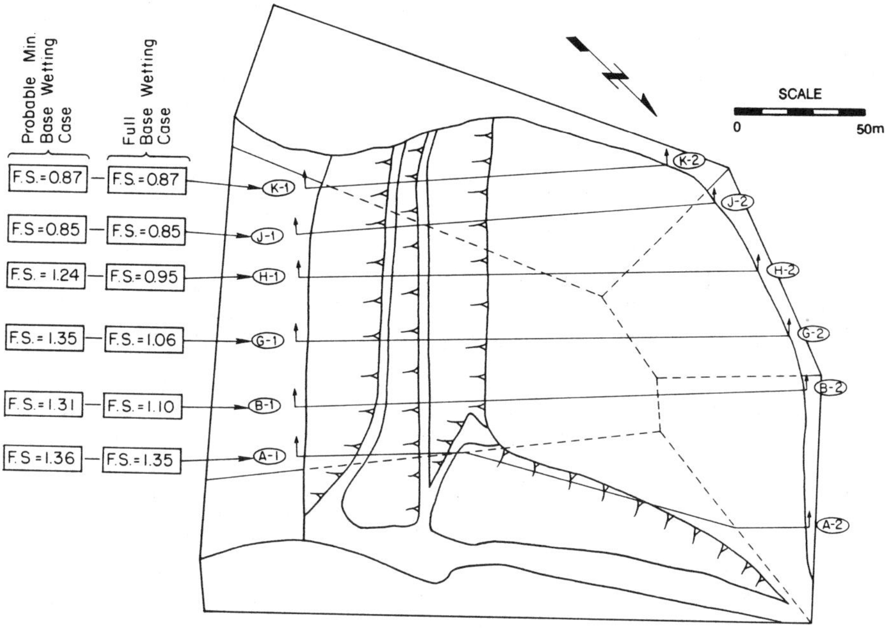

FIG. 6 -- Plane section factors of safety: Kettleman Hills Unit B-19, Phase I-A landfill

the sides must rely on the resistance provided by the fill on the base for support. Any component of this downslope force that acts in the direction of potential sliding of the mass on the base will contribute to instability. Accordingly, consideration must be given to three dimensional effects.

No generally applicable methods for the three-dimensional stability analysis of systems such as the Kettleman Hills repository have been developed and verified. Assumptions must be made concerning failure blocks, interblock stresses and kinematics, just as is the case for two-dimensional analyses. Two approaches were used (Seed et al.,[1]).

In the first, a five block system as shown in Fig. 7 was analyzed. Block boundaries were assumed vertical, side forces acting between blocks were assumed to act horizontally, and the vertical equilibrium of each block was considered along with overall translational equilibrium of all five blocks for some given direction of sliding. The overall potential for sliding in any given direction could then be determined. This force-equilibrium analysis indicated the most critical potential sliding direction to agree closely with the actual observed direction of sliding. The resulting calculated factors of safety were 1.14 and 1.06 for the minimum base wetting and the full base wetting cases, respectively. Although this approach satisfies classical force equilibrium requirements, it is not clear that this "rigid block" analysis satisfies all of the requirements of kinematics and compatibility necessary to reproduce the actual failure motions. Some out of plane movements and progressive failure are likely to have occurred in the actual field case.

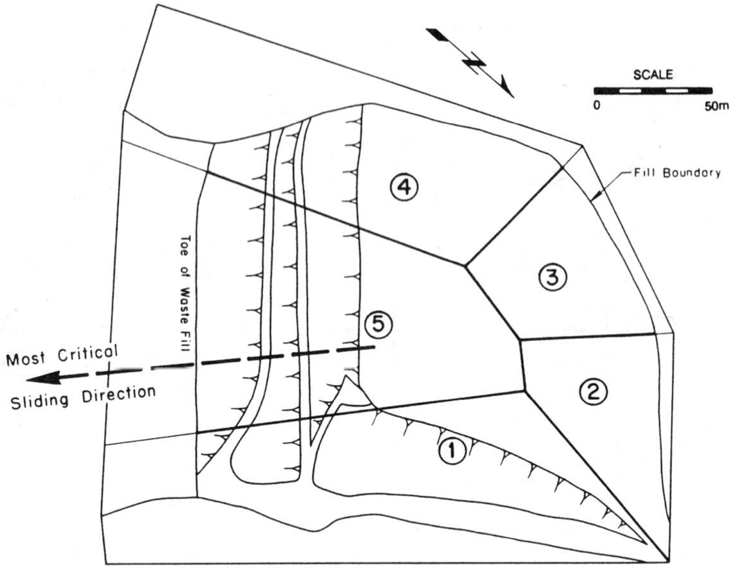

FIG. 7 -- Five block analysis and critical sliding direction

Approximate analyses were made to take these possibilities into account. Active driving forces and passive resisting forces were considered as shown schematically in Fig. 8. The convergence of the active forces on the central resisting block illustrate why the three-dimensional condition is potentially more critical than the two-dimensional case.

An approximate three-dimensional analysis for this situation was made by subdividing the mass by vertical planes into blocks for consideration of representative cross sections which were then weighted in proportion to their masses (Seed, et al., 1988). The overall factors of safety calculated by this method were about 20 percent lower than for the two-dimensional case and about 15 percent lower than for the three-dimensional force equilibrium case, but this unconventional analysis did not fully satisfy all conditions of translational equilibrium. After consideration of all factors, it was decided that the 5-block "rigid block" analysis of Fig. 7 missed some out of plane motions and progressive failure mechanisms likely to have occurred in the field (and evident in subsequent scale model studies of the Kettleman Hills slide), and the factors of safety calculated by this method were reduced by 5% to give "best estimates" of the overall F.S. by three-dimensional analyses.

After consideration of the uncertainties in the analysis methods, and in the properties of the system, overall best estimates of the factors of safety and their ranges were made as indicated in Table 2. These results show that use of the measured interface strength properties and three-dimensional stability analyses yield factors of safety at the time of failure close to unity, in satisfactory agreement with the observed field behavior.

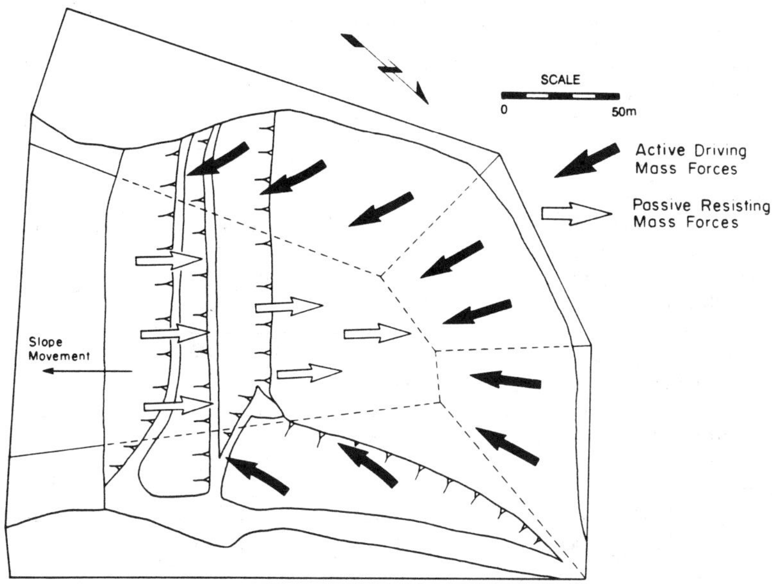

FIG. 8 -- Schematic illustration of three-dimensional failure mechanism

DISCUSSION

The Kettleman Hills Landfill failure and subsequent evaluation of its causes has focused attention on several issues relating to the safe design and operation of lined waste repositories. In the development of liner systems to date, primary emphasis has been concentrated on the collection of leachate and prevention of passage of any leachate through the liner system into the ground and groundwater below. It is now clear that the safe filling of landfills also requires careful consideration of the stability of slopes within the fill that are bounded by the liner system.

At the time of the Kettleman Hills failure little published information was available on interface shear strength values. Additional data developed by the authors for other interface combinations and conditions are listed in Table 5. Also, as shown on this table, new geomembranes with "textured" faces are now available. It may be seen that a wide range of interface shear strengths have been obtained, and that some of these for "non-textured" geomembrane to geotextile and geomembrane to compacted clay combinations are very low.

The data in Tables 3, 4, and 5 show clearly that geosynthetic material type, soil type, and test conditions must all be taken into account in selection of appropriate values for analysis and design. Interface wetting effects, consolidation conditions, grid orientations,and the surface texture and cleanliness of geomembranes may all be important. Although the tests on materials from the Kettleman Hills liner system indicated that the values of interface friction were not significantly influenced by the magnitude of normal stress, this does not appear to be the case for different material combinations.

TABLE 5 -- Additional Liner Interface Shear Strength Data

Interface Material Combination and Testing Conditions	Range of Normal Stresses	Interface Shear Strength Parameters
Non-Textured HDPE/Compacted Soil-Bentonite[(1)]		
(a) Sheared "As-Compacted"	250 to 500 kPa	$\phi_r \approx 5°$ to $24°$ [(2)]
(b) Pre-Soaked Under 300 psf Surcharge, U-U Sheared	250 to 500 kPa	$\phi_r \approx 5°$ to $13°$
Non-Textured HDPE/Compacted Soil-Bentonite[(3)]		
(a) Sheared "As-Compacted"	250 to 500 kPa	$\phi_r \approx 8°$ to $21°$ [(2)]
(b) Pre-Soaked Under 300 psf Surcharge, U-U Sheared	250 to 500 kPa	$\phi_r \approx 7°$ to $15°$
Non-Textured HDPE/Compacted Soil-Bentonite[(4)] Pre-Soaked Under 300 psf Surcharge, U-U Sheared	150 to 200 kPa	$\tau_r \approx 24$ to 28 kPa
Textured HDPE/Compacted Soil-Bentonite[(3)] (Pre-Soaked Under 300 psf Surcharge, U-U Sheared)	150 to 450 kPa	$c_r \approx 1{,}000$ psf, $\phi_r \approx 10.5°$
Textured HDPE/HDPE Geonet	150 to 450 kPa	$\phi_r \approx 14°$ to $21°$
Geotextile[(5)]/HDPE Geonet	150 to 450 kPa	$\phi_r \approx 15°$ to $21°$
Geotextile/Compacted Clay (Pre-Soaked Under 300 psf)	150 to 450 kPa	$c_r \approx 15$ kPa, $\phi_r \approx 28°$
Geotextile[(5)]/Compacted Soil-Bentonite[(3)] (Pre-Soaked Under 300 psf)	150 to 450 kPa	$c_r \approx 20$ kPa, $\phi_r \approx 25°$

1. Sandy clay till +5% bentonite: LL = 37, PI = 19.
2. Large variation as a function of compaction conditions.
3. Crushed claystone +5% bentonite: LL = 82, PI = 48.
4. Silty clay + 5% bentonite: Atterberg limits not available.
5. Nylon spun geotextile filter fabric.

Interfaces between geomembranes and compacted clay may be very critical, and their shearing resistance may also be extremely sensitive to the compaction conditions. This is illustrated by the data in Fig. 9, where strength values for compaction to "as compacted" moisture contents and densities wet of the line of optimums are significantly less than those for compaction dry of the line of optimums. The wide range of interface strengths and the many factors that influence the specific values in any case mean that project specific values should be determined by rigorous testing under anticipated "worst-case" field conditions for use in evaluation of landfill stability.

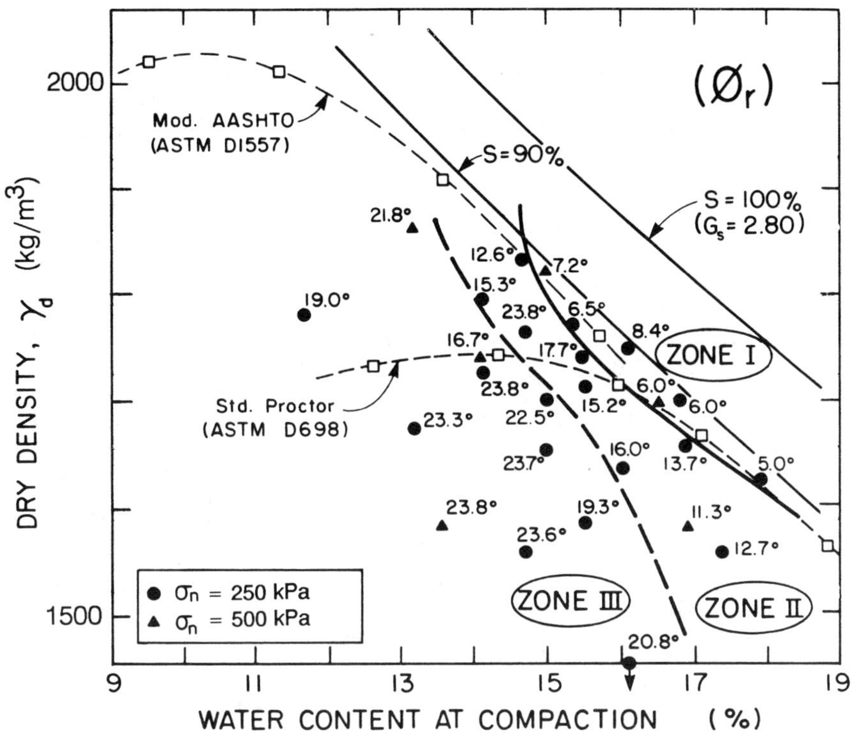

FIG. 9 -- Smooth HDPE/compacted soil-bentonite interface residual friction angles (samples sheared "as-compacted")

It can be argued that stability problems could easily be avoided by filling repositories uniformly from the bottom, thus avoiding large elevation differentials. This is impractical in most cases, however, for economic and operational reasons. The use of flatter side slopes in a repository would also reduce the risk of slope failures for any given fill height. This, however, reduces the available storage volume for a given landfill area.

These considerations mean that careful stability analyses should be made to enable the development of safe filling operational plans. The investigation of the Kettleman Hills repository slide suggests that three-dimensional effects should be considered in analyzing the stability of lined waste repositories. Further research is needed, however, to develop and verify suitable generalized methods for making these three-dimensional analyses.

Finally, insufficient attention has thus far been given to the seismic stability of waste landfills. It is not clear what levels of seismicity must be considered "during construction," and what levels of performance must be provided for long-term seismic stability under maximum credible earthquake loading. Seismic response characteristics of typically heterogeneous waste fill masses are, thus far, largely unknown. Similarly, there is little information currently available regarding dynamic strength and stress-deformation behavior of liner interfaces.

CONCLUSIONS

The results of the investigation of the 1988 slope stability failure at the Kettleman Hills Class I hazardous waste repository teach valuable lessons about the safe design and construction of lined waste landfills. These include:

(1) Multi-layer liner systems which have been devised for the safe containment and removal of landfill leachate may contain liner interfaces with low shear strength, including friction angles as low as 8 degrees or less.

(2) Simple direct shear tests can be used to determine reliable values for interface strength properties.

(3) Because of the variability in interface strengths that is associated with different geosynthetic liner system components and compacted clays used in composite double liner systems, values of interface strength should be determined specifically for each project using samples of the actual materials and representative placement, loading, and wetting conditions.

(4) In situations involving low shear strengths such as may occur in liner systems, three-dimensional effects may be important in evaluating stability.

(5) The repository filling operations should be planned in such a way that an adequate factor of safety can be maintained at all times and for all fill heights. This can be done by means of systematic analyses of stability for different fill geometries.

(6) While the concept embodied in (5) is simple in principle, it is presently difficult in detail owing to (1) the lack of a suitable generalized method for doing three-dimensional stability analyses, and (2) uncertainties about the effects of seismic loadings on the response and stability of lined landfills. Both of these issues need further study.

ACKNOWLEDGMENTS

The authors thank Clarence K. Chan, Richard C. Sisson, Makram Jaber, Mu Hsiung Chang, and Peter G. Nicholson of the University of California, Berkeley Geotechnical Group who assisted in the testing.

REFERENCES

[1] Seed, R. B., Mitchell, J. K. and Seed, H. B., "Slope Stability Failure Investigation: Landfill Unit B-19, Phase I-A, Kettleman Hills, California," Geotechnical Engineering Report No. UCB/GT/88-01, University of California, Berkeley, CA, July, 1988, 96 pp.

[2] Martin, J. P., Koerner, R. M. and Whitty, J. E., "Experimental Friction Evaluation of Slippage Between Geomembranes, Geotextiles and Soils," Proceedings International Conference on Geomembranes, Denver, Colorado, Jun 20-23, 1984, pp. 191-196.

Tuncer B. Edil[1], Valeri J. Ranguette[2] and William W. Wuellner[3]

SETTLEMENT OF MUNICIPAL REFUSE

REFERENCE: Edil, T. B., Ranguette, V. J., and Wuellner, W. W., **"Settlement of Municipal Refuse,"** Geotechnics of Waste Fills - Thoery and Practice, ASTM STP 1070, Arvid Landva and G. David Knowles, Eds., American Society for Testing and Materials, Philadelphia, 1990.

ABSTRACT: Refuse settlement in sanitary landfills is a complex process, which is dominated by secondary compression. Two mathematical models are used to model refuse settlement at four different sites. A data bank of empirical parameters was obtained and the trends observed.

KEYWORDS: refuse, settlement, settlement model, empirical settlement parameters, compression, rate of compression, landfill

INTRODUCTION

Refuse settlement in sanitary landfills, though frequently a troublesome and unpredictable problem offers at the same time a significant opportunity to the landfill operator for increasing the potential disposal capacity. Although filled to design grades, after closure landfill surfaces rapidly settle below the approved final grades. While this phenomenon is understood, a suitable model to predict this behavior has not been available or generally accepted. Consequently, valuable air space (volume), previously approved by regulatory agencies, may not be fully used. More accurate prediction of this settlement may permit a range of opportunities including better estimates of the remaining life of existing landfills, and improved predictability of surficial refuse movement and its impact on cover integrity, future vertical expansions, and ultimate use of the landfill surface.

The mechanics of compression of refuse are many and complex. Settlement-time curves from sanitary landfills differ from those for typical clay settlement curves; however, they are similar to those from organic soils and peats. This paper outlines an analysis approach using two different mathematical functions for the refuse settlement-time relationship. These relationships, which have previously been used to predict peat and soil settlement incorporate the significant factors controlling refuse settlement.

[1]Professor of Civil and Environmental Engineering, University of Wisconsin-Madison, Madison, Wisconsin
[2]Geotechnical Engineer, Warzyn Engineering Inc., Madison, Wisconsin
[3]Group Manager, Construction Services, EWI Engineering Associates, Madison, Wisconsin

The mathematical models were applied to the settlement data obtained from four different landfills and the parameters characterizing the compression of refuse were calculated. The dependency of the refuse settlement parameters on the stress level and the strain rate was evaluated for a range of conditions such as age of refuse, active filling or post-closure and leachate extraction. The limited data preclude broad generalization of the models for a variety of conditions; however, their versatility and utility as a tool for estimating the settlement of refuse, especially during active filling or post-closure, are demonstrated.

MECHANICS OF REFUSE SETTLEMENT

The mechanics of refuse settlement are many and complex, even more so than for a soil due to the extreme heterogeneity of, and large voids present in, the refuse fill. The main mechanisms involved in refuse settlement are the following:

1. Mechanical (distortion, bending, crushing and reorientation; similar to consolidation of organic soils);

2. Ravelling (movement of fines into large voids);

3. Physical-chemical change (corrosion, oxidation and combustion); and

4. Bio-chemical decomposition (fermentation and decay, both aerobic and anaerobic processes).

The factors affecting the magnitude of settlement are many and are influenced by each other. These factors include: 1) initial refuse density or void ratio; 2) content of the decomposable materials in the refuse; 3) fill height; 4) stress history; 5) leachate level and fluctuations thereof; and 6) environmental factors (such as moisture content, temperature and gases present or generated within the landfill).

It should be noted that refuse settles substantially both under its own self weight as well as under the weight of a new load (for example, the placement of new refuse over existing refuse). A factor complicating the computation of stress changes due to these weights is the introduction of cover soil to refuse fill. The addition of cover soil makes the measurement and interpretation of unit weight values more difficult. As a result, two types of refuse unit weight can be defined: 1) Actual refuse unit weight (weight of refuse per unit volume of refuse); and 2) Effective refuse unit weight (weight of refuse plus cover per unit volume of landfill), [1].

In themselves, actual refuse unit weights are highly erratic. Within a landfill, refuse unit weights typically vary from 5 to 11 kN/cu m. Moisture contents typically range from 10 to 50 percent, on a percent of dry weight basis [1-3].

Settlement of refuse fill is characteristically irregular. Initially, there is a large settlement within one or two months of completing construction, followed by a substantial amount of secondary compression over an extended period of time. The magnitude of settlement decreases over time and with increasing depth below the surface of the fill. Under its own weight, refuse settlement typically ranges from 5 to 30 percent of the original thickness, with most of the settlement occurring in the first year or two.

PREVIOUS REFUSE SETTLEMENT STUDIES

Various methods of analysis and prediction of post-construction refuse fill settlement are reported by several investigators [2-5]. Settlement due to compression of refuse fill under external surface load can be plotted in terms of strain (ratio of settlement to initial fill height) versus the logarithm of effective stress (pressure). Settlement magnitude can be predicted based on the settlement coefficient, the slope of the straight line connecting two selected stresses [2, 4]. The problems with this method include: 1) for older fills, the initial fill height is usually unknown; 2) effective stress is a function of refuse density, which usually cannot be determined accurately; and 3) the strain-log stress relationship is not a straight-line relationship; therefore, the settlement coefficient, which is proportional to settlement magnitude, varies as the stresses (pressures) within the fill change.

Another approach, is to calculate settlement rate as settlement magnitude per unit time interval. Yen and Scanlon [5] collected settlement platform data from several landfill sites and calculated settlement rate as the ratio of change in platform elevation to elapsed time between surveys. Since the settlement platform data for these sites covered periods of up to nine years following the end of construction, they were able to plot settlement rate versus log time and determine the best-fit linear relationship by the least-squares method.

Yen and Scanlon compared their data with Sowers' field observations and noted that the rate of settlement decreases with time logarithmically [5]. Sowers noted the time-dependent secondary compression of refuse and reported values of the coefficient of secondary compression, α (based on Buisman's definition for soils) for some sanitary landfills. Sowers noted that the α values for refuse were comparable to those of peat and organic soils and dependent on how favorable the conditions were for decomposition [2, 3].

PROPOSED REFUSE SETTLEMENT MODELS USED

The conventional approach to soil compression requires a separation of primary and secondary compression and treatment of each with different mathematical expressions. In the long term, secondary compression of refuse is larger than other compression, and it is often difficult to make a distinction between primary and secondary compression. Therefore, a simple model combining all stages of compression is needed. Two such models are investigated in this study.

Gibson and Lo Model

The rheological model proposed by Gibson and Lo [6] for the long-term (secondary) compression of soils was found to be rather useful in predicting the settlement of peats [7]. Peat, like refuse, involves mechanisms of compression different than those in inorganic clays. Both peat and refuse have relatively large void spaces that compress quickly during initial and primary settlement, but by far the largest compression is due to the slow and continuous process of secondary settlement, where the particle structures begins to break down. Encouraged by the simplicity and usefulness of the rheological model proposed by Gibson and Lo we decided to apply the same model to field refuse settlement records.

This rheological model is shown in Fig. 1a, and it represents the average compression characteristics in the one-dimensional compression of the refuse fill shown in Fig. 1b. The applied increment of stress can be either the self-weight of the refuse or it may be imposed on the refuse surface.

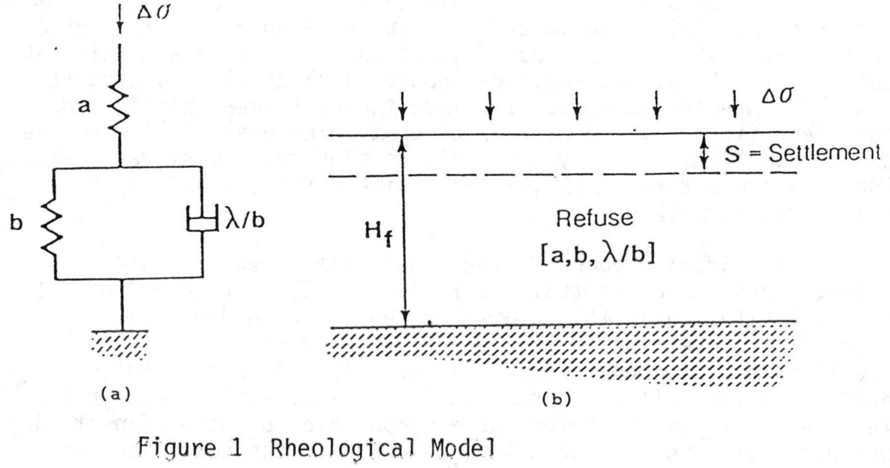

Figure 1 Rheological Model

When a stress increment, $\Delta\sigma$, acts on the model, the Hookean spring, with a spring constant of a, compresses instantaneously. This is analogous to primary compression. The compression of the Kelvin element, with a spring (spring constant of b) parallel to a dashpot (viscosity of λ/b) is retarded by the Newtonian (linear) dashpot. This is similar to the continuous process of secondary compression under sustained effective stress. The sustained load is transferred progressively to the Hookean spring from the Newtonian dashpot. After a long time, (i.e., in the secondary compression range), the full effective stress will be taken by the two springs, thus the dashpot will sustain no load. The time-dependent settlement can be expressed as:

$$S(t) = H\,\epsilon(t) = H\,\Delta\sigma\,\{a + b(1 - \exp[-(\lambda/b)\,t]\} \qquad (1)$$

where

- S = settlement
- H = initial height of refuse
- ϵ = strain (settlement divided by the layer thickness, i.e., S/H)
- $\Delta\sigma$ = compressive stress
- a = primary compressibility parameter
- b = secondary compressibility parameter
- λ/b = rate of secondary compression
- t = time since load application

Power Creep Law

One of the simplest forms of a relation for time-dependent deformation under constant stress and one that has been extensively used in representing the transient creep behavior of many engineering materials is the power creep law. According to this law, the time-dependent settlement can be expressed as:

$$S\,(t) = H\,\epsilon(t) = H\,\Delta\sigma\,m(t/t_r)^n \qquad (2)$$

where

- m = reference compressibility
- n = rate of compression
- t_r = reference time introduced into the equation to make time dimensionless (t_r = 1 day in this study)

Other terms are as defined before.

Determination of Model Parameters

An interactive spread sheet program was developed at the University of Wisconsin-Madison on a personal computer in fitting Eqs 1 and 2 to the settlement-time records from various fills. The program uses the method of the logarithm of strain rate ($\Delta\epsilon/\Delta t$) versus time in determining the parameters, a, b, and λ [7, 8] and the method of the logarithm of strain (ϵ) versus logarithm of time in determining parameters m and n from the settlement-time record.

Accordingly, the settlement-time record is incrementalized and the operator chooses the range over which these functions give a linear plot. From this portion of the plots, model parameters are calculated. The program provides a plot of the actual strain versus time, along with the calculated strain (from the calculated model parameters) versus time so that the quality of curve-fitting can be visually evaluated.

SITES MODELED

Data from four different existing refuse fills were analyzed. All four of these sites are municipal landfills. It was assumed that the refuse in each of these four sites is about the same composition. The sites are in northern climates, thus there is sufficient rainfall to promote the degradation of the refuse which affects settlement. Table 1 summarizes the known values of refuse thickness, settlement, data collection duration, and refuse placement conditions.

Site A

This refuse site is in southeastern Wisconsin. Placement of the refuse fill at this site began in the early 1970's. The settlement data was collected using settlement platforms surveyed periodically from 1984 to 1986. The age of the refuse fill below each platform varied but was estimated to be between 0 to 4 years at the time the data were obtained. The data collection at this site continued for approximately 1.8 years. For this study two categories of loading conditions were considered. The first category is called "minimal filling". This category represents a condition of settlement under essentially self weight during data collection. The second category is called "active filling", since additional refuse and daily cover were added during data collection. Thus, the second category represents a condition of settlement under both self weight and the placement of additional fill above the platforms [9]. The leachate level was about 7.6 m above the base of the landfill during data collection. Near Platforms 7 and 9, there was occasional leachate extraction.

Site B

This refuse site is in southern Michigan. The refuse fill has been placed in the landfill since 1969. In 1985 an expansion area was constructed on top of the existing fill. The settlement data collection began in 1985, during placement of the additional fill in the expansion area. Again, two differing conditions existed during this 1.2 year study. The first was the old refuse that was already in place below the settlement platform. After the placement of new refuse, settlement was monitored at varying horizontal distances of between 50 to 250 feet from the expansion

TABLE 1 -- REFUSE SETTLEMENT DATA

Platform Number	Refuse Thickness (m)	Settlement (m)	Time Duration (yr)	Placement Condition
SITE A				
1	13.73	0.52	1.6	Fresh Refuse: No filling
2	8.01	0.59	1.6	Fresh Refuse: No filling
3	9.25	1.11	1.6	Fresh Refuse: No filling
4	9.84	1.19	1.8	Fresh Refuse: No filling
15	26.74	0.37	1.5	Fresh Refuse: Minimal filling < 1m
16	25.15	0.43	1.3	Fresh Refuse: Minimal filling < 1m
7	36.28	1.89	1.7	Fresh Refuse: Minimal filling < 1m
9	34.72	1.12	1.5	Fresh Refuse: Minimal filling < 1m
8	36.75	3.20	1.5	Fresh Refuse: Active filling > 6m
10	37.38	2.99	1.7	Fresh Refuse: Active filling < 6m
11	19.66	2.10	1.4	Fresh Refuse: Active filling < 6m
12	27.94	1.94	0.7	Fresh Refuse: Active filling < 6m
13	23.46	0.72	1.1	Fresh Refuse: Active filling > 6m
14	19.86	1.62	1.3	Fresh Refuse: Active filling > 6m
17	22.28	2.74	1.1	Fresh Refuse: Active filling > 6m
SITE B				
S-4	15.24	0.09	1.2	Old Refuse: No filling
S-5	15.24	0.21	1.2	Old Refuse: No filling
S-6	15.24	0.94	1.2	Fresh Refuse: Active filling
SITE C				
84-2	10.06	0.65	4.0	Old Refuse: Relocated/Compacted
84-3	10.06	0.58	4.0	Old Refuse: Relocated/Compacted
84-4	11.58	0.60	4.1	Old Refuse: Relocated/Compacted
84-5	5.49	0.33	4.1	Old Refuse: Relocated/Compacted
84-6	10.06	0.61	3.9	Old Refuse: Relocated/Compacted
84-7	11.58	0.52	3.3	Old Refuse: Relocated/Compacted
SITE D				
SP1	3.05	0.38	0.9	Old Refuse: Surcharge
SP2	3.05	0.64	0.9	Old Refuse: Surcharge
SP3	3.05	0.44	0.9	Old Refuse: Surcharge

area. The second condition was the active filling of fresh refuse directly above the settlement platform [10].

Site C

This refuse site is in western Connecticut. A 40 to 50 year old "town dump" was excavated and relocated at a new site. This site was monitored for 5 years. The only settlement that occurred was due to the self weight of the compacted refuse [11].

Site D

Site D consisted of (a) experimental cells that were constructed to monitor settlement and (b) an area of old refuse below the settlement platform when an embankment load was added. The duration for this study was 1 year [12].

MODELING

The settlement platforms were placed as shown in Fig. 2. The refuse thickness (H_o) is the initial thickness of the refuse below the settlement plate. Additional refuse was placed above the settlement platform (h). The times of placement of the additional refuse above the platforms were unavailable for the four sites. it was assumed that the additional refuse above the platform was placed at time zero, when the data collection was started. The average applied stress ($\Delta\sigma$) in the layer of refuse below the platform was calculated as follows:

$$\Delta\sigma = \gamma h + 1/2(\gamma H_o) - 1/2(\gamma + \gamma_w - \gamma_{sat})(H_w^2/H_o) \quad (3)$$

where:

$\Delta\sigma$ = average applied stress
γ = moist unit weight of refuse
γ_{sat} = saturated unit weight of refuse
γ_w = unit weight of leachate (water)
H_o = initial height of refuse below platform
h = height of refuse above platform
H_w = height of leachate above base liner

A moist unit weight of 10.7 kN/m^3 and a saturated unit weight of 14.6 kN/m^3 were used in computing the average applied stress at each of the sites.

The programs that were used to model the Gibson and Lo model and the power creep law plotted predicted strain versus log time, along with the actual strain versus log time for the data that was input. Emphasis was placed on the later portion of the curve in the case of the Gibson and Lo model, where secondary compression occurs, and there is a constant effective stress. In the case of the power creep law, the whole range of the data was considered in

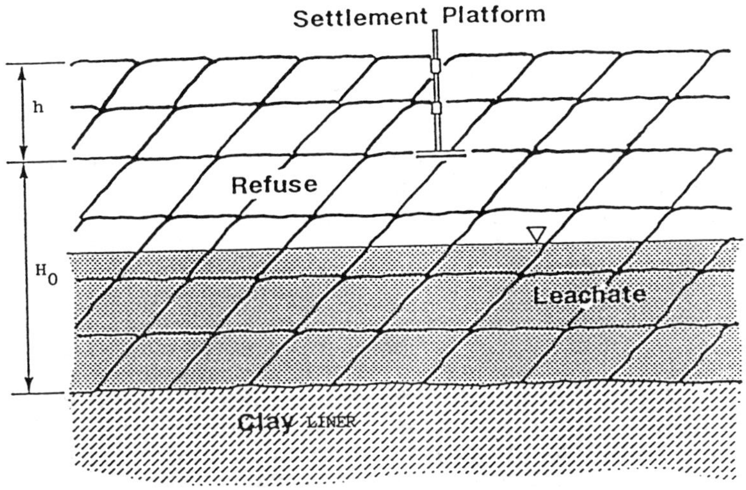

Figure 2 Typical Settlement Platform

curve-fitting.The curve fitting predictions were generally accurate, with a few exceptions of the early data not fitting the computed curve, especially for the Gibson and Lo model. Fig. 3 shows typical curves fitted to the data using the two methods.

A few of the settlement records could not be analyzed with these models. It is believed that these records violated the assumption of constant stress change. Overall, the power creep law gave a better representation of the data in 65% of the cases than the Gibson and Lo model. For the remaining 35% of the cases, it was comparable to the Gibson and Lo model, except only one case.

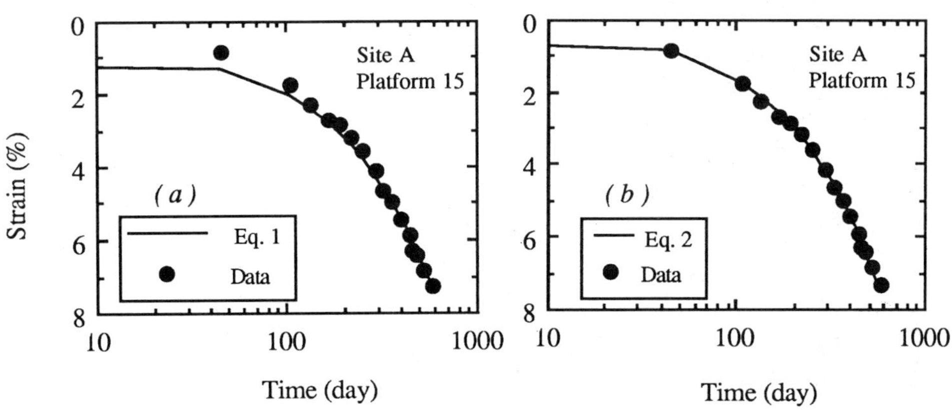

Figure 3 Strain-time curves by a) the rheological model and b) the power creep law fitted to the measured data

REFUSE COMPRESSION MODEL PARAMETERS

Gibson and Lo Model Parameters

The three empirical parameters of the Gibson and Lo model derived for the four sites are summarized in Table 2 and plotted in Figs. 4, 5 and 6. The curves in these figures are trends not actual relations. Fig. 4 plots a, the primary compressibility as a function of stress. The amount of primary compression decreases with an increase in stress. For the "active" filling in Site A, a higher value was obtained indicating more primary settlement was occurring during placement of the fresh refuse. In Fig. 5 the secondary compressibility, b, is shown to decrease with increasing stress. Generally, the "active" filling sites show lower amounts of secondary settlement than for the "minimal" filling sites. This is due to the fact that the "active" filling cases were still experiencing substantial primary settlement. Undisturbed old refuse from sites B and C has a lower secondary compressibility compared to fresh refuse or old refuse recently surcharged (Sites A and D and Platform S6 from Site B). The rate of secondary compression, λ/b, as a function of average strain rate is illustrated in Fig. 6. Average strain rate is defined as total strain divided by elapsed time during data collection. As expected, as the average strain rate increases, so does the rate of secondary compression. This behavior was also observed for peat soils [7] and indicates that the dashpot is essentially nonlinear in the model. The implification of this is that the parameters obtained from a fill must be extrapolated with care to another fill with different refuse thickness and applied stress even if the composition and location are similar. There was not any observable effect of leachate extraction on the parameters, perhaps because of its limited scope.

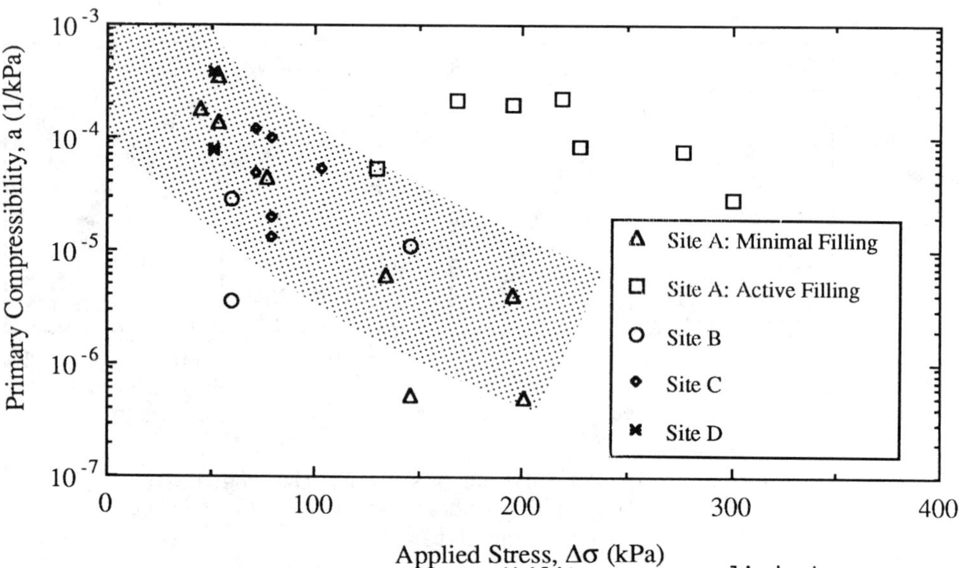

Figure 4 Primary compressibility verus applied stress

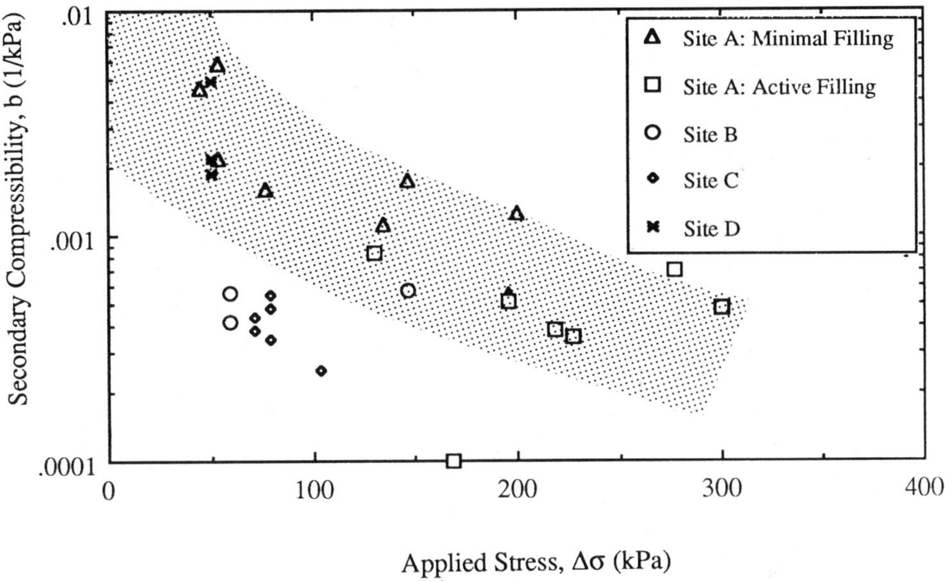

Figure 5 Secondary compressibility versus applied stress

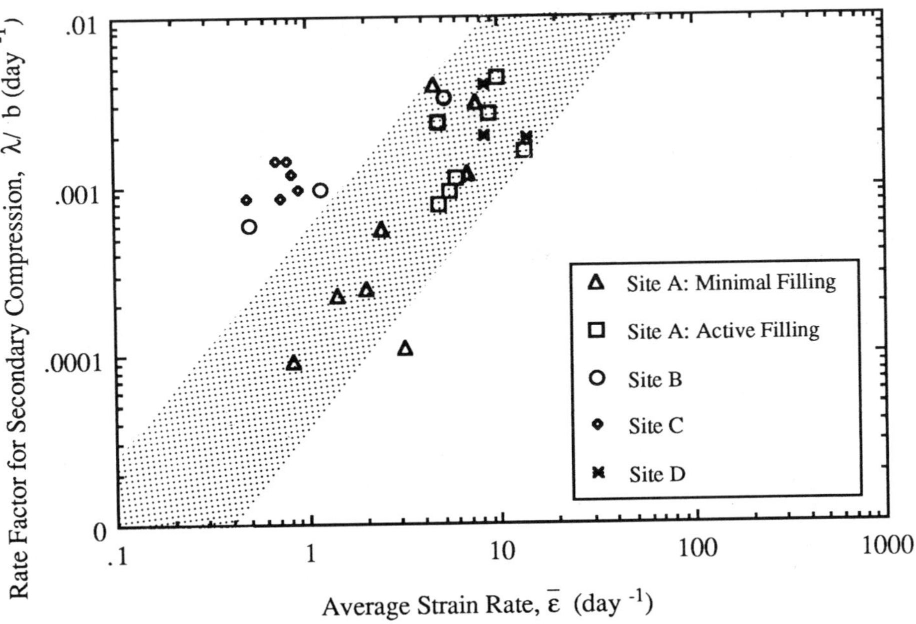

Figure 6 Rate of secondary compression versus average strain rate

TABLE 2 -- EMPIRICAL MODEL PARAMETERS

Platform Number	Applied Stress (kPa)	Average Strain (%/yr)	Gibson & Lo a (1/kPa)	Gibson & Lo b (1/kPa)	Gibson & Lo λ/b (1/day)	Power Creep m (1/kPa)	Power Creep n (t_r = 1 day)
SITE A							
1	77.21	2.37	4.42e-5	1.62e-3	5.60e-4	5.48e-6	0.702
2	54.09	4.59	1.40e-4	5.87e-3	4.00e-3	5.75e-6	0.862
3	53.58	7.51	3.52e-4	2.18e-3	3.10e-3	1.38e-4	0.438
4	45.00	6.81	1.78e-4	4.58e-3	1.20e-3	1.18e-5	0.850
15	146.27	0.83	5.32e-7	1.77e-3	9.20e-5	7.52e-8	1.131
16	134.12	1.42	6.11e-6	1.13e-3	2.30e-4	9.00e-8	1.170
7	195.65	3.14	4.10e-6	5.49e-4	1.10e-4	1.61e-6	0.804
9	200.16	2.01	5.11e-7	1.24e-3	2.50e-4	3.15e-7	0.980
8	276.40	5.50	7.76e-5	6.01e-4	9.40e-4	3.10e-6	0.744
10	227.76	4.84	8.35e-5	3.54e-4	2.40e-3	3.40e-6	0.746
11	168.01	13.58	2.12e-4	1.00e-4	1.60e-3	1.67e-5	0.619
12	195.32	4.74	1.99e-4	5.05e-4	7.70e-4	5.48e-5	0.297
13	219.07	5.89	2.30e-4	3.75e-4	1.10e-3	5.89e-5	0.302
14	130.12	8.98	5.34e-5	8.40e-4	2.70e-3	1.30e-5	0.670
17	300.29	9.82	2.86e-5	4.74e-4	4.30e-3	1.16e-6	1.005
SITE B							
S-4	59.88	0.50	3.60e-6	4.10e-4	6.00e-4	7.85e-7	0.779
S-5	59.88	1.17	2.80e-5	5.60e-4	9.70e-4	2.25e-6	0.759
S-6	146.10	5.17	1.10e-5	5.70e-4	3.30e-3	8.83e-6	0.648
SITE C							
84-2	79.42	0.90	1.00e-4	4.70e-4	9.70e-4	6.48e-5	0.264
84-3	79.42	0.48	1.30e-5	3.50e-4	8.40e-4	1.10e-5	0.409
84-4	71.66	0.83	1.20e-4	4.30e-4	1.20e-3	5.14e-5	0.304
84-5	102.79	0.68	5.20e-5	2.50e-4	1.40e-3	2.75e-5	0.314
84-6	79.42	0.72	2.00e-5	5.40e-4	8.40e-4	1.40e-5	0.465
84-7	71.66	0.79	4.90e-5	3.80e-4	1.40e-3	1.67e-5	0.443
SITE D							
SP1	50.97	8.33	7.50e-5	1.90e-3	4.00e-3	4.69e-5	0.593
SP2	50.97	14.00	8.00e-5	4.90e-3	1.90e-3	4.85e-5	0.666
SP3	50.97	8.44	3.80e-4	2.20e-3	2.00e-3	8.57e-5	0.486

Power Creep Law Parameters

The two empirical parameters of the power creep law derived for the four sites are given in Table 2. These parameters did not indicate any discernible trends with the respect to applied stress or average strain in each site within the range of variation of these factors. Reference compressibility, m has an average value of about 2.5 x 10^{-5} 1/kPa and it is about 1.7 times higher for old refuse (3.4 x 10^{-5} 1/kPa) than fresh refuse (2.0 x 10^{-5} 1/kPa). It shows no discernible patterns with respect to placement conditions of the refuse. However, it is quite variable, especially in Sites A and B. Rate of compression, n has an average of 0.65 and indicates some patterns with respect to age and placement conditions of the refuse. For instance, old relocated refuse from Site C that was compacted during placement had the lowest average n = 0.37 and, in general, fresh refuse had an average n value of nearly 1.5 times as that of old refuse. The variability of n is not as great as that of m; however, it is more variable in Site A than the other three sites.

COMPARISON OF THE MODELS

For Site A, the first year of data obtained was used to predict the amount of settlement that could be expected at the end of the data collection period which was about two years. The results obtained using both models are compared with the actual measurements in Table 3. The Gibson and Lo model predicted the amount of settlement at the end of two years within 2 to 18% of the actual settlement that occurred for minimal filling and 4 to 21% for active filling. The power creep law predictions for the same conditions were 0 to 6% and 0 to 14%, respectively.

TABLE 3 -- COMPARISON OF PREDICTED SETTLEMENT

Platform Number	Settlement (m)			Percent Diviation (%)	
	Actual	Gibson & Lo	Power Creep	Gibson & Lo	Power Creep
Minimal Filling					
1	0.52	0.43	0.53	-17	2
2	0.59	0.59	0.59	0	0
3	1.11	1.09	1.06	-2	-4
4	1.19	1.23	1.24	4	5
7	1.88	1.54	2.00	-18	6
Active Filling					
8	3.34	3.19	3.38	-4	1
10	2.99	2.93	3.18	-2	6
12	1.94	1.91	1.94	-1	0
13	2.03	2.00	1.97	-2	-3
14	2.95	2.32	2.53	-21	-14

CONCLUSIONS

From the limited data, the following conclusions can be made:

1. Refuse settlement can be modeled satisfactorily with either a rheological model as presented in the Gibson and Lo theory or the power creep law.

2. Power creep law provides a better representation of the settlement data than the rheological model. However, the rheological model has parameters that can be assigned physical meaning and reflect the effects of certain refuse placement conditions.

3. While active filling is ongoing, primary compression is significant compared to secondary compression; and

4. Once filling has stopped, secondary compression is more evident.

Further landfill sites need to be instrumented for data collection and analyzed to develop a data bank of ranges for the empirical parameters for different stages of the landfill life.

ACKNOWLEDGEMENTS

The authors would like to thank Waste Management of Wisconsin, Inc. and Debra L. Nelson of Warzyn Engineering Inc., for supplying some of the initial data used in preparation of this paper.

REFERENCES

[1] Ham, R. K., Reinhardt, J. J. and Sevick, G. W., "Density of Milled and Unprocessed Refuse," in Journal of the Environmental Engineering Division, A.S.C.E., Vol. 104, No. EE1, 1978.

[2] Sowers, G. F., "Settlement of Waste Disposal Fills," Proceedings, 8th International Conference on Soil Mechanics and Foundation Engineering, Moscow, 1973.

[3] Sowers, G. F., "Foundation Problems in Sanitary Landfills," A.S.C.E. Journal of the Sanitary Engineering Division, Vol. 94, No. SA1, 1968.

[4] Rao, S. K., Moulton, L. K. and Seals, R. K. "Settlement of Refuse Landfills," Geotechnical Practice for Disposal of Solid Waste Materials, A.S.C.E. Geotechnical Engineering Division, 1977.

[5] Yen, B. C. and Scanlon, B., "Sanitary Landfill Settlement Rates," A.S.C.E. Journal of the Geotechnical Engineering Division, Vol. 101, No. GT5, 1975.

[6] Gibson, R.E. and Lo, K. Y., "A Theory of Soils exhibiting Secondary Compression," Acta Polytechnica Scandinavica, C; 10 296, 1961, p. 1-15.
[7] Edil, T. B., and Mochtar, N. E., "Prediction of Peat Settlement," A.S.C.E. Preceedings of Symposium: Sedimentation/Consolidation Models, San Francisco, 1981.
[8] Edil, T. B., and Dhowian, A. W., "Analysis of Long-Term Compression of Peats", A.S.C.E. Geotechnical Journal, Vol. 10, 1979.
[9] Edil, T. B., Ranguette, V. J. and Otzelberger D. G, "Comparison of Refuse Settlement Prediction Models," Proceedings of the Twelfth Annual Madison Waste Conference, Madison, Wisconsin, 1989.
[10] Dodt, M. E., Sweatman, M. B. and Bergstom, W. R. "Field Measurement of Landfill Surface Settlements," Geotechnical Practice for Waste Disposal '87, A.S.C.E Geotechnical Special Publication No. 13, 1987.
[11] Keene, P. F., "Sanitary Landfill Treatment, Interstate Highway 84", Geotechnical Practice for Disposal of Solid Waste Materials, A.S.C.E., Geotechnical Engineering Division, 1977.
[12] Rao, S. K., Moulton, L. K. and Seals, R. K., "Settlement of Refuse Landfills", Geotechnical Practice for Disposal of Solid Waste Materials, A.S.C.E, Geotechnical Engineering Division, 1977.

Sukhmander Singh and Bruce J. Murphy

EVALUATION OF THE STABILITY OF SANITARY LANDFILLS

REFERENCE: Singh, S., and Murphy, Bruce, **Evaluation of the Stability of Sanitary Landfills,"** Geotechnics of Waste Fills - Theory and Practice, ASTM STP 1070, Arvid Landva, G. David Knowles, editors, American Society for Testing and Materials, Philadelphia, 1990.

ABSTRACT: In most of the major U.S. cities, both the capacity and availability of solid waste landfill sites are declining. One option for increasing landfill capacity has been to build landfills to greater heights and new sites are being planned to store refuse to unprecedented heights. This situation has raised concerns by many state and federal regulators regarding the stability of high refuse fill under static and dynamic loading conditions. This paper includes:

- a critical evaluation of the published and unpublished studies on the shear strength properties of refuse and settlement characteristics of refuse fills.
- a discussion of the inadequacy of the Mohr-Coulomb theory to account for the large, yet non catastrophic deformations that refuse undergoes.
- a presentation of a new approach to stability analysis based on the bearing capacity and settlement criteria of landfills
- an examination of the dynamic strength properties of refuse fills, including recently reported field shear wave data and a deformation analysis approach for evaluating stability when earthquake motions pass through a high landfill.

KEYWORDS: sanitary landfill, engineering properties, stability, bearing capacity, settlement, earthquake response.

Dr. Sukhmander Singh is an Associate Professor of Civil Engineering at Santa Clara University, Santa Clara, California 95053
Mr. Bruce Murphy is Director, Waste Management Services at Purcell, Rhoades and Associates, 2504 Technology Drive, Hayward, California 94545

INTRODUCTION

The long arduous journey, back and forth from New York to Florida, of a barge loaded with New York refuse has brought to public attention the critically inadequate space for accommodating the ever increasing mounds of refuse. Nobody wants a landfill in their backyard, yet in almost all major cities of the United States, solid waste landfill capacity and available new landfill sites are declining. The demand for increased capacity has been achieved in some cases by building landfills to greater heights. New sites in planning are being designed to store refuse to unprecedented heights.

These situations have raised concerns among regulators regarding the stability of high refuse fill under static and earthquake loading conditions. In the State of California, regulatory agencies such as the Regional Water Quality Control Board (RWQCB) require the evaluation of structural stability in addition to traditional site evaluations like leachate control and removal . Consequently, slope stability studies have recently been presented in reports prepared by various consultants. Although these reports were considered adequate by the various State of California regulatory agencies, basic questions regarding the strength and cohesion values of the refuse material and the applicability of certain slope stability techniques for the evaluation of landfill sites still remain.

The realization, shared by almost all consultants and industry representatives in this field, is that there is a lack of published data on both static and dynamic strength properties of sanitary landfills. This lack of data is understandable because landfills have a complex and heterogeneous structure and it is not easy to adapt it into conventional laboratory testing methods.

Accordingly, the following questions must be resolved to further the progress in the analysis, design and construction of safe, large-capacity sanitary landfills:

(1) what are the reliable strength properties of the refuse?

(2) which slope stability method is applicable for analyzing slopes made of refuse?

(3) is it a reasonable or useful assumption that refuse slopes behave like a soil slopes?

This paper describes and presents the results of the research efforts undertaken to find answers to the above questions.

EXISTING DOCUMENTS ON STRENGTH PROPERTIES OF REFUSE

Efforts have been made to collect and examine existing data on the strength properties of sanitary landfills. Among those documents reporting strength data obtained through laboratory tests are: Stoll (1), Fang (2), Landva et.al. (3), Landva & Clark (4), Los Angeles County Sanitation District (5), Cooper & Clark (6), Saarela (7), Cooper engineers (8), Earth Technology, Inc. (9), and

Klienfelder (10). Another set of documents uses field performance data and includes: Volpe (11), Woodward-Clyde (12), Purcell, Rhoades & Associates (13, 14, 15, 16), EMCON (17, 18, 19), Dames and Moore (20). The performance record of a field load test carried out by Converse et.al. (21) and performance records of existing fills under earthquake shaking were used by these authors to estimate strength characteristics. Other techniques include estimating strength values by analogy to observed slope stability. Recently, attempts have been made to obtain strength values by using in-situ techniques such as SPT and Vane Shear (Earth Tech, 9).

In summary, it appears that estimates on refuse strength have been established using three approaches:

(1) laboratory testing;

(2) back-calculations of field test and operational records; and

(3) in-situ testing.

However, in each case, because of the difficulties and the complexities in estimating material properties of refuse, only limited data has been obtained. Nevertheless, it is important to examine the limited data in an attempt to more reliably predict and thus plan for landfill stability.

STRENGTH ESTIMATES BASED ON LABORATORY TEST DATA

The data obtained from laboratory testing of refuse samples has been plotted in Figure 1. The Los Angeles Sanitary District (1984) tested simulated refuse samples in which various substitutes were made. Substitutions included sand for stones, ceramic for bones, crushed glass for glass containers, lint for cloth, rubber threads for rubber, etc. These materials were constituted

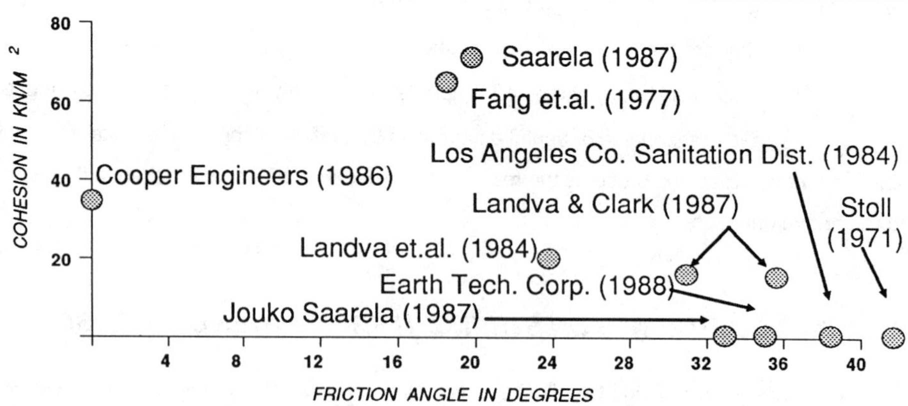

Fig. 1 - Strength Parameters from Laboratory Shear Tests

into a sample of 6.15 cm (2.42 inches) in diameter and 2.54 cm (1 inch) thick for use in a direct shear machine. Fang (2) tested compacted bales of refuse at Fritz Engineering Laboratory of Lehigh University. Cooper Engineers (6) tested Shelby tube samples of refuse in a triaxial testing machine. Saarela (7) has reported test data from Finland. Landva et.al. (3) tested large sized (28 cm by 43 cm) samples of refuse in direct shear. One year later, retesting again in direct shear, Landva and Clark (4) reported a decrease in strength. More recently, Earth Technology Corporation (9) used a California drive sampler to obtain samples of refuse and reported a multistage triaxial test on one of these samples.

Clearly, there is a large scatter in the laboratory test results (see Figure 1). Several factors could have contributed to the wide variations, such as

- the highly heterogeneous composition of the refuse,
- the method by which a sample was obtained,
- too small a sample size to be representative of the in-place refuse with its different unit weights and dissimilar composition.

It is interesting to note that most of the laboratory investigators have treated the refuse as cohesionless material and have reported the results accordingly. However, the users of the back-calculating approach described in the next section have indicated the refuse to possess both cohesion and frictional properties. Landva and Clark(4) indicated the need for more data to establish the loss of strength with time due to decomposition of refuse.

STRENGTH ESTIMATES BASED ON BACK-CALCULATIONS OF FIELD TESTS AND OPERATIONAL RECORDS

The back-calculation approach is chiefly based on the field load test made in the Los Angeles area by Converse, et al (21). The Los Angeles area landfill is in Monterey Park, and was field tested by loading and monitoring the deformation of the fill. Many of the slope stability studies have used strength parameters obtained from this test study, such as Kirby Canyon Landfill (Volpe, 11), Zanker Road Landfill (Cooper, 6; Woodward-Clyde, 12), Sunnyvale Landfill (Cooper, 6), Newby Island Landfill (Purcell, Rhoades & Associates, 13), Corinda Los Trancos Landfill (Purcell, Rhoades & Associates, 14), Acme Landfill (Harding-Lawson & Associates, 22) and Sunnyvale Landfill (Dames & Moore, 20). These studies estimated the back-calculated values in various combinations of cohesion and angle of friction, and are plotted in Figure 2.

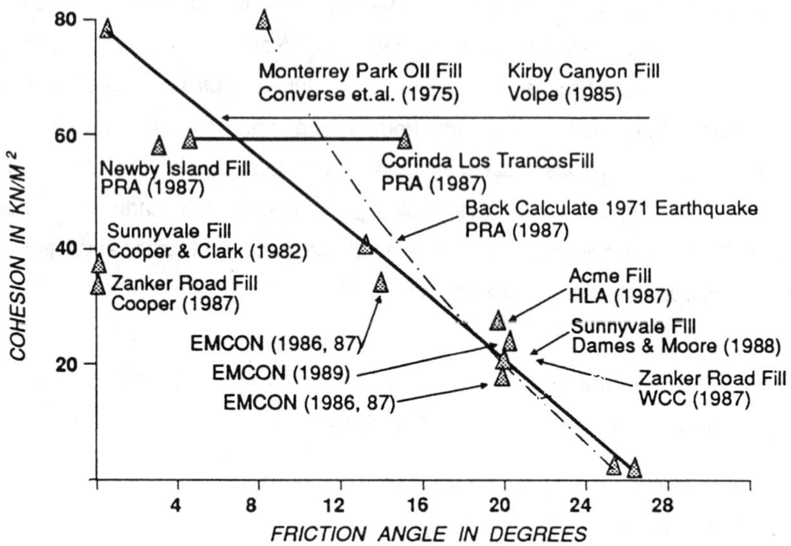

Fig. 2 - Strength Parameters Estimated by Back Calculation from a Field Load Test and from Performance Records

The main justification in using these values by various authors in their studies was that these values represent the lower boundary of the available strength and therefore are conservative. As also pointed out by Dames & Moore (20), because these combinations of strength parameters are estimated based on the results of a field test by Converse, et al (21), they may not represent independent estimates of the shear strength properties of the refuse.

Back-calculated strength data has also been obtained on the basis of the satisfactory performance of the numerous landfill slopes in southern California during the San Fernando earthquake of 1971 and more recently, during the Whittier earthquake of 1987. Purcell, Rhoades & Associates, (15) and Earth Tech, (9) observed the stability of the relatively steep slopes, including nearly vertical cuts. Again, the basis for justifying the use of this approach was that the back-calculated values represent minimum available strength of refuse and are therefore conservative.

STRENGTH ESTIMATES BASED ON IN-SITU TESTING OF REFUSE

Recently, attempts were made to evaluate the shear strength of the refuse by in-situ testing. Cooper-Clark and Associates (6) and EMCON (18) obtained standard penetration test (SPT) data on refuse material during field exploration of the Sunnyvale Landfill. An average blow count value of 15 was used by Dames & Moore (20) after rejecting values larger than 50 that may represent the encounter of obstructions. Earth Tech Corporation (9) reported the results of a vane shear test and a standard penetration test. These results are shown in Figure 3.

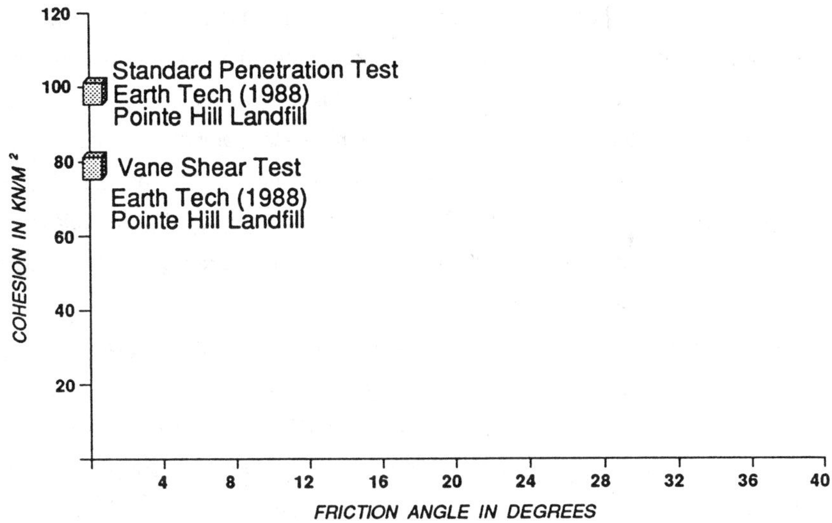

Fig. 3 - Strength Parameters from In-Situ Testing

Finally, the results of all the foregoing tests are plotted in Figure 4. Because of the scatter and scarcity of the data, it is difficult to draw any definitive conclusions on the shear strength characteristics of sanitary fill material.

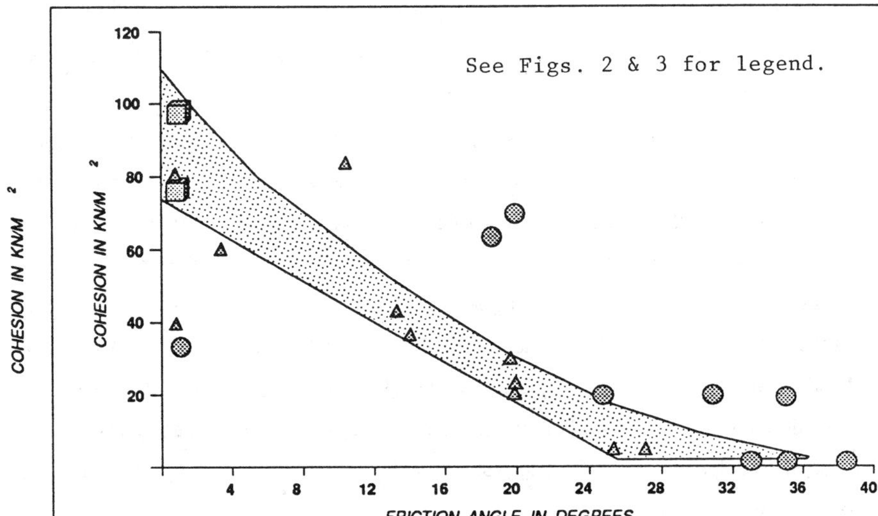

Fig. 4 - Summary Plot for All Tests with Recommended Parameters Shaded

A CRITICAL EVALUATION OF STRENGTH DATA

The in-situ testing, involving shearing of the refuse by a vane shear, has been reported by Earth Tech (9) with the following statement, "A review of the boring log indicates that solid inclusions (asphalt and solid wood) were present in many of the vane shear tests. The size of the vane shear device is relatively small 6.35 cm (2-1/2-inch diameter) as compared to the inclusions and, therefore, the vane shear data may not be representative of the refuse strength." This is understandable considering the physical makeup of the refuse fill material which includes plastics, tires, carpets, etc. and hence, the 6.35 cm size vane can be completely inadequate if it shears across a carpet or wood. Dames & Moore (20) argued that since no published correlation between refuse strength and blow count is found in the literature, assigning a strength value to the average blow count would be subjective in nature.

Cooper Engineers (6) performed triaxial tests on Shelby tube samples of refuse. The samples were compressed to a strain of over 30 percent, with load capacity still increasing and no failure in sight. Similarly, in a field load test at the Operating Industries, Inc. landfill in Monterey Park, California, Converse, et al (21) surcharged the refuse slope and the slope underwent a large deformation but no failure plane was evident. Converse then assumed a failure plane typical of a soil slope with a conservative factor of safety, and back-calculated a range of values for cohesion and friction angle. No information on the deformation for tests reported by Fang (2) or from Finland reported by Saarela (7) are available. Multistage triaxial test results on one California drive sample reported by Earth Tech (9) does indicate a leveling of the load capacity with an increase in strain. However, the development of a shear failure plane or large lateral strain development has not been reported. The simulated direct shear sample tested by the Los Angeles County Sanitation District (5) appears to be the only test where a typical shearing would have taken place.

In view of the foregoing, two significant questions arise about the application of soil mechanic principles to refuse material strength and stability evaluation. The Mohr-Coulomb theory may not adequately account for why refuse material undergoes large deformation without failure. Secondly, the incompatibility of strains that produce shear failure in soils and those that would produce shear failure in refuse, suggests that stability analysis of a refuse fill may be related more to its settlement and foundation bearing capacity than to its slope failure. Satisfactory performance of relatively steep slopes (1-3/4:1, H:V) of high refuse fills in southern California during earthquakes and observations of no slope instability of nearly vertical cuts made in sanitary fills (Volpe, 11), seem to suggest that a slope failure may not be the most critical aspect of a sanitary landfill.

It is important to note the following observations made by Converse, et al (21) at the time of the well known field test made at Monterey Park, California. Converse, et al (21) states that, "Movements observed in the test fill were primarily caused by vertical settlement due to compression of the refuse material, with secondary movement out of the slope resulting from lateral spreading of the foundation of the test fill." This could be related to a bearing capacity problem. Converse, et al, further states that, "It is our opinion that the sanitary landfill test slope remained stable during and after the placement of the soil test fill."

In view of the foregoing discussion, it appears that the characteristics:

- settlement of the fill and the
- bearing capacity of the foundation

might be the more significant parameters than the slope stability in evaluating the structural integrity of sanitary landfills.

SETTLEMENT CHARACTERISTICS

Settlements and settlement monitoring of sanitary landfills have been discussed by Eliassen (23), Sowers (24, 25), and Rao, et al (26). These investigators primarily examined the load-bearing characteristics of the refuse fill for building foundation support. Sowers (25) cited five mechanisms causing settlements of a sanitary landfill. These factors include mechanical reorientation of materials, infilling of void areas by finer materials, "physico" chemical, and "bio" chemical changes.

Recently, Huitric (27) has presented a comprehensive treatment of the subject of the sanitary landfill settlement rates. Huitric defines three possible modes of settlement:

- consolidation
- shrinkage
- compaction.

The term consolidation refers to settlement resulting from the dewatering of the saturated materials. Shrinkage is the process by which organic solids and moisture are microbially converted to carbon dioxide and methane, resulting in a corresponding decrease in the volume of the fill. Compaction is defined as the re-orientation of solids into a more dense configuration due to the gradual loss of rigidity in solids from the creep of solids under high stress or from decomposition.

Huitric believes that such highly stressed solids may initially "bridge" across voids, but eventually collapse and this may be judged to be potentially the most significant feature of

It is important to note the following observations made by Converse, et al (21) at the time of the well known field test made at Monterey Park, California. Converse, et al (21) states that, "Movements observed in the test fill were primarily caused by vertical settlement due to compression of the refuse material, with secondary movement out of the slope resulting from lateral spreading of the foundation of the test fill." This could be related to a bearing capacity problem. Converse, et al, further states that, "It is our opinion that the sanitary landfill test slope remained stable during and after the placement of the soil test fill."

In view of the foregoing discussion, it appears that the characteristics:

- settlement of the fill and the
- bearing capacity of the foundation

might be the more significant parameters than the slope stability in evaluating the structural integrity of sanitary landfills.

SETTLEMENT CHARACTERISTICS

Settlements and settlement monitoring of sanitary landfills have been discussed by Eliassen (23), Sowers (24, 25), and Rao, et al (26). These investigators primarily examined the load-bearing characteristics of the refuse fill for building foundation support. Sowers (25) cited five mechanisms causing settlements of a sanitary landfill. These factors include mechanical reorientation of materials, infilling of void areas by finer materials, "physico" chemical, and "bio" chemical changes.

Recently, Huitric (27) has presented a comprehensive treatment of the subject of the sanitary landfill settlement rates. Huitric defines three possible modes of settlement:

- consolidation
- shrinkage
- compaction.

The term consolidation refers to settlement resulting from the dewatering of the saturated materials. Shrinkage is the process by which organic solids and moisture are microbially converted to carbon dioxide and methane, resulting in a corresponding decrease in the volume of the fill. Compaction is defined as the re-orientation of solids into a more dense configuration due to the gradual loss of rigidity in solids from the creep of solids under high stress or from decomposition.

Huitric believes that such highly stressed solids may initially "bridge" across voids, but eventually collapse and this may be judged to be potentially the most significant feature of settlement of sanitary fills. Huitric (27) cites on the basis of settlement measurements at Canyon 3 Fill, that the "percent settlement varies from 15 to 20 percent in the central portions of the fill and decreases rapidly at points along the canyon sides. On the whole, the variations in the settlement contours expressed in meters are more uniform". This suggests that "the fill settles as an integral unit and is not overly sensitive to local variations in depth." This phenomena seems quite possible, because fills of a young age settle as a unit and are relatively insensitive to local variations in unit weight and composition. With age, as the decomposition proceeds and becomes a prevailing factor in settlement, then a wide range in the rate and the magnitude of settlements can be expected.

BEARING CAPACITY CHARACTERISTICS

In view of the foregoing discussions, it is suggested that correlations of lateral movements of foundation soils of the fill should be made with the vertical movements due to the settlement/compression of the fill. Such correlations can provide clues to the bearing capacity problem. For instance, the foundation of the Operating Industries, Inc. (OII) landfill in Monterey Park (Converse, et al, 21) is reported to have undergone a movement of about 43 cm (17 inches) on its upper portion. Data from slope inclinometers indicated that movement accelerated with the fill placement. When the fill placement ceased, the rate of movement markedly decreased.

In order to further gain insight into the bearing capacity problem associated with the foundation soils, data from the load test at the OII landfill was used to estimate the bearing capacity of foundation soils. The data was obtained from the report documenting the field program and prepared by Converse, et al (21).

The field program consisted of a full-scale field load test performed on one portion of an existing exterior slope approximately 100 feet high with a 1.6:1 slope. Compacted earth fill was placed over the existing refuse to a maximum height of 38 feet in a period of 24 days. The overburden pressure was calculated using as an average height of 28 to 30 feet for the compacted soil and unit weight for the refuse and the compacted soil as 0.8 gm/cm^3 (50 pcf) and 2.05 gm/cm^3 (128 pcf), respectively. The overburden pressure was calculated to be 685 kN/m^2 (14,300 psf.)

For a clay or silt as foundation soils with an average shear strength of 95.8 kN/m^2 (2000 psf) under an undrained condition, the estimated ultimate bearing capacity (5.14 Su) equals about488.4 kN/m^2 (10,200 psf). An undrained shear strength of 95.8 kN/m^2 (2000 psf) for soils of marginal lands is not uncommon. However, it is significant to note that even for a strength of 143.6 kN/m^2 (3000 psf) for the foundation soils at OII fill, the load intensity due to the additional soil fill brought the factor of safety against bearing capacity failure close to unity. Unfortunately, the information on the shear strength characteristics of the foundation soils at the OII fill are not available. However, the following statement from the Converse, et al (21) report is worth noting. "Movements observed in the test fill were primarily caused by vertical settlement due to compression of the refuse material, with secondary movement out of the slope resulting from lateral spreading of the foundation of the test fill."

In the case of the New Jersey global landfill failure, the lateral movements of the foundation resulted from increased refuse filling on soft silt. However, similar deep-seated movements in foundation soils of the OII test fill did not cause noticeable cracks along the toe or on the top. Converse, et al believes that the foundation soils could be on the verge of failure before any cracks become noticeable in a landfill. The New Jersey failure opened a chasm 18 m (60 feet) wide, 122 m (400 feet) deep and 183 m (600) feet long over a period of a few days. A nearby toe dike was raised up several feet in places and an adjacent tidal marsh experienced cracking and lifting. Clearly, the bearing capacity of foundation soils at landfills - especially the high fills on marginal lands - should be seriously investigated.

RECOMMENDED APPROACH FOR STABILITY ANALYSIS

The following section presents the authors' approach to the analysis of landfill stability. This approach considers both stability and bearing capacity factors in analyzing the structural integrity of a landfill.

For a typical refuse fill of moderate height (about 61 m) with a 3:1 (horizontal to vertical) slope and sitting on a relatively strong foundation soils ($S_u \geq$ 192 kN/m^2), a classical soil slope stability analysis can be performed by using shear strength parameters for refuse from Figure 4. The shaded zone in figure 4 is the recommended range of strength parameters for use in stability studies. Results of such analysis should be interpreted judgmentally in favor of least conservatism because large, moderately steep sanitary landfills have had no slope failures even when they were shaken by relatively strong ground motions during earthquakes.

For a relatively high refuse fill (height greater than 68 m or 200 feet) with moderately steep slopes (1.5-2H to 1V) and placed on soft, marginal or relatively weak foundation soils ($S_u \leq$ 96 kN/m^2), the potential mode of failure as shown in Figure 5 should be considered. To analyze such a case, two approaches can be used; one based on slope stability analysis and the other based on bearing capacity analysis. As may be seen in Figure 5, the stability analysis would involve calculating the overturning movement and estimating the resisting movement. There is one important difference recommended in estimating the resisting movement. Authors believe that the contribution to the resisting force by the landfill should be neglected as a conservative approach. This belief assumes that any disruptive force large enough to tear the landfill would have already compromised the foundation. The foundation soils would have already undergone considerable movement and shear resistance of the foundation soils would have been completely mobilized before the refuse could tear along the potential failure surface shown in Figure 5.

Fig. 5 - Typical Failure for High Fills on Soft Foundations

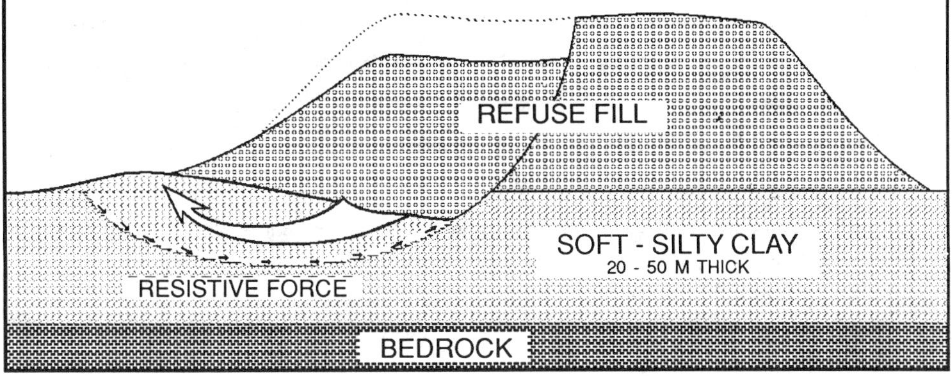

This approach would yield highly conservative results for predicting slope stability involving a slip surface through the fill. However, these results may not be conservative for predicting deformation of foundation soils, which may threaten the integrity of leachate collection or other drainage systems including the liner under the fill.

The method for evaluating bearing capacity is simple, and involves estimating the ultimate bearing capacity of the foundation soils and comparing it with the overburden pressure of the refuse fill. The bearing capacity for saturated clay or silty foundation soils can be readily calculated from the simple relationship: $Q_{ult} = 5.14\ S_u$, where S_u is the shear strength under undrained loading. For soils other than soft clay or silt, the bearing capacity is likely to be quite high and can be estimated using bearing capacity equations. When such an approach is used to analyze the failure of New Jersey landfill which was placed on weak foundation soils, it may not be surprising to find that it failed. A reasonable estimate of the weight of the fill is important if the bearing capacity of a relatively weak foundation soils is in question. Unit weight of a sanitary landfill is also an important parameter in addition to its shear strength for evaluating the stability conditions of the fill. Within a given refuse fill, the unit weight of the refuse can widely vary because of the difference in composition, state of decomposition, amount of compactive effort and settlement. It is generally believed that the unit weight of the fill somewhat increases with depth (Converse, et al, 21; Volpe, 11: Purcell, Rhoades & Associates, 15; Dames & Moore, 20). The average unit weight, according to the above authors, varies from 0.8 gm/cm^3 (50 pcf) to 0.96 gm/cm^3 (60 pcf.) On the other hand, studies by Earth Tech Corporation (9) have used a nearly constant unit weight with depth.

DYNAMIC STABILITY ANALYSIS

Analysis of soil slope stability under earthquake loading has been carried out using pseudo-static and deformational analysis. The deformational analysis approach is based on the works of Newmark (28), Seed (29) and Makdisi & Seed (30). This approach has also been used for analyzing the dynamic stability of sanitary landfills. According to these analyses, both the simplified and the rigorous method of estimating time history of accelerations and shear stress requires representative data on the dynamic strength properties of the material at different points within the fill. Obtaining dynamic strength properties of soil through equivalent linear or non-linear models has not been easy (Finn (31)), and there is hardly any test data on the dynamic strength properties of refuse material.

Until recently, the shear modulus and damping characteristics of refuse were assumed to be similar to peat because of the low unit weights, high void ratios and high compressibilities of both peat and refuse materials. Seed and Idriss (32) developed shear modulus and damping curves for peat on the basis of static and dynamic laboratory tests. This data was used by several authors (Volpe, 11, EMCON, 17; Purcell, Rhoades & Associates, 14, 15; Earth Tech, Inc., 9) in the response analyses made for sanitary landfills in California. Volpe (11) estimated shear wave velocity for refuse based on the static load settlement results from the full scale load test performed by Converse, et al (21) at the Operating Industries Landfill in Monterey Park, California. The computed average of shear wave velocity was 26 m/sec (85 ft/sec). The range of shear wave velocity for peat reported by Seed & Idriss (32) was 21.6 to 43.3 m/sec (71 to 142 ft/sec).

More recently, test data based on field shear wave velocity tests have been reported by The Earth Technology, Inc. (9). An average shear wave velocity of about 274 m/sec (900 ft/sec) has been estimated by Earth Technology, Inc. on the basis of geophysical cross-hole and downhole shear wave velocity tests. Results of seismic survey by downhole shear wave velocity tests carried out for EMCON (19) by Redpath Geophysics and by Portola Geophysics at West Richmond Fill and at Redwood Fill indicate respectively, gave average shear wave velocities of 213 m/sec (700 ft/sec) and 91 m/sec (300 ft/sec). A value as low as 31.4 m/sec (103 ft/sec) has also been reported at the Redwood Refuse Fill. These values are much higher than those estimated on the basis of tests on peat or field load tests. Apparently, as more test data on refuse will accumulate, representative values for shear modulus and its damping for the refuse material at various stages of decomposition should emerge. In the meanwhile, the use of

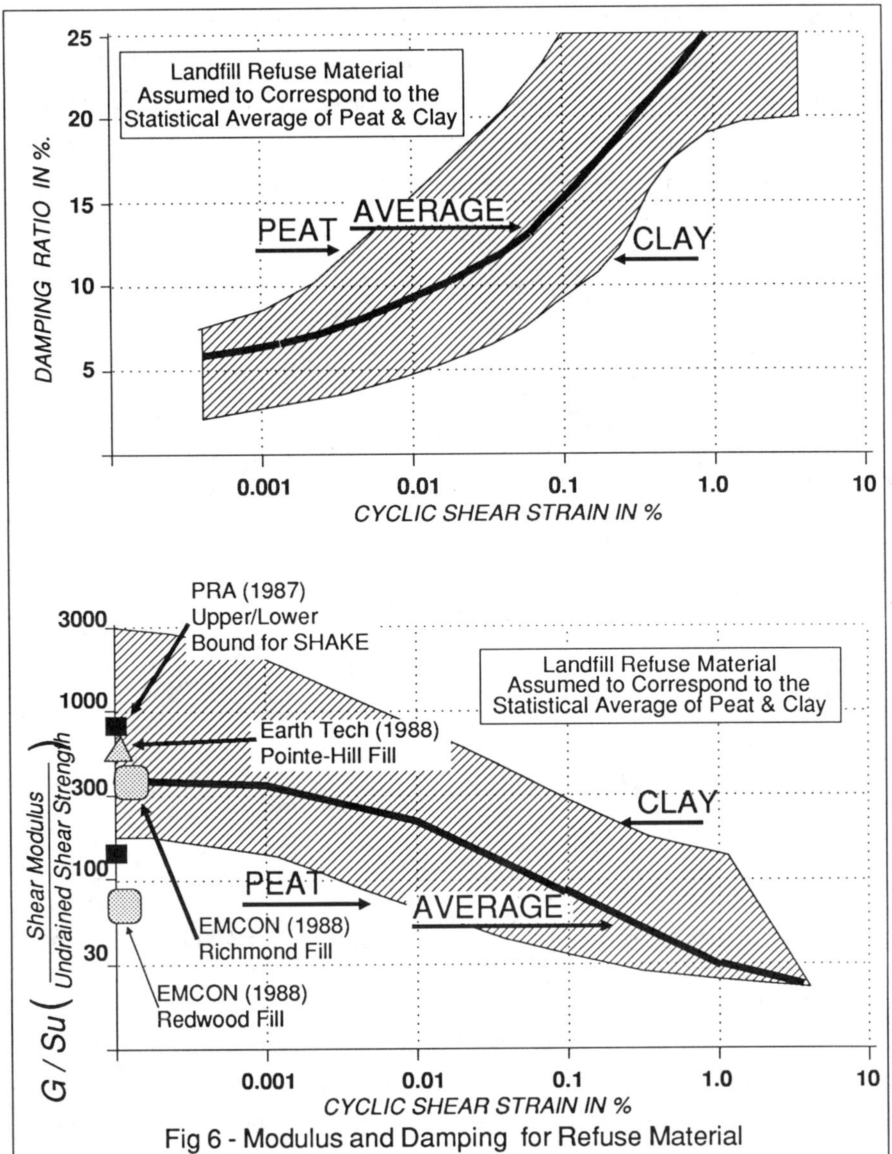

Fig 6 - Modulus and Damping for Refuse Material

Note: Data Points are estimated average values from shear wave data for Su=143 KN/m^2

static stress-strain data or the downhole shear wave velocity data to estimate dynamic shear modulus of refuse may be used with caution because of the highly compressive nature of the refuse and its non-soil like strength deformation characteristics.

Because of the foregoing reasons, authors carried out parametric studies on the dynamic response of landfills using a range of values for shear modulus and damping characteristics of the refuse material. Based on the assumption that the strength properties of the refuse are more cohesive than frictional, the range of values chosen were similar to that of peat and clay. Figure 6 shows the modulus and damping curves used in the SHAKE analyses made to evaluate the response of a 122 m (400-foot) high sanitary landfill.

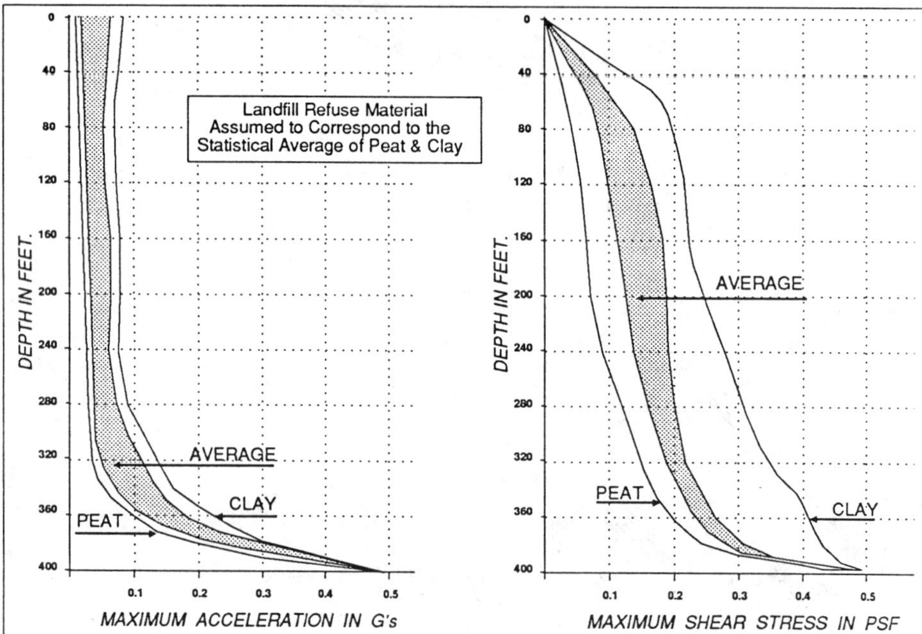

Fig 7 - Results of Response Analysis by SHAKE for 400 ft. High Landfill

Significant attenuation of baserock motions is evident in each case (see Figure 7). Apparently, the influence of varying the shear modulus and the damping values is not significant for the range of values considered. Because of the absence of recorded data on earthquake motion attenuation on refuse fills and the excellent performance of sanitary landfills to relatively strong earthquake motions, it can be argued on the basis of the physical makeup of the refuse that the refuse inherently has strong energy absorption mechanisms.

The results of SHAKE analyses are most significant in showing that the maximum bedrock accelerations were considerably reduced as they propagated up through the height of the landfill. This is quite different from what one would expect from the response of an earth dam subjected to similar motions. The results showed that the maximum bedrock acceleration was reduced from

0.5g at the bedrock level to average values of 0.06g or less at crest level. This is due to the low modulus, high damping and light weight of the fill.

These factors were probably also responsible for the natural period estimated by SHAKE analysis to be relatively high (12 to 16 seconds). Accordingly, amplification of energy may only be expected for frequencies less than one Hz. The effect of damping on motions as they propagate upward was noted in the filtering out of the high frequencies. However, more data on the dynamic response of refuse fills is needed to confirm these results.

CONCLUSIONS

As a result of the studies presented in this paper, the following conclusions can be drawn:

1.Shear strength characteristics of the refuse material are not yet adequately defined. There is a large scatter on the shear strength data obtained by laboratory testing, in-situ testing or estimates based on field performance records. The Mohr-Coulomb theory may not adequately account for the large deformations a refuse material undergoes without failure.

2.The application of the soil mechanics principles to refuse material strength and stability evaluation should be viewed with caution because of the incompatibility of strains that produce a shear failure in soils and those that would produce shear failure in refuse.

3.Satisfactory performance of relatively steep slopes of high refuse fills during earthquakes, and observations of no slope instability of nearly vertical cuts made in sanitary landfills and the large deformations a refuse material undergoes without failure, seem to suggest that a slope failure may not be the most critical aspect of a sanitary landfill.

4.In view of the observed lateral spreading of foundation soils, it appears that the settlement of the fill and the bearing capacity of the foundation soils might be the more significant parameters than slope stability in evaluating the structural integrity of sanitary landfills. The simplified approach proposed by the authors may be used to evaluate bearing capacity as well as stability characteristics.

5.Settlement characteristics are sensitive to the age of the refuse fill and become complex and more non-uniform with age.

6.Because of the complex and heterogeneous structure of a refuse material and the lack of test data, very little is known about its dynamic strength characteristics. Results from recent but limited shear wave velocity tests indicate shear moduli values to be somewhere between that of clay and peat.

7.Results of SHAKE analysis indicate significant attenuation of bedrock motion as they propagate up through the refuse fill. The use of the pseudostatic and deformation analysis approach for dynamic stability analysis should consider the strong energy absorption mechanism of the refuse material.

REFERENCES

1. Stoll, U.W., "Mechanical Properties of Milled Domestic Trash," American Society of Civil Engineers, presented at the National Water Resources Engineering Meeting, Pheonix, AZ, January 1971.

2. Fang, et al, "Strength Testing of Bales of Sanitary Landfill, Fritz Engineering Laboratory, Lehigh University, Department of Civil Engineering, 1977.

3. Landva, A.O., Clark, J.I., Weisner, W.R., and Burwash, W.J., "Geotechnical Engineering and Refuse Landfills," Sixth National Conference on Waste Management in Canada, Vancouver, British Columbia, 1984.

4. Landva, A.O., and Clark, J.I., Geotechnical Testing of Wastefill, Report submitted to the London Institute of Civil Engineers, 1987.

5. Los Angeles County Sanitation District, "Puente Hills Landfill Draft EIR", Appendix IV-A, 1984.

6. Cooper & Clark Consulting Engineers, "Stability Evaluation of Sunnyvale sanitary Landfill, Report to the City of Sunnyvale, CA, April, 1982

7. Saarela Jouko, "Some facts about the Sanitary Landfill Investigations in Helsinki City", an attachment to a communication to Prof. S. Singh from Jouko Saarela of National Board of Waters and Environment, Helsinki, Finland, 1987.

8. Cooper Engineers, "Unconsolidated-Undrained Triaxial Shear Test Results", Job No. 2659-A3, Zanker Road Resource Recovery, May 1986.

9. Earth Technology Corporation, "Instability of Landfill Slope, Puente Hills Landfill, Los Angeles County, California", submitted to Los Angeles County Sanitation District, 1988.

10. Kleinfelder, "Slope Stability Evaluation, Eastern Alameda County Landfill, Livermore, CA" Report to DePaolia Equipment Company Inc., Livermore, CA 94550, July,1988

11. Volpe, R.L. & Associates, "Slope Stability Investigation, Intermediate and Final Slopes Static and Dynamic Loading Conditions, Kirby Canyon Sanitary Landfill, San Jose, California, prepared for EMCON,1985

12. Woodward-Clyde Consultants, "Geotechnical Engineering Study, anker Road Landfill, Proposed Expansion, San Jose, California", Report to Zanker Road Resources Recovery, Inc., San Jose, California, April 1987.

13. Purcell, Rhoades & Associates, "Stability Evaluation, Newby Island Sanitary Landfill", Report to Browning-Ferris Industries, March 1983.

14. Purcell, Rhoades & Associates, "Stability Analysis, Static and Dynamic Loading Conditions, Corinda Los Trancos Sanitary Landfill, Ox Mountain, Half Moon Bay, California", Report to California Regional Water Quality Control Board, August 1987.

15. Purcell, Rhoades & Associates, "Slope Stability Analyses, Apanolio Canyon Expansion Site, Ox Mountain, San Mateo County, California", Report to Browning-Ferris Industries, 1987.

16. Purcell, Rhoades, and Associates, "Stability Analysis - Static and Dynamic Loading Conditions, Sunshine Canyon Extension Site, Sylmer, California," Report to Browning-Ferris Industries of California, April, 1988

17. EMCON, "Slope Stability Analysis - City of Lompoc Landfill- Santa Barbara Co. CA", Report to City of Lompoc, California, 1986.

18. EMCON, "Stability Studies - Weldon Canyon Landfill", Ventura County, California, Report to Waste Management Inc. of California, 1987.

19. EMCON, "Analysis of Deformations under Static & Seismic Loading - West Contra Costa Sanitary Landfill, Richmond, CA" Report to Richmond Sanitary Service, Richmond CA, January, 1989

20. Dames & Moore, "Landfill Slope Stability Analysis, Sunnyvale Sanitary Landfill, Sunnyvale, California," Report to City of Sunnyvale, California, 94086, February, 1988.

21. Converse, Davis, Dixon Associates, "Slope Stability Investigation, Proposed Final Slope Adjacent to the Pomona Freeway", Operating Industries Disposal Site, 1975.

22. Harding-Lawson Associates, "Seismic Stability Evaluation, ACME Landfill, Martinez, California", Report to ACME Fill Corporations, July 1987.

23. Eliassen, R., "Load Bearing Characteristics of Landfills", Engineering News Record, Volume P. 361, pp. 103-105, 1042.

24. Sowers, G.F., "Foundation Problems in Sanitary Landfills", Journal of Sanitary Engineering Division, Proceedings of the ASCE, Vol. 94, No. SA1, pp. 103-116, 1968.

25. Sowers, G.F., "Settlement of Waste Disposal Fills", Proceedings of the Eighth International Conference on Soil Mechanics and foundation Engineering, Moscow, Vol. 2, pp. 207-210, 1973.

26. Rao, S.K., K.L. Moulton and R.K. Seals, "Settlement of Refuse Landfills", Proceedings of the Conference on Geotechnical Practice for Disposal of Solid Waste Materials, Ann Arbor, Michigan, pp. 574-598, 1977.

27. Huitric, R., "Sanitary Landfill Settlement Rates", Los Angeles County Sanitation District, paper presented at Technische Universitat, Berlin, 1981.

28. Newmark, N.M., Effects of Earthquakes on Dams and Embankments", Rankine Lecture, Geotechnique No. 15, 1965.

29. Seed, H.B., "Considerations in the Earthquake-Resistant Design of Earth and Rockfill Dams, The Rankin Lecture, Geotechnique Vol 29 no.3 pp215-263, 1979.

30. Makdisi, F.I. and Seed, H.B., "Simplified Procedure for Estimating Dam and Embankment Earthquake-Induced Deformations", Journal of the Geotechnical Engineering Division of ASCE, July 1978.

31. Finn, W.D. Liam, "Dynamic Analysis in Geotechnical Engineering, State-of-the-Art Report, Proceeding of the Specialty Conference on Earthquake Engineering and Soil Dynamics - II", sponsored by Geotechnical Engineering Division of ASCE, Park City, Utah, pp. 523-592, June 27-30, 1988.

32. Seed H.B. and Idriss, I.M., "Analyses of Ground Motions at Union Bay, Seattle During Earthquakes and Distance Nuclear Blasts", Bulletin of the Seismological Society of America, Vol. 60, No. 1, pp. 125-136, 1970.

Ronald A. Siegel, Richard J. Robertson, and Donald G. Anderson

SLOPE STABILITY INVESTIGATIONS AT A LANDFILL IN SOUTHERN CALIFORNIA

REFERENCE: Siegel, R.A., Robertson, R. J., Anderson, D. G., "**Slope Stability Investigations at a Landfill in Southern California,**" Geotechnics of Waste Fills-Theory and Practice, ASTM STP 1070, Arvid Landva, G. David Knowles, editors, American Society for Testing and Materials, Philadelphia, 1990.

ABSTRACT: The results of geotechnical investigations and slope stability studies of an existing municipal/commercial landfill in southern California are presented and discussed. Investigations included cone penetration testing, installation and monitoring of inclinometers, piezometers, surface monuments, and laboratory classification and direct shear testing of refuse. The landfill was also monitored for dynamic response to earthquake-induced ground motions.

KEYWORDS: municipal refuse, engineering properties, municipal landfill, slope stability, shear strength, seismic response

Geotechnical investigations and slope stability studies [1,2] have been carried out for the South Parcel of the Operating Industries, Inc. (OII) Landfill in Monterey Park, California (Figure 1). The program objectives were to determine the magnitude and direction of current slope movements and to assess the potential for slope failure under both gravity (static) and seismic loading.

Slopes of the OII landfill are municipal and industrial/commercial refuse covered with soil of varying thickness (Figure 2). The slopes range up to 70 m (230 ft) high and have narrow benches at several levels. The average steepness over the total slope height ranges from approximately 2:1 to 3:1 (horizontal:vertical). Intermediate slopes are up to 30 m (100 ft) high and up to 1.4:1 in steepness.

Mr. Siegel is a senior geotechnical engineer at the Rocky Mountain regional office of CH2M HILL, P.O. Box 22508, Denver, CO 80222-0508; Mr. Robertson is a project engineer assigned to CH2M HILL's Southern California regional office, 2510 Red Hill Ave., Suite A, Santa Ana, CA 92705; Dr. Anderson is a senior geotechnical engineer at CH2M HILL's Seattle regional office, P.O. Box 91500, Bellevue, WA 98009-2050.

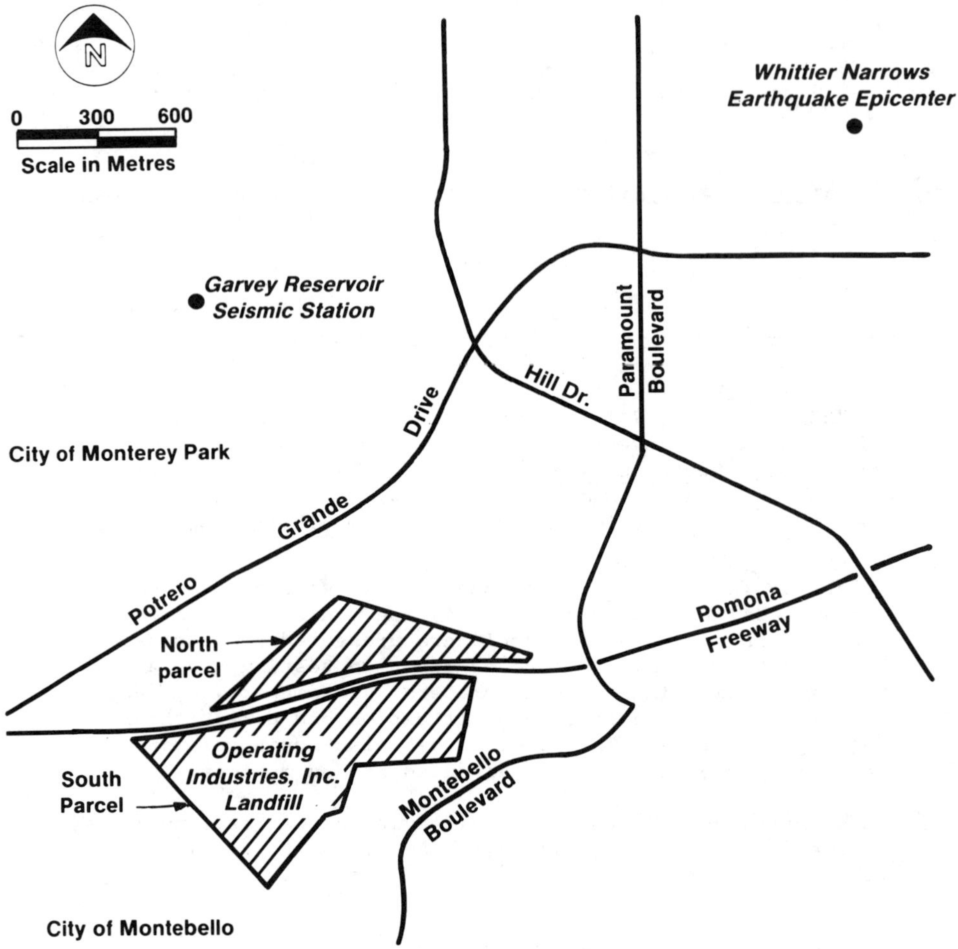

FIG. 1-- Site Location Map

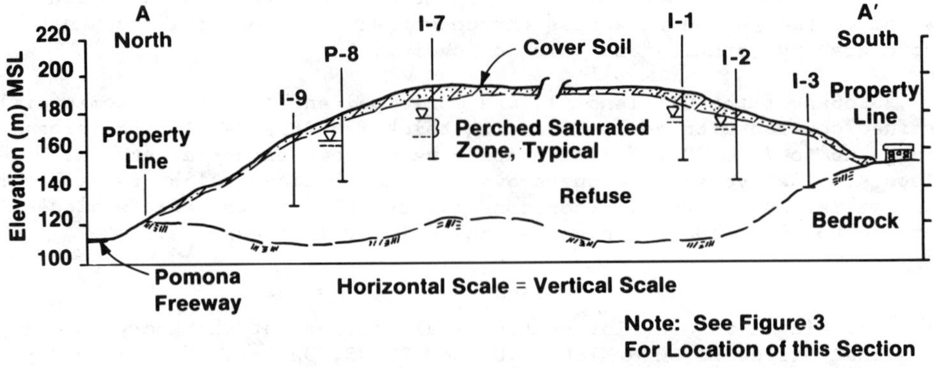

FIG. 2-- Generalized Landfill Cross Section A-A′

The landfill was active from 1948 until 1984. Some of the steeper slopes have been standing nearly full height for 20 years. During the facility's life, no slopes have failed; however, there are many signs of slope movement, such as surface cracking, bulging, and slumping. The stability of these slopes is of concern because of the potential impacts to the Pomona Freeway and nearby residential areas.

This paper discusses field and laboratory investigations and the data collected. Also discussed are the results of preliminary slope stability analyses and seismic response data collected during several earthquakes.

CONE PENETRATION TESTS

Cone penetration tests (CPT) were performed to help delineate stratigraphy and saturated zones within the landfill [3]. An 18-Mg (20-ton) truck transported and housed the CPT equipment and provided the reaction to push the cone into the landfill. The CPTs were conducted in general accordance with ASTM D3441-86 [4].

CPTs were conducted at nine locations (Figure 3) with depths ranging from 4.8 to 37.5 m (16 to 123 ft). Planned depths were 45.7 m (150 ft). Each test ended whenever the angular deflection of the probe or the penetration resistance was excessive. Two or three attempts were made at most locations to penetrate to greater depths. In 18 attempts, only half penetrated more than 6.7 m (22 ft). Two cone instruments broke off in the landfill.

A piezocone penetrometer was used for the first two tests in an attempt to identify saturated zones within the landfill. Because the probe's porous stone element was crushed when pushed against hard objects in the refuse and because of low piezometric response, the pore pressure transducer system was abandoned for subsequent tests. The low piezometric response was a result of loss of saturation in the measuring system as the piezocone penetrated unsaturated material.

The CPT results indicated that the cone frequently encountered stiff objects, which produced sharp peaks in the tip resistance measurements. This resulted in highly variable readings (Figure 4). However, a trend of increasing lower bound tip resistance with depth, about 0.8 $kg/cm^2/m$ (0.25 tsf/ft), was apparent in most of the tests. Because of the relatively erratic readings from the CPT probes, daily or interim cover soil could not be distinguished from the refuse.

Soil types were interpreted from soil behavior classification charts [5] to be primarily sandy or clayey silt. This classification generally agreed with the types of cover soil encountered in borings. Although cover soils at the surface could be identified from the test records, their thickness was sometimes hard to determine.

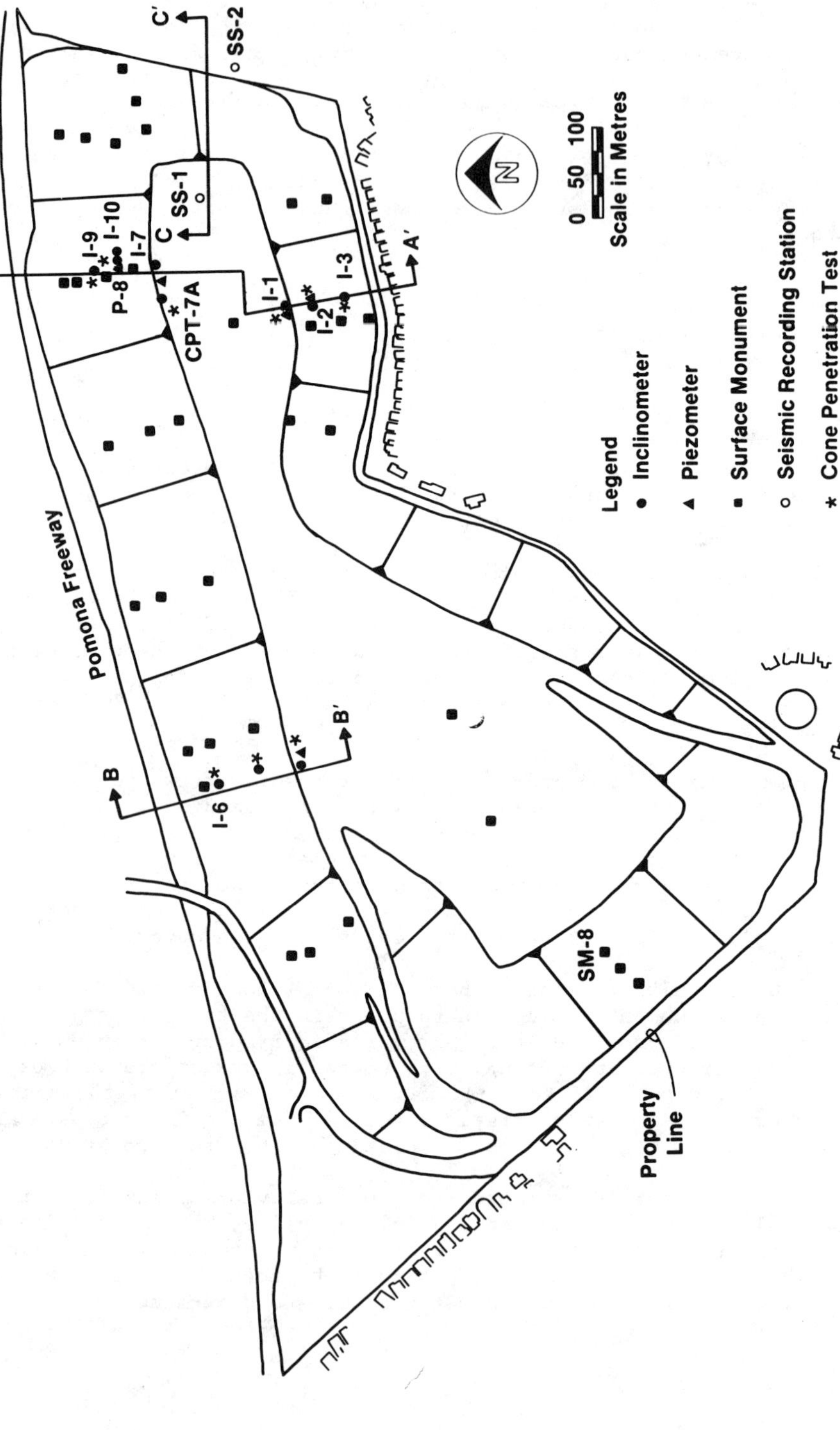

FIG 3-- Instrumentation and CPT Locations

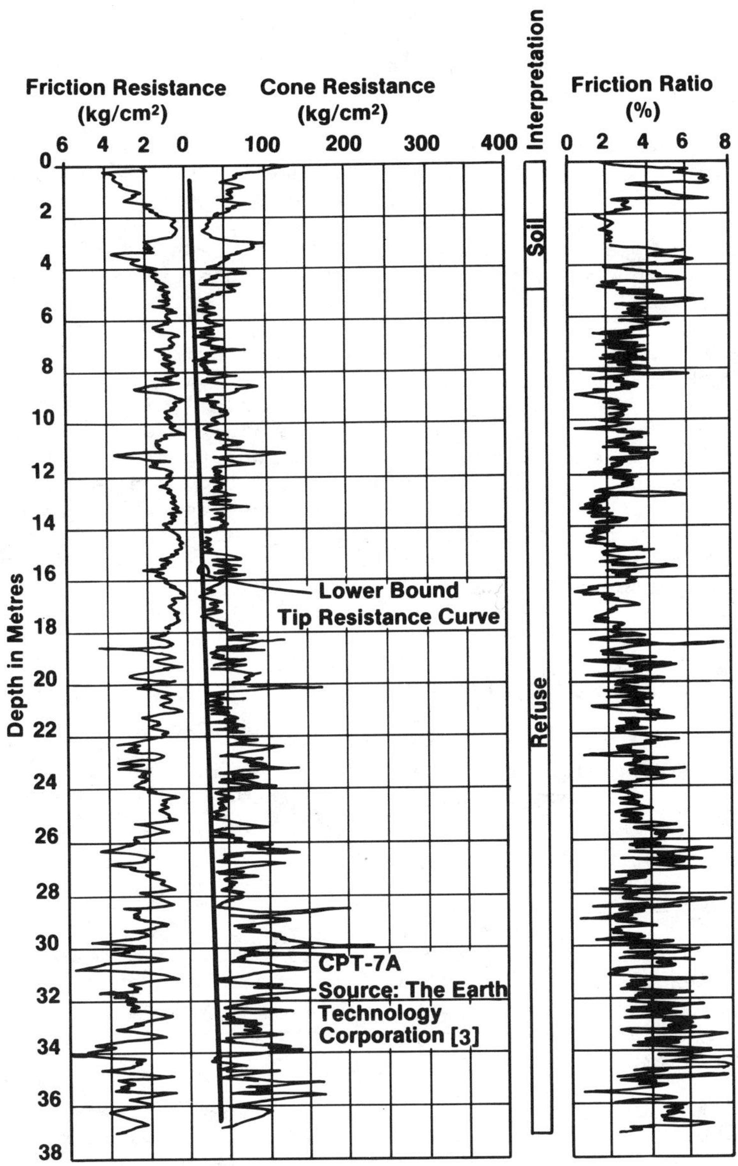

FIG. 4-- Example of CPT Results

DRILLING AND SAMPLING

To install inclinometers and piezometers, sixteen borings were drilled using 16-cm (6-1/4-in.)-inside-diameter (ID) hollow-stem augers rotated by CME-75 drill rigs, which are rated to deliver nearly 11,500 N-m (8450 ft-lb) of torque. Inclinometer and piezometer locations are shown in Figure 3. All 11 inclinometer borings were advanced to their planned 37-m (120-ft) depths, except for one that met refusal at 29.7 m (97.5 ft) and another that was terminated at 29.4 m (96.5 ft) where bedrock was encountered. Only one boring was too crooked for inclinometer installation; it was partially backfilled and completed as a piezometer.

Refuse samples were obtained using standard penetration test (ASTM D 1586-84), 7.62-cm (3-in.) outside-diameter (OD) ring-lined barrel (ASTM D 3550-84), and 13-cm (5-1/8-in.) ID acrylic tube-lined split-core barrel sampling techniques. Recovery was generally good. Several attempts to sample refuse using 7.62-cm (3-in.) Shelby tubes (ASTM D 1587-83) were unsuccessful because the tube crimped when it was pushed into refuse.

Saturated zones were detected during drilling by observing moisture in recovered soil samples and cuttings. Six of the boreholes remained dry during drilling and inclinometer installation. Shallower companion borings were drilled and completed as piezometers near inclinometer borings that encountered free fluid. Unsaturated material was found to underlie the saturated material. Saturated zones were less than 9 m (30 ft) thick. Saturated zones may have been substantially thinner than sample moisture indicated, as free liquid could have flowed from saturated zones down the annular space and wetted unsatura-ted material at the bottom of the borehole.

GEOTECHNICAL INSTRUMENTATION

Geotechnical instrumentation included 11 inclinometers, 5 piezometers, and 36 surface monuments. Figure 3 shows the locations of all geotechnical instrumentation.

Surface Movements

The location and elevation of each surface monument, inclinometer, and piezometer were monitored for 1 year using both ground and aerial methods. Aerial survey data consistently varied more than ground survey data, but the overall movement trends were essentially the same for both methods.

The lateral component of surface movement based on ground survey data ranged from about 1 to 23 cm (0.4 to 9 in.) over the 1-year monitoring period (Figure 5). Even though trends of generally consistent out-of-slope movement are apparent, the lateral movements do not necessarily indicate slope creep or incipient slope failure. This is because the lateral movement may be related to settlement rather than slope instability. Some locations show transverse and upslope movements. These may be influenced by localized deposits of more rapidly decomposing refuse in-slope or transverse from these monitored locations.

The instrumented location that moved laterally the most (SM-8) is on a relatively flat slope (approximately 3:1). This slope also

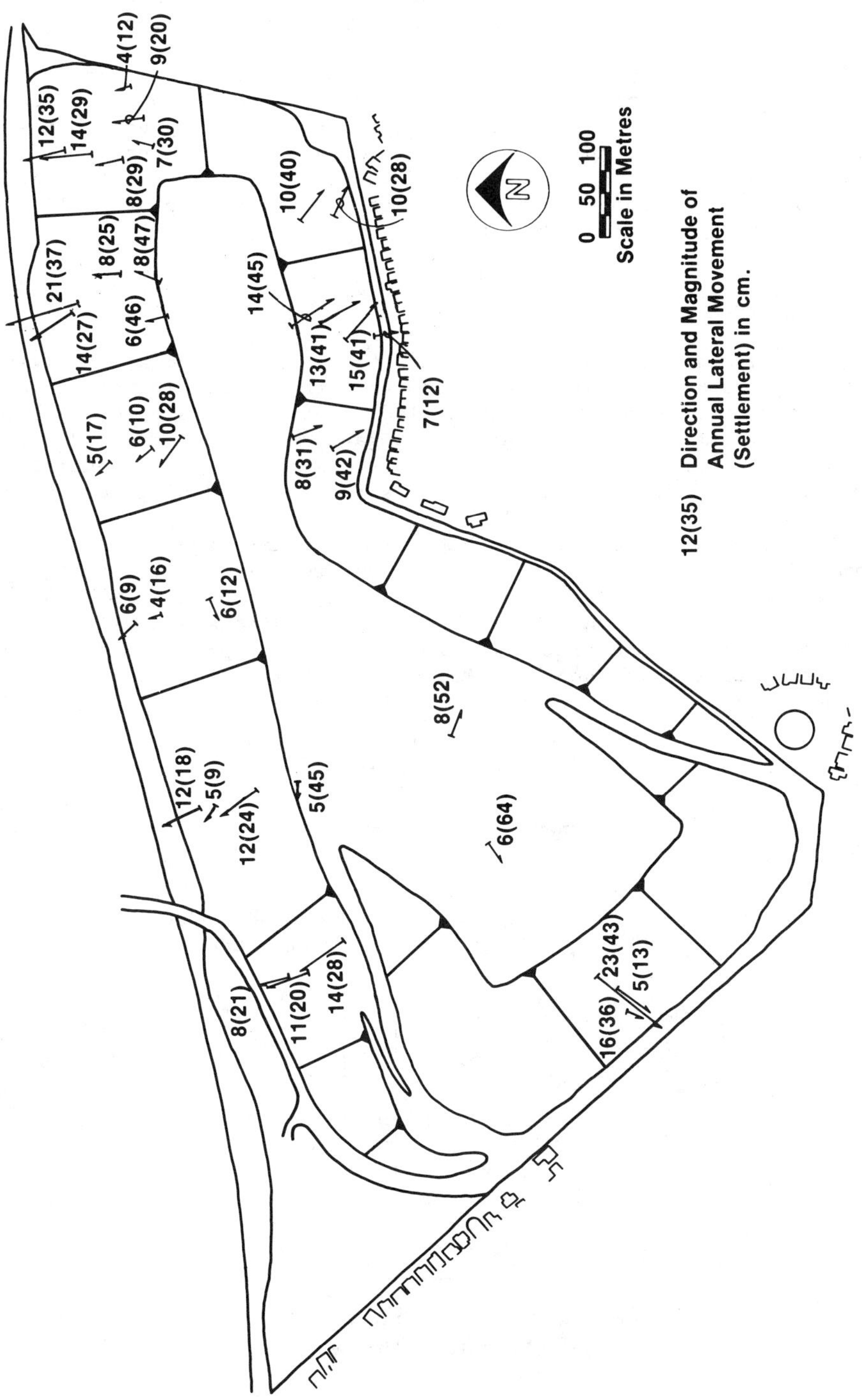

FIG. 5-- Annualized Surface Movements

consists of the youngest refuse. The north-central slopes, which are the oldest, steepest, and tallest, generally moved the least.

The vertical component of surface movements, based on ground survey data, ranged from approximately 9 to 64 cm (3.5 to 25 in.) per year (Figure 5). Generally, settlements of locations on top and on the south slopes of the landfill were the greatest. These settlement rates are significantly lower than rates of 90 to 120 cm (3 to 4 ft) per year determined for reference points on aerial photographs of the OII landfill from 1974 until 1983 [6].

Inclinometers

Six inclinometers were installed with Sondex (corrugated polyethylene tubing) sleeves; four were installed with telescoping sections; and one standard inclinometer was installed. The Sondex sleeves or telescoping sections were installed to mitigate downdrag forces and avoid buckling that might otherwise develop from landfill settlement; the standard inclinometer had no provisions for accommodating settlement.

Lateral Displacements: Over the 1-year monitoring period, the maximum lateral displacement that occurred relative to the bottom of each casing (only one extended into bedrock) was about 9 cm (3.5 in.). Displacements are generally out of slope, and displacement-depth profiles suggest slope creep (Figure 6). Several profiles also suggest slippage at particular depths. Movement of the standard inclinometer has been similar to that of the other inclinometers.

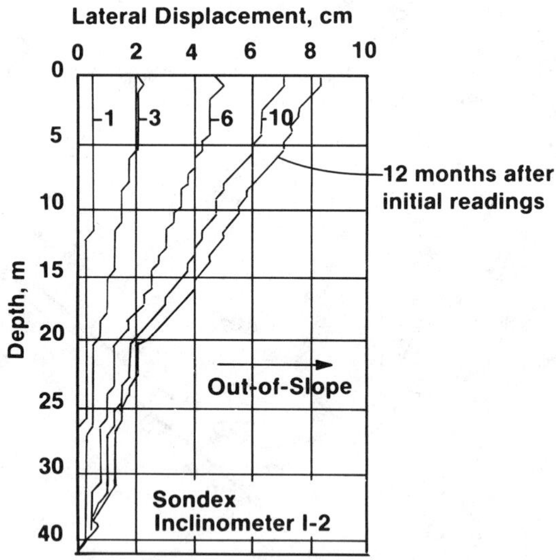

FIG. 6-- Example of Inclinometer Displacement Profile

Over 4 months, the upper 21.3 m (70 ft) of one inclinometer moved a couple of inches in-slope, rather than out-of-slope. The grout surrounding this inclinometer had subsided approximately a metre, which may have caused the casing to shift within the borehole. However, the in-slope movements may also have been caused by a localized landfill shift, which resulted from differential refuse decomposition.

All inclinometers, except one, were still functional 15 months after installation. The exception, a telescoping inclinometer, collapsed for an unknown reason, 2 m (6 ft) below the landfill surface 6 to 9 months after installation. The plastic casing of telescoping inclinometers softened and swelled, and their cross sections changed from circular to nearly square. This was apparently caused by solvents within the landfill. A few months after installation, the reaction appeared to stop, and the plastic hardened. The deformed cross section did not interfere with inclinometer monitoring, nor has it resulted in questionable readings. The casings of the standard and Sondex inclinometers did not appear to have been subjected to chemical attack. This probably resulted from the protection that the sealed joints and Sondex sleeves provided these casings.

Subsurface Settlements: The sleeved (Sondex) and telescoping inclinometers were monitored for vertical as well as lateral subsurface movement. Figure 7 shows the subsurface settlement for one inclinometer. Settlements are generally larger at the surface than at depth. Approximately 36 cm (14 in.) of settlement occurred at the landfill surface; 15 cm (6 in.), at the inclinometer's base. About 24 m (80 ft) of refuse is estimated to underlie the base of this inclinometer.

Subsurface settlement for sleeved inclinometers was monitored by determining changes in the elevation of metal rings attached to the sleeves at 3-m (10-ft) intervals. A Sondex probe lowered down the casing senses the location of the metal rings by electrical induction. The probe's electronic circuitry can be damaged by temperatures that exceed 50°C (120°F). To avoid damage from landfill temperatures that rise to 65°C (148°F), each casing was flushed with water to reduce the temperature before the ring depth was measured.

The depths of telescoping joints were measured with a latch hook attached to the end of a steel measuring tape. The tape was moved up and down until the hook latched onto the bottom edge of the upper slip joint of each telescoping section. Locating the telescoping sections was difficult using the latch-hook. This difficulty was probably aggravated by deformation, swelling, and softening of the casing.

Piezometers

Five piezometers were constructed using 5-cm (2-in.) OD, flush-jointed, Schedule 80 PVC pipe with 3- or 6-m (10- or 20-ft)-long screened sections at the bottom having 0.5-mm (0.02-in.) wide slots. Top caps are vented. All piezometers are packed with sand extending approximately a metre above the uppermost perforations of each screen. Bentonite pellet seals were placed above the sand. The remaining annular space was filled with a cement-bentonite grout.

Liquid levels that were measured using an electrical well sounder remained essentially constant throughout the year-long monitoring

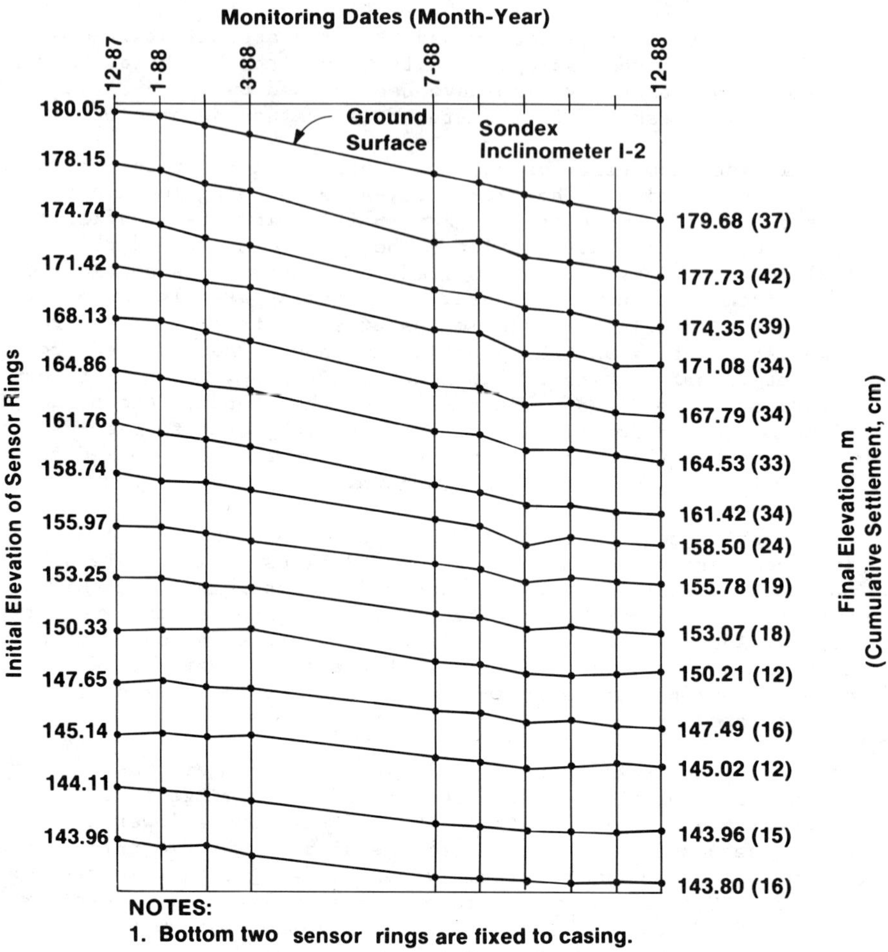

FIG. 7-- Example of Subsurface Settlement Versus Time

period, with the exception of two piezometers. The level in one piezometer dropped substantially throughout the monitoring period. Determining liquid levels in this piezometer was particularly difficult because foam, sludge, and other foreign matter in the casing may have influenced measurements. The liquid level in the second piezometer dropped 3 m (10 ft) during the first month of monitoring and then remained essentially constant. A drop in liquid level may be attributed to the puncturing of an underlying layer of relatively impervious soil or refuse.

LABORATORY TESTING

The 13-cm (5-1/8-in.) acrylic-tube samples of refuse from depths ranging from 4.6 to 25 m (15 to 82 ft) were selected for testing.

Testing [7] included determining direct shear strength, moisture content and unit weight, and percentages by weight of the different types of materials comprising each sample. Samples were x-rayed to identify characteristics such as voids, metal objects, density differences, and sample disturbances. Although the original plan was to select samples for direct shear testing based on the review of the x-rays, the samples were so heterogeneous that x-rays were not particularly helpful for sample selection.

Direct Shear Strength

The direct shear strength was determined in general accordance with ASTM D 3080-72. The tests were conducted under consolidated drained conditions on 7.6- to 10.2-cm (3- to 4-in.) high, 13-cm (5-1/8-in.) diameter specimens of refuse.

For all test specimens, the peak or maximum shear strength corresponded to shear displacements substantially exceeding 10 percent of the sample diameter; that is, 16 to 39 percent. Figure 8 shows shear stress, at 10 percent shear displacement, versus normal stress for each specimen of the five samples tested. Individual test specimens of each sample were dissimilar in both composition and behavior. Shear stress versus shear displacement curves were grossly different. Given the refuse variability, deriving Mohr-Coulomb angles of internal friction and cohesion intercepts for individual samples was deemed inappropriate.

Two interpretations of lower bound Mohr-Coulomb envelopes are shown in Figure 8. One interpretation considers all test specimens, and the other neglects specimens that contained relatively large soil percentages. Both interpretations simplistically assume no cohesion; interpretations with cohesion and smaller friction angles can be made as well.

The friction angle of 53 degrees is significantly higher than previously published for refuse. Landva, et.al. [8], performed several large-scale, 28- by 43-cm (11- by 17-in.), direct shear tests on refuse samples. Friction angles were found to vary from 24 to 42 degrees; cohesion varied from 16 to 23 kPa (335 to 480 psf). Stoll [9] performed triaxial shear strength tests on anisotropically consolidated specimens of 2-year-old milled domestic refuse. The effective friction angle was 44 degrees, and there was no cohesion.

Landva and Clark [10] found that refuse had lower strength when retested in direct shear after a year of decomposition. The friction angle that had originally ranged from 38 to 42 degrees had decreased to 33 degrees, and the cohesion intercept that had ranged from 16 to 19 kPa (330 to 400 psf) was 16 kPa (330 psf). Landva and Clark cautioned that more data would be required to confirm that strength loss occurs with refuse decomposition.

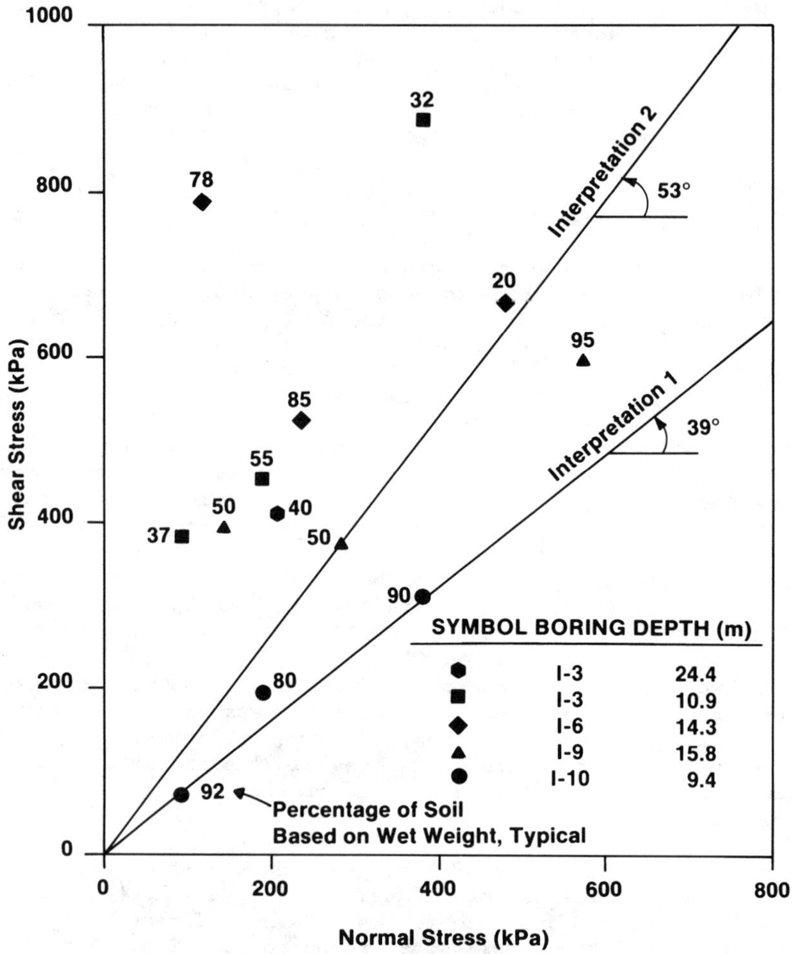

FIG. 8-- Direct Shear Test Results

Unit Weight and Moisture Content

Moisture contents were determined in accordance with ASTM D 2216-80. The samples were dried at 60°C (140°F) to avoid burning the organic content of the samples. Volumes for unit weight determinations were calculated based on specimen dimension measurements.

Moisture content of the refuse samples was found to range from 10 to 45 percent. In one sample, the moisture content varied from 13.5 to 44.5 percent within a distance of about 8 cm (3 in.). Dry unit weights ranged from 0.96 to 1.73 gm/cm^3 (60 to 108 pcf). Samples containing relatively high soil percentages had higher unit weights.

Classification

Percentages of different types of material in each sample were estimated by manually separating each sample into the following components and then weighing the components:

- Metal (0-11%)
- Wood (0-20%)
- Soil (20-95%)
- Paper (0-46%)
- Glass (0-5%)
- Rock and brick (0-15%)
- Rubber and plastic (0-35%)
- Miscellaneous (0-13%)

The individual components were weighed moist. The miscellaneous category included canvas, rags, cloth, and decomposed matter. Portions classified as soil also included decomposed material that could not be readily separated from soil. The samples contained a large percentage of soil (20 to 95 percent by moist weight). On a volume basis, the percentages of soil were substantially less. Large amounts of paper, wood, and rock were found in the samples with smaller amounts of metal, rubber, and glass.

SLOPE STABILITY ANALYSES

A slope stability study was performed to evaluate the effect of refuse shear strength on slope stability under both static and seismic (pseudostatic) conditions [2]. Analysis was limited to circular-arc shear surfaces that pass through the soil cover and into the underlying refuse and was based on the simplified Janbu method of slices [11]. The effect of refuse saturation was also evaluated.

Slope Geometry

The analysis was limited to Section B-B' (Figure 9) that is located in the north-central portion of the landfill (Figure 3). The overall slope height is approximately 76 m (250 ft), and the average slope steepness is 2:1. Intermediate slopes are as high as 23 m (75 ft) with a maximum steepness of 1.6:1.

Refuse and Bedrock Properties

Material properties for the refuse and underlying bedrock are summarized in Figure 9. Moist refuse unit weights are based on existing site data [12] and are supported by typical in-place values reported in the literature [10,13]. Bedrock properties are average values based on existing data obtained from various reports addressing bedrock properties at or near the site [14-19].

Results of Static Analyses

Figure 9 shows the critical shear surface (the lowest factor of safety), assuming an unsaturated slope and static loading, for several combinations of cohesion and friction angle. As cohesion increases,

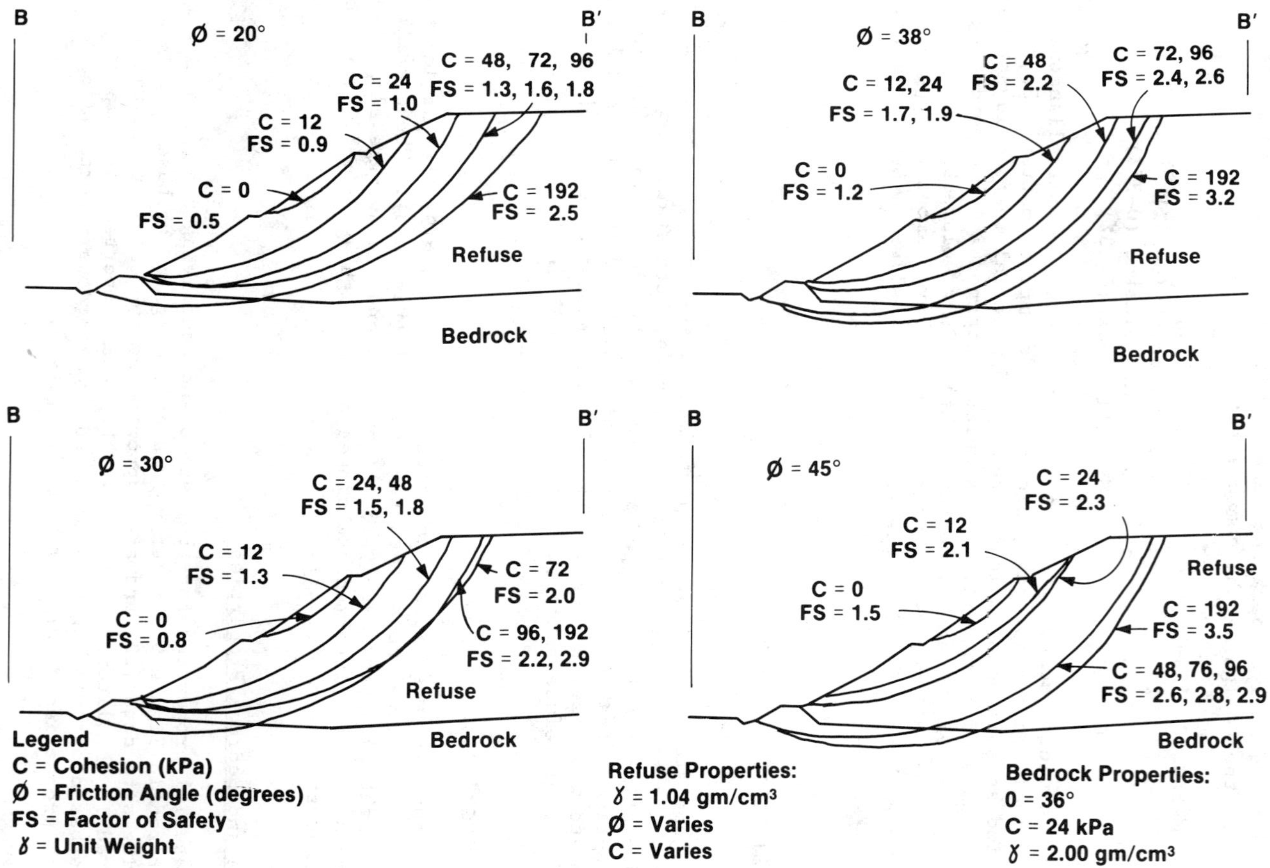

FIG. 9-- Critical Shear Surfaces for Unsaturated Condition and Static Loading for Section B-B′

the critical shear surface becomes deeper. For relatively high values of cohesion, the critical shear surface passes through bedrock. However, a slope failure involving bedrock is unlikely as the factors of safety corresponding to these deep shear surfaces are quite high.

Figure 10 shows the relationship between the factor of safety (FS) and cohesion for various friction angles for the slope under unsaturated conditions and static loading. This figure also shows several points representing refuse shear strength data reported in the literature (Points A, B, C, D, and E). Two interpretations of the results of direct shear tests performed on samples of OII landfill (Figure 8) are also shown (Points F and G). The data suggest a factor of safety greater than 1.2.

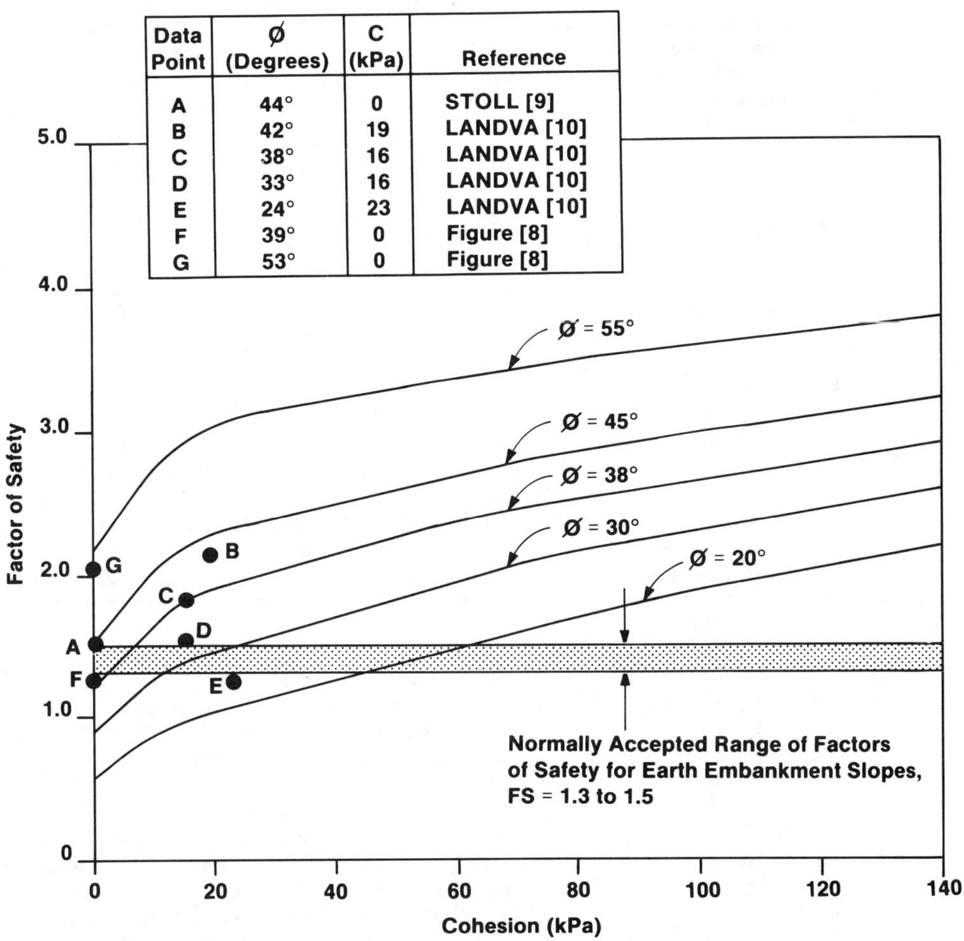

Data Point	Ø (Degrees)	C (kPa)	Reference
A	44°	0	STOLL [9]
B	42°	19	LANDVA [10]
C	38°	16	LANDVA [10]
D	33°	16	LANDVA [10]
E	24°	23	LANDVA [10]
F	39°	0	Figure [8]
G	53°	0	Figure [8]

FIG. 10-- Factor of Safety for Unsaturated Condition and Static Loading

Although saturated zones of liquids were identified within the landfill, unsaturated conditions were considered because, in general, the landfill is believed to be unsaturated. Saturated zones at the investigated locations are relatively few and thin and are of limited lateral extent compared to the overall mass of the landfill.

A perched, 9-m (30-ft)-thick saturated zone located 12.2 m (40 ft) below the top of the landfill was also considered. The saturated unit weight of refuse was assumed to be 1.44 gm/cm^3 (90 pcf). Results indicate a relatively small reduction in the factor of safety for static loading compared to that for the unsaturated condition. The reduction is less than 5 percent and is greatest for refuse without cohesion.

A saturated zone at the bottom of the landfill was also considered. This condition is suspected but has not yet been confirmed. The elevation of the top of this zone was assumed to be 7.6 m (25 ft) above the bottom of the landfill. Although results indicate somewhat larger reductions in the factor of safety for static loading compared to that for the perched saturated zone, the reduction is less than 10 percent for combinations of friction angle and cohesion resulting in factors of safety less than 2.

SEISMIC RESPONSE

The OII landfill is in an area of high seismicity. Figure 11 shows the epicenters of earthquakes with magnitudes greater than 4 that have occurred in the general Los Angeles area between 1932 and February 1989.

The San Andreas fault lies approximately 50 km (31 miles) northeast of the OII site and is believed capable of producing up to a magnitude (M) 8.5 earthquake. The Whittier fault zone is located nearest to the site, about 5 km (3 miles) southeast. It is believed to be capable of producing up to an M 7.0 earthquake. At the OII site, peak accelerations associated with these earthquakes are estimated to be from 0.25 to 0.50 g (g is the acceleration of gravity) for the San Andreas and Whittier faults, respectively [20].

During a seismic event, ground motions will propagate from the base of the landfill to its top. Accelerations could be amplified or attenuated depending on the duration and frequency content of the earthquake motion and the response to the ground motions of the materials comprising the landfill.

Ground motions in the landfill will result in additional vertical and horizontal forces in the slope due to the inertial response of the landfill material. If the inertial forces from the earthquake in combination with gravity-induced shearing stresses exceed the strength of the landfill material, the slope will deform. The deformation could vary from negligible to movements of tens of metres, depending on the earthquake characteristics and the dynamic strength of the landfill material.

Source: The Earth Technology Corporation [21]

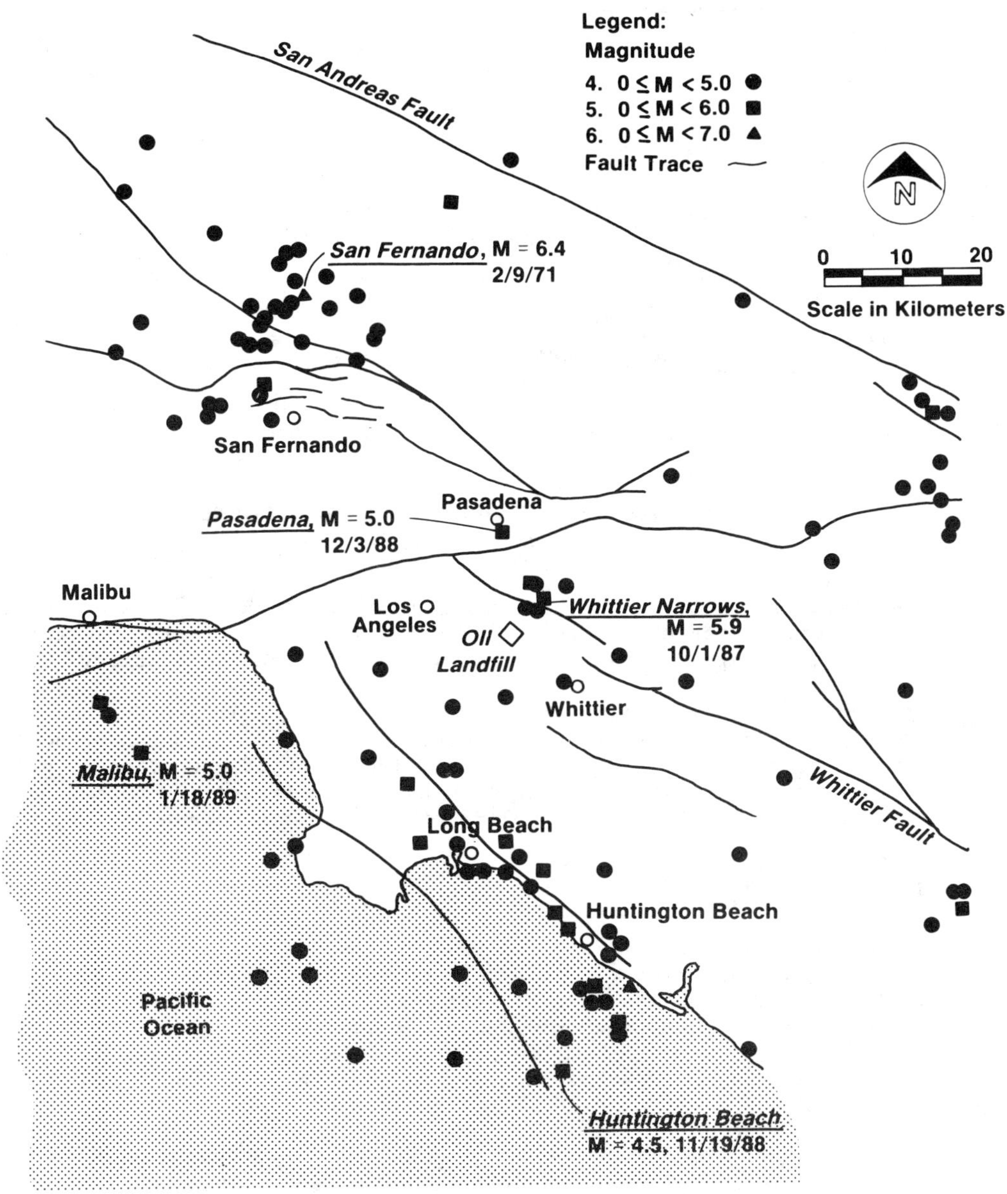

FIG. 11-- Seismicity in Los Angeles Region 1932 to February 1989

Seismic Response Monitoring

Two seismic stations were installed at OII in September 1988 to monitor the response of the landfill during earthquakes [21]. One of

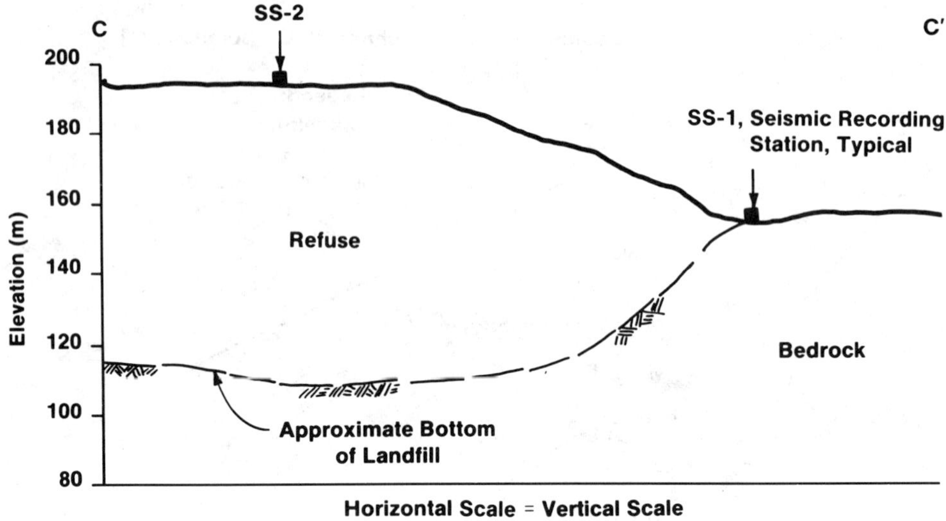

FIG. 12-- Seismic Recording Stations in Cross Section C-C′

the stations is located on bedrock east of the landfill; the other is located on the top of the landfill approximately 34 m (110 ft) from its east edge (Figure 3). Figure 12 shows the stations in section.

Each seismic recording station consists of a solid-state accelerograph. The accelerograph unit contains three force-balance accelerometers to monitor three orthogonal directions of vibration (longitudinal, transverse, and vertical). Each unit triggers independently when acceleration from any accelerometer exceeds a preset level.

Seismic Records for OII

Several earthquakes have occurred in the Los Angeles area since the the seismic monitoring stations were installed. Seismic records of the first three earthquakes that triggered one or both accelerographs at the OII landfill were analyzed and are discussed:

- Huntington Beach earthquake on November 19, 1988. This was an M 4.5 event with an epicenter located 60 km (37 miles) south of the landfill.

- Pasadena earthquake on December 3, 1988. This was an M 5.0 event with an epicenter located 15 km (9.3 miles) north of the landfill.

- Malibu earthquake on January 18, 1989. This was an M 5.0 event with an epicenter located 50 km (31 miles) west of the landfill.

The accelerograph at the base of the landfill triggered for all three earthquakes. However, acceleration levels at the top of the landfill were too low during the Huntington Beach event to trigger the accelerograph. The trigger level was lowered following the Huntington Beach event to improve the chances of recording data at the top of the landfill. Since then, accelerations for both the Pasadena and Malibu earthquakes were recorded.

Huntington Beach Earthquake: The Huntington Beach earthquake resulted in a peak acceleration of 0.01 g at the base of the landfill. Results of Fourier Analyses indicate that most of the energy of this earthquake was between 3 to 4 Hz. As noted above, the accelerograph at the top of the landfill failed to trigger. The trigger level of the unit at the top of the landfill was 0.008 g. This indicates that by the time the acceleration propagated to the top of the landfill, the acceleration had attenuated to less than 0.008 g. This attenuation is consistent with results of an ambient vibration survey [22] that also suggests attenuation of energy with frequencies in excess of 3 Hz.

Pasadena Earthquake: The Pasadena earthquake caused a longitudinal peak acceleration of 0.22 g at the base of the landfill. At the top of the landfill, the longitudinal peak acceleration had decreased to 0.10 g. Similar results were recorded in the transverse and vertical directions, although the peak amplitudes were less at the base (0.14 g for transverse and 0.10 g for vertical) and at the top of the landfill (0.11 g for transverse and 0.07 g for vertical).

The peak acceleration attenuated because of the large amount of energy at frequencies in excess of 2 Hz. Amplification occurred only at frequencies of less than 2 Hz. However, the amount of energy at less than 2 Hz was small compared to the overall frequency content, which resulted in a net attenuation of the input motion. Strong peaks in the transfer function plots occurred at frequencies of about 0.8 to 1 Hz. This suggests that the fundamental modes of vibration in the longitudinal, transverse, and vertical directions of the landfill are in this range.

Malibu Earthquake: Accelerations associated with the Malibu event were very small, typically ranging from 0.01 to 0.011 g at the base and the top of the landfill, respectively. A change in frequency content occurred between the base and the top. Most energy at the base was in the frequency range of 2 to 4 Hz; at the top, the frequency range decreased to 0.6 to 2.7 Hz. Although the energy attenuation above 3 Hz and amplification below 3 Hz resulted in little change in peak acceleration, the predominant frequency changed.

Response During Recent Earthquakes

Important information about the performance of the OII landfill can also be obtained by observing whether or not the landfill underwent permanent deformation during previous earthquakes. By correlating observed response to the level of acceleration within the slope, information about refuse properties during a seismic event can be deduced.

As part of an earlier seismic stability study of the OII landfill, Woodward-Clyde Consultants [20] reviewed the performance of the OII landfill after the 1971 M 6.4 San Fernando earthquake. The OII landfill was 39 km (24 miles) from the epicenter of this earthquake. At that time the footprint of the landfill was smaller in plan, and the slopes were about 9 m (30 ft) shorter than their present height. The firm-ground acceleration at the base of the landfill was estimated to be about 0.09 g. Under this loading, it was reported that three parallel cracks, up to 0.6 m (2 ft) wide developed on the east side of the landfill during the earthquake. The Woodward-Clyde report expressed surprise over the damage, given the low acceleration level, and stated further that the cracks may have existed before the earthquake.

Several earthquakes have occurred in the Los Angeles area since the landfill stopped accepting waste in 1984. The most important event was the Whittier Narrows earthquake, which occurred on October 1, 1987. The magnitude for this event was 5.9, and its epicenter [23] was about 3.7 km (2.3 miles) northeast of the center of the landfill (Figure 1). This earthquake was followed by significant aftershocks on October 4, 1987, (M 5.5) and on February 11, 1988, (M 4.7).

Unfortunately, there were no onsite seismic monitoring stations at the time of the Whittier Narrows earthquake and its aftershocks. The acceleration level on firm ground associated with the main shock of the Whittier Narrows earthquake is estimated to be as high as 0.47 g [24], based on acceleration measurements recorded at Garvey Reservoir located 3.1 km (1.9 miles) from the epicenter (Figure 1). Because geological and topographic conditions at the OII Landfill and the Garvey Reservoir sites, as well as the distances from the epicenter for the two locations, are similar, the base of the landfill probably experienced shaking similar to that recorded at Garvey Reservoir.

A field reconnaissance of the South Parcel in the afternoon after the main Whittier Narrows event identified several areas of ground cracking [1]. The most significant surface cracking was observed on the north-central and northeastern portions of the landfill along the lower bench, along the southeastern corner of the top bench of the landfill, and along the southeastern area along the middle bench. Cracks were typically 2.5 to 8 cm (1 to 3 in.) wide and up to 90 m (300 ft) long and ran transverse to the slopes. No additional distress was observed after either of the two aftershocks.

Only one area of open-ground cracking was apparently caused by the Pasadena earthquake in December of 1988. Cracks were on the northeastern lower bench, in the same vicinity as previous cracking from the Whittier Narrows earthquake. The cracks were typically 1 to 2 cm (1/2 to 1 in.) wide, 5 to 10 cm (2 to 4 in.) deep, and 45 m (150 ft) long. The western edge or beginning of the cracking displayed a 15- to 30-cm (6- to 12-in.)-wide, 30- to 60-cm (1- to 2-ft)-deep, 3-m (10-ft)-long crack. No cracking was observed after the other recent earthquakes.

Preliminary Interpretations

The results of the landfill monitoring program can be used with the results of pseudostatic slope stability analyses to make some preliminary observations about the static strength properties of landfill materials. Yield acceleration (horizontal acceleration at which the factor of safety equals unity) was determined for different combinations of refuse cohesion and friction angle.

Figure 13 shows yield acceleration as a function of the friction angle and the cohesion intercept used in the pseudostatic slope stability analyses. This figure also shows peak ground accelerations measured at the base and the top of the landfill for the Pasadena earthquake, as well as the base acceleration for the Whittier Narrows earthquake at the Garvey Reservoir.

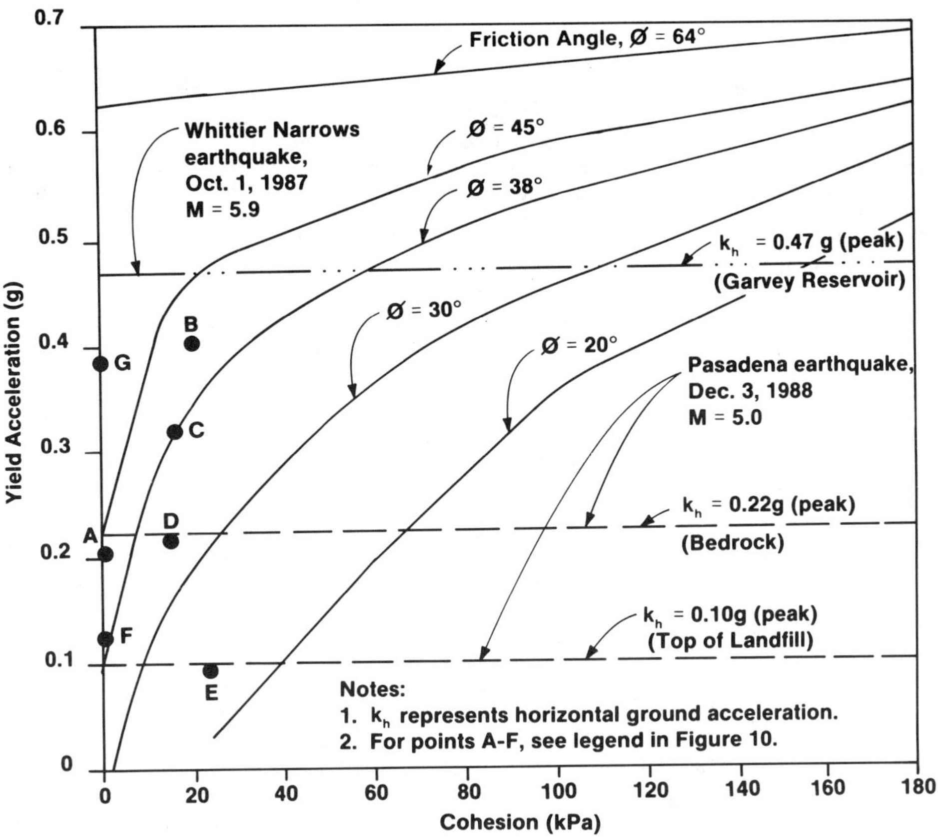

FIG. 13-- Yield Acceleration for Unsaturated Condition and Seismic (Pseudostatic) Loading

For both earthquakes, either minor or no slope movement was observed during postearthquake damage surveys. This suggests that factors of safety during these events were likely at or above 1.0. By assuming that the factor of safety was 1.0, it can be concluded that the landfill strength properties had to be equal to or greater than the intercepts with the horizontal acceleration values for each record.

For example, during the Pasadena event, accelerations in the landfill were between 0.10 and 0.22 g. If the average acceleration is conservatively assumed to be 0.10 g, then the strength of the landfill material could have been one of the following or a number of similar combinations:

Friction Angle (degrees)	Cohesion Intercept (kPa/psf)
38	0
30	10/200
20	40/800

These combinations of refuse strengths for the analyzed slope that sustained a yield acceleration of 0.1 g suggest a factor of safety greater than 1.2 for unsaturated static slopes (Figure 10).

These interpretations are believed to be conservative from the standpoint that Pasadena accelerations plotted in Figure 13 are the peak values recorded at the top and bottom of the landfill. These values are instantaneous occurrences. A single peak equal to the yield will normally cause very little, if any, deformation. For noticeable deformations to occur, the peak usually has to be at least twice the yield acceleration [25]. This suggests that, in the absence of any noticeable deformation on most slopes, higher strength properties than given above must have existed within the landfill.

CONCLUSIONS

In general, other than identifying cover soil types, CPTs were not particularly helpful for investigating the OII landfill. However, CPTs might still be useful to identify relatively weak zones within the landfill, or to qualitatively evaluate whether refuse shear strength changes over time.

Direct shear tests results should be used with caution. In general, the direct shear testing device used in the OII testing program was not well suited for the large shear displacements that were required to fully mobilize the shear strength of refuse. Also, large pieces of relatively strong refuse may have become wedged between the shear boxes and produced artificially high shearing resistance. Finally, sample sizes tested were significantly smaller than many particles of in situ refuse.

With respect to the stability of OII landfill slopes, the following conclusions can be made:

- Stability analyses for unsaturated slopes suggest that for likely combinations of friction angle and cohesion, the factor of safety of the slope section analyzed is greater than 1.2. However, in view of the uncertainties associated with determining refuse strength and the potential for refuse strength to change with time, additional studies are required to support this conclusion.

- For earth embankment design, the minimum acceptable factor of safety typically ranges from 1.3 to 1.5 for static conditions, which suggests that OII slopes may be marginally acceptable.

However, a factor of safety that is reasonable for earth slopes eventually may not prove reasonable for slopes composed of refuse. Observed slope movements, if due to creep, also suggest marginally stable conditions.

- Results of pseudostatic slope stability analyses determined that yield accelerations (acceleration at which the factor of safety is 1.0) will range widely depending on strength properties of refuse. During the 1987 Whittier Narrows earthquake and the 1988 Pasadena earthquake, slopes withstood accelerations in excess of 0.1 g with little or no slope movement. This suggests that the yield acceleration was at least 0.1 g and combinations of shear strength parameters correspond to a static factor of safety greater than 1.2.

The following conclusions can be drawn from interpreting the earthquake records:

- Peak accelerations attenuated or remained the same for the two earthquakes that triggered both the top and base seismic recording units. During the Pasadena earthquake, the base acceleration in the longitudinal direction decreased from 0.22 g at the base to 0.10 g at the top of the landfill. Lower peak accelerations were recorded in the transverse and vertical directions. The amount of attenuation in these directions was also less. Negligible change was recorded from the base to the top during the Malibu earthquake.

- The input motions at the base of the landfill for the three records included a significant amount of energy above 2 to 3 Hz. Results of transfer function studies of these records suggest that energy above 3 Hz will attenuate as it propagates to the top of the landfill.

- Strong amplification of energy occurred at frequencies less than 2 to 3 Hz as the motion at the base of the landfill propagated to the top. The three recorded motions had low amounts of energy in this frequency range, which resulted in a net attenuation of the ground motion.

- Little to no damage was observed after the three recorded earthquakes. Only minor damage was observed after the Whittier Narrows earthquake that occurred before the seismic stations were installed. These results suggest that the landfill can withstand moderate earthquakes with only need of minor repair following the earthquake.

For final evaluation of the static and dynamic stability of the OII landfill slopes, the shear strength of refuse must be known with more certainty. To improve present knowledge of the shear strength of refuse, large-scale laboratory and/or in situ shear tests on refuse are recommended. Additional seismic response data should also be collected.

ACKNOWLEDGEMENTS

The work discussed in this paper was performed by CH2M HILL for the United States Environmental Protection Agency (EPA) as part of the Phase 2 Remedial Investigation/ Feasibility Study of the OII landfill site. Although the work was funded by the EPA, this paper has not been subject to the Agency's review and, therefore, does not necessarily reflect the Agency's views, and no official endorsement should be inferred.

The Earth Technology Corporation (TETC) of Huntington Beach and Long Beach, California, under subcontract to CH2M HILL, provided cone penetration testing, laboratory testing, and seismic response monitoring services. TETC also performed the analysis of ambient vibration survey and earthquake records.

REFERENCES

[1] U.S. Environmental Protection Agency, Agency Review Draft Technical Memorandum, Data for the Geotechnical Site Assessment, RI2/FS South Parcel, Operating Industries, Inc., Monterey Park, California, prepared by CH2M HILL, Santa Ana, California, February 6, 1989.

[2] U.S. Environmental Protection Agency, Agency Review Draft Technical Memorandum, Slope Stability and Seismic Response Studies, RI2/FS South Parcel, Operating Industries, Inc., Monterey Park, California, prepared by CH2M HILL, Santa Ana, California, March 17, 1989.

[3] The Earth Technology Corporation, Cone Penetration Tests, Operating Industries, Inc., South Parcel Landfill, Monterey Park, California (report prepared for Layne Environmental Services of Fontana, California), Long Beach, California, December 1987.

[4] American Society of Testing and Materials (ASTM), Annual Book of ASTM Standards, Volume 4.08 Soil and Rock; Building Stones; Geotextiles, Section 4, Construction, Philadelphia, 1987.

[5] Douglas, B. J. and Olsen, R. S., "Soil Classification Using Electric Cone Penetrometer," Symposium on Cone Penetration Testing and Experience, Geotechnical Engineering Division, American Society of Civil Engineers, St. Louis, Missouri, October 1981, pp. 209-227.

[6] Converse Consultants, Seismic and Slope Stability, Operating Industries Landfill, Monterey Park, California (report prepared for Operating Industries Inc.), Pasadena, California, December 12, 1985.

[7] The Earth Technology Corporation, Report, Laboratory Testing of Soils and Refuse, Operating Industries Landfill (prepared for CH2M HILL, Santa Ana, California), Long Beach, California, December 19, 1988.

[8] Landva, A. O., Clark, J. I., Weisner, W. R., and Burwash, W. J., "Geotechnical Engineering and Refuse Landfills," Sixth National Conference on Waste Management in Canada, Vancouver, British Columbia, 1984.

[9] Stoll, U. W., "Mechanical Properties of Milled Domestic Trash," American Society of Civil Engineers," presented at the National Water Resources Engineering Meeting, Phoenix, Arizona, January 1971.

[10] Landva, A. O. and Clark, J. I., Geotechnical Testing of Wastefill, Report submitted to the London Institution of Civil Engineers, 1987.

[11] Janbu, N., "Slope Stability Computations," Embankment-Dam Engineering, Casagrande Volume, Hirschfeld, R. C. and Poulos, S. J., Editors, John Wiley and Sons, New York, 1972.

[12] Converse Davis Dixon Associates, Slope Stability Investigation, Proposed Final Slope Adjacent to the Pomona Freeway, Operating Industries Disposal Site, Monterey Park, California (report prepared for Operating Industries, Inc.), Pasadena, California, October 15, 1975.

[13] Sowers, G. F., "Foundation Problems in Sanitary Landfills," Proceedings of the American Society of Civil Engineers, Journal of the Sanitary Engineering Division, Vol. 94, No. SA1, New York, February, 1973.

[14] Converse Davis Dixon Associates, Supplemental Geotechnical Studies, Operating Industries Disposal Site, Monterey Park, California (report prepared for Lockman and Associates, Monterey Park, California), Pasadena, California, July 15, 1976.

[15] Gorian and Associates, Soils Engineering Investigation of Monterey Park Facility Nu FUEL Co. (report prepared for the NGR Nu FUEL Company, Newport Beach, California), Westlake Village, California, February 10, 1975.

[16] Leighton-Yen & Associates, Geotechnical Investigation of Stability of Cut-Slope on Southwestern Perimeter of Cut and Cover Dump Site, Monterey Park, California (report prepared for Operating Industries, Monterey Park, California), La Habra, California, August 31, 1973.

[17] Woodward-Clyde Consultants, Geotechnical Investigation, Montebello Reservoir, Montebello, California (report prepared for the City of Montebello, California), Orange, California, January 22, 1979.

[18] Pacific Soils Engineering, Preliminary Soils Engineering and Geologic Investigation, Proposed Gas Recovery Facility, City of Monterey Park, California (report prepared for Lockman and Associates, Monterey Park, California), Harbor City, California, October 5, 1978.

[19] Medall, Aragon, Worswick & Associates, Geotechnical Investigation Report, Proposed Commercial Development, Northeasterly of Greenwood Avenue - Pomona Freeway Grade Separation, Monterey Park, California (prepared for the Bennett Development Company, Newport Beach, California), Los Angeles, California, September 26, 1983.

[20] U.S. Environmental Protection Agency, Seismic Stability Evaluation, Operating Industries, Inc. Landfill, Monterey Park, California, Report prepared by Woodward-Clyde Consultants, Santa Ana, California, May 27, 1987.

[21] The Earth Technology Corporation, Draft Landfill Response to Seismic Events, Seismic Evaluation Study, Operating Industries, Inc. (report prepared for CH2M HILL, Santa Ana, California), Long Beach, California, March, 1989.

[22] The Earth Technology Corporation, Ambient Vibration Survey Seismic Evaluation Study, Operating Industries, Inc. (report prepared for CH2M HILL, Santa Ana, California) Long Beach, California, February 13, 1989.

[23] Jones, L. and Hauksson, E., "The Whittier Narrows, California Earthquake of October 1, 1987 - Seismology," Earthquake Spectra, Vol. 4, No. 1, Earthquake Engineering Research Institute, El Cerrito, California, 1988.

[24] Brady, A. G., Etheredge, E. C., and Porcella, R. L., "The Whittier Narrows, California Earthquake of October 1, 1987, Preliminary Assessment of Strong Ground Motion Records," Earthquake Spectra, Vol. 4, No. 1, Earthquake Engineering Research Institute, El Cerrito, California, 1988.

[25] Makdisi, F. I. and Seed, B. H., A Simplified Procedure for Estimating Earthquake-Induced Deformations in Dams and Embankments, Report No. UCB/EERC-77/19, Earthquake Engineering Research Center, University of California, Berkeley, California, August 1977.

Gregory E. Tieman, Gregory W. Druback, Kenneth A. Davis, and Charles H. Weidner

STABILITY CONSIDERATION OF VERTICAL LANDFILL EXPANSIONS

REFERENCE: Tieman, G.E., Druback, G.W., Davis, K.A., and Weidner, C.H., "Stability of Vertical Piggyback Landfill Expansions", Geotechnics of Waste Fills - Theory and Practice, ASTM STP 1070, Arvid Landva, G. David Knowles, eds., American Society for Testing and Materials, Philadelphia, 1990.

ABSTRACT: It should be expected that vertical expansions of ash residue and municipal solid waste landfills in the future will be required to demonstrate an acceptable level of structural stability in addition to meeting other federal and state regulations. In particular, the liner systems of the vertical piggyback expansions should not suffer damage due to settlement of the underlying existing waste material or due to side slope instability.

In May 1987, prior to the development of any New York state or federal regulations addressing the issue of structural stability relative to vertical piggyback expansions, separate ash residue and municipal solid waste vertical expansions were designed by Malcolm Pirnie, Inc. for the Town of Islip, New York. A fundamental design consideration was the evaluation of long-term piggyback structural stability. Construction of both piggyback expansions began in September 1987, with substantial completion in August 1988.

The design of both piggybacks evaluated structural stability in the following modes: (1) failure between the various natural and geosynthetic materials comprising the piggyback liner system, (2) deep seated circular failure down through the underlying existing waste material and subgrade and (3) subgrade reinforcement with polyethylene geogrids. Malcolm Pirnie's analyses indicated that several geotechnical parameters were of critical concern for the stability of the Town of Islip's vertical piggyback expansions: (1) bottom liner and existing side slope grades, (2) interface friction angles, (3) type and orientation of the geosynthetics relative

Gregory Tieman is a Project Engineer, Gregory Druback, P.E. is a Senior Associate, and Kenneth Davis, P.E. is a Senior Project Engineer with Malcolm Pirnie, Inc., 2 Corporate Park Drive, White Plains, NY 10602; Charles Weidner, P.E. is Executive Vice President of the Islip Resource Recovery Agency, 40 Nassau Avenue, Islip, NY 11751.

to grade, (4) depth of leachate in the drainage layer above the piggyback liner system, and (5) utilization of geogrids for subgrade reinforcement.

The stability analyses which assessed design conditions (1) through (4) utilized the computer program STABL, written at Purdue University for the general solution of slope stability problems. Calculation of factor of safety values against incipient instability of the slope was performed for the sliding block and deep-seated circular failure analysis, using the Modified Bishop's and Simplified Janbu methods, respectively. The results of the structural analyses indicated that a combination of geogrid reinforcement and appropriate design would yield vertical piggyback expansions which would have acceptable structural stabilities over their operational and post-closure lifetimes.

KEYWORDS: piggyback, structural stability, uniaxial geogrid, textured HDPE liner, factor of safety

INTRODUCTION

The Town of Islip, New York, encompasses 285 square kilometers (110 square miles) along the southern shore of Suffolk County, Long Island. The Town operates recycling and yard waste and leaf composting programs and has placed into operation a 363 tonnes per day (400 ton per day) resource recovery facility. Although these facilities are in place, the Town has continued to rely upon the Blydenburgh Landfill, for municipal solid waste (MSW) disposal. The success of this integrated system has allowed the Town to process and dispose of its estimated 290 kilotonnes per year (320,000 tons per year) of MSW in the Town without having to contract for off-Long Island hauling and disposal.

In the mid 1980's, the viability of this integrated program was in jeopardy. Passage of the Long Island Landfill Law in 1983, which effectively banned landfilling of raw garbage on most of Long Island after December 18, 1990, precluded the economic development of new landfill capacity in the Town. Furthermore, Blydenburgh Landfill was projected to reach capacity in early 1987 and close. To conserve the landfill's dwindling capacity for residential use, the Town prohibited commercial haulers from the landfill in the fall of 1986. In response, on March 22, 1987, the Mobro barge departed Long Island City, New York loaded with commercial MSW which could no longer be landfilled at the Blydenburgh site.

Partly in response to significant negative publicity surrounding the Mobro barge and the continuing need to provide for waste disposal, the New York State Department of Environmental Conservation (NYSDEC) and the Town's Resource Recovery Agency entered into an Order on Consent on May 12, 1987. This allowed for continued operations and an 81,000 square meter (20.0 acre) vertical MSW "piggyback" expansion on top of a closed and capped portion of the existing 181,000 square meter (44.8 acre) landfill mound. In addition, the Order on Consent permitted construction of a separate 12,000 square meter (3.0 acre) ash residue vertical piggyback expansion adjacent to the MSW piggyback expansion. During design and construction of the vertical piggybacks, residential MSW disposal continued on the remaining 61,000 square meter (15.1 acre) portion of the landfill mound and commercial MSW disposal recommenced.

Both expansions have been designed for and constructed on top of existing landfilled MSW. Reflecting the similarity in their design, and construction, this paper will concentrate on the MSW vertical piggyback expansion. Design and construction of the vertical MSW piggyback expansion represented a unique geotechnical challenge, due to its area, need to protect existing landfill environmental systems, thickness of existing refuse, and the need to construct the expansion on top of side slopes as steep as 2.5:1 (horizontal to vertical).

Substantial completion of construction in August 1988 gave the Town an estimated 841,000 cubic meters (1,100,000 cubic yards) of additional MSW capacity and an estimated 57,415 cubic meters (75,100 cubic yards) of ash residue capacity. Based on waste inflow rates, this additional MSW and ash residue capacity should be more than sufficient to allow landfilling to continue until both landfills' required closures on December 18, 1990.

EXISTING SITE CONDITIONS

The Blydenburgh Landfill is located in an area characterized by hilly and irregular topography typical of a glacial terminal moraine. The land surface elevation at the southerly end of the site is approximately 58 meters msl (140 feet) and dips toward the north, where the elevation is approximately 24 meters msl (82 feet). The surface soils within the landfill area prior to excavation and landfilling activities were primarily the Plymouth loamy sand and the Riverhead sandy loam. In this area the subsoils appear to be composed predominantly of coarse sand to fine gravel, with lenses of finer material. These sandy soils have been used extensively as cover soil over the refuse and for sand drainage layers in the on-site landfills.

Landfilling commenced on the site in the 1950's. As of May 1987, approximately 120,000 square meters (29.7 acres) of the landfill had been closed to a peak elevation of 76 meters msl (250 feet). This portion had been capped with a 20 mil polyvinyl chloride (PVC) flexible membrane liner, 0.61 meters (2.0 feet) of sand, and 0.15 meters (6 inches) of topsoil. Much of the topsoil had eroded off the slope and as part of piggyback construction, the remaining portion was removed. The landfill had a thickness of as much as 58 meters (190 feet) of MSW by the time this portion of the landfill was capped. Side slopes of the capped portion of the landfill ranged from 2.5:1 to 9:1. Active landfill operations were occurring on a 61,000 square meter (15.1 acre) section of the existing mound but would cease in 1987 at the 72 meter msl (250 foot) elevation. The 61,000 square meter (15.1 acre) section was underlain by a bottom 30 mil PVC liner, but the remaining 120,000 square meter (29.7 acre) closed and capped portion of the landfill was unlined.

The landfill has historically operated a perimeter landfill gas collection system to inhibit the lateral migration of gas off-site. This system consists of a series of vertical collection wells utilizing negative pressure, header piping, and two flares. Prior to construction of the vertical MSW piggyback, vertical gas wells had been installed under another contract through the 20 mil PVC cap into the underlying waste. As part of the vertical MSW piggyback design, the integrity of these wells had to be maintained. In the fall of 1989, the Town began generating electricity from the gas collected by these wells beneath the 20 mil PVC cap; while perimeter gases continued to be flared. Storm water runoff is collected in a series of perimeter drainage swales and directed to the north end of the site with discharge into an existing ground water recharge basin. Storm water recharge is a general construction and development requirement of Suffolk County for the artificial recharge of the underlying Magothy Aquifer.

PIGGYBACK DESIGN

The MSW piggyback was designed for and constructed on top of a 81,000 square meter (20.0 acre) unlined portion of the Blydenburgh Landfill. It was projected that the final height of the entire 181,000 square meter (44.8 acre) landfill would increase by an additional 15.1 meters (50 feet) to elevation 91 meters msl (300 feet). The general design concept for both of the vertical expansions was developed by the White Plains, New York office of Malcolm Pirnie, with review and input by the NYSDEC and the Town of Islip's engineering staff. The primary components of the MSW piggyback design are:

- Polyethylene (PE) geogrids to provide subgrade support;

- Geomembrane liner system utilizing the existing 20 mil PVC cap with an upgraded leachate collection system and an overlying 80 mil high density polyethylene (HDPE) liner with a leachate collection system;

- A PE drainage net overlying the HDPE liner to provide lateral leachate drainage; and

- Gravity flow leachate collection system utilizing perforated PE piping inside the perimeter of the piggyback to collect leachate. Solid piping penetrates perimeter berms, intersecting a header PE pipe in junction manholes, and then to a pump station. MSW leachate is then pumped into one of three 833 cubic meter (220,000 gallon) storage tanks.

The overall geomembrane liner system for the MSW piggyback expansion consists of two separate flexible membrane liner systems. But it should be noted that in 1987, the NYSDEC was only requiring a single liner system for the piggyback. The existing 20 mil PVC

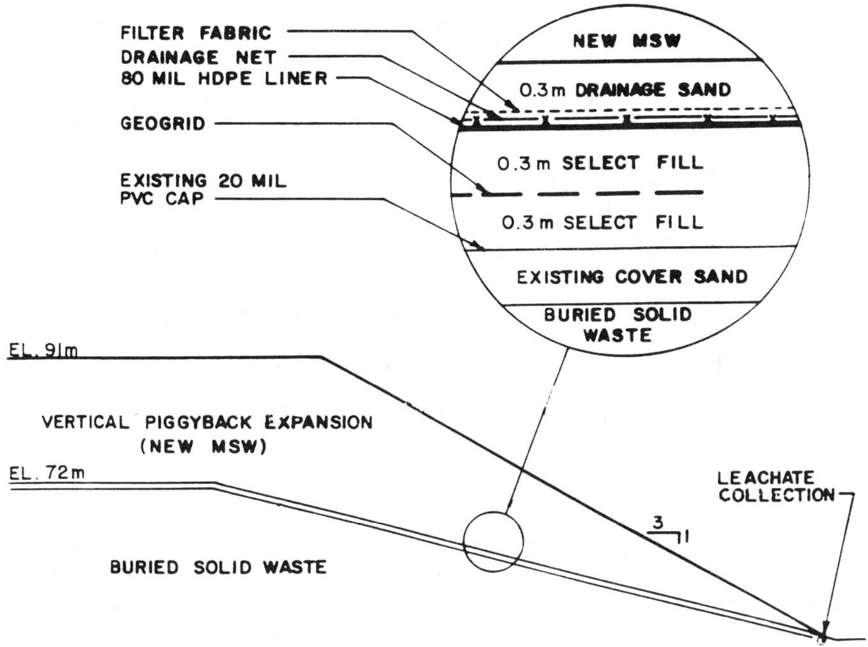

FIGURE 1. Liner System Design.

cap was upgraded as part of the construction of the MSW piggyback's HDPE liner system to operate in a sense as a lower liner system. A minimum thickness of 0.15 meters (6 inches) of sand was maintained on top of the 20 mil PVC cap during construction. A perimeter HDPE leachate collection piping system was constructed to collect any leachate which could collect in this sand layer above the 20 mil PVC. The intent of this upgrade was not to collect leachate which might migrate down from the overlying piggyback liner, but instead to collect mounded leachate which historically had seeped from beneath and through the 20 mil PVC cap out onto the side slope surfaces.

Once the 20 mil PVC cap upgrade had been completed, the MSW piggyback's single geomembrane liner system was constructed. The intent of this liner system was to hydraulically isolate the vertical piggyback from the underlying capped landfill and ground and surface water. To accomplish this goal, a 80 mil HDPE synthetic liner with leachate collection and transmission was designed and constructed. This liner system consisted of the following material, from top to bottom, as shown in Figure 1:

- Design thickness of 0.30 meters (12 inches) of compacted, select on-site sand for liner protection; 0.38 to 0.45 meters (15 to 18 inches) actually placed by the Contractor. During operations there have been no reported incidents of mechanical damage to the liner system from equipment operations or refuse placement;

- 0.30 kilograms/square meter (8 ounce/square yard) nonwoven, needle-punched polypropylene (PP) filter fabric;

- PE drainage net to provide primary leachate drainage off the liner. Calculations indicate that the PE drainage net, when placed on a 3:1 slope, has a transmissivity equivalent to 12.9 meters (42.3 feet) of 5.0 X E-5 meters per second on-site drainage sand.

- Smooth and textured 80 mil HDPE flexible membrane liner;

- 0.30 meters (12 inches) of compacted, select fill;

- Two PE, uniaxial geogrid reinforcement layers; and

- 0.30 meters (12 inches) of compacted, select fill.

SUBGRADE STABILITY AND GEOGRIDS

A major consideration in the design of a piggyback type landfill expansion is the survival of the landfill's environmental protection systems. Specifically, the piggyback's geomembrane liner system must be able to accept potential settlement of the refuse in the lower, existing capped landfill. This includes maintaining the integrity of the piggyback's leachate collection system (whether drainage net or piping), filter fabric, and sand drainage layers. Furthermore, the expansion should not interfere with other peripheral systems such as gas collection, storm water management, ground water monitoring, utilities, roads, etc. Peripheral systems represent a unique economic opportunity. The piggyback should be able to utilize these systems with a resultant capital cost savings. Realizing this savings is dependent upon a stable design. Therefore, an engineering analysis was performed prior to piggyback construction at the Blydenburgh Landfill to design for the structural integrity of the piggyback. Structural integrity was assessed via settlement and side slope stability analysis.

The analysis of potential settlement in the refuse underlying the 20 mil PVC cap was broken down into three components. First, the relationship between the additional MSW piggyback loading and soils settlement was reviewed. The site soils consist of very dense and compact sands with blow counts ranging from 16 to greater than 100 blows per 0.15 meters (6 inches) from 3 meters (10 feet) to borings termination at 12 meters (40 feet) depth. Near surface values ranged from 8 to 40 blows per 0.15 meters (6 inches). Because of the nature of these sands, the likelihood of further soil settlement of any consequence was considered unlikely and not analyzed further. Second, the integrity of the perimeter leachate collection piping systems was assessed. Where the piping systems were placed on sand there would be no additional refuse loading to induce settlement. Where the piping systems were underlain by existing refuse, the degree of strain in the HDPE piping which could potentially be induced by uniform settlement would be well within the allowable design values for PE pipe. Additionally, the positive flexibility characteristic of the piping would serve to ensure its integrity.

The third component of the settlement analysis assessed consolidation of the existing MSW. Two types of settlement occur in landfills. Primary consolidation occurs rapidly during the placement of the MSW. Secondary consolidation occurs over longer periods of time and is a function of environmental conditions and composition of the refuse. This material presumably had completed its primary consolidation and much of its secondary consolidation prior to construction of the piggyback. This reflects the age of the refuse, much of which had been landfilled in the 1950's and 1960's. Using a methodology developed specifically for estimating settlement in MSW landfills due to new applied loads, calculations were performed which indicated that the piggyback had the potential to induce further settlement of this refuse. (References 1 - 2). The methodology utilized a relative height-log pressure relationship (relative layer thickness versus applied stress) which was based on laboratory and field experiments to estimate the magnitude of settlement in discrete waste layers within the existing landfill. The discrete settlements were then summed to provide an estimate of the total magnitude of the resultant settlement.

The potential for additional settlement was dependent on the estimated thickness of old refuse which could be as much as 58 meters (190 feet). Although the magnitude of the calculated potential settlement due to the additional MSW piggyback loading could be as much as 3.9 meters (12.8 feet), this value would only be applicable where the potential thickness of both the old refuse and the new piggyback loading would both coincide and be greatest. The values for settlement would then decrease uniformly from this location at the center of the expansion to zero at the perimeter. The resultant strain which could develop in the HDPE flexible membrane liner from uniform settlement was then calculated to be 0.13 percent, compressive. This value is significantly less than 13 percent, elongation, which represents the yield point for the 80 mil HDPE liner utilized for the piggyback and would theoretically result in strain relief or relaxation. Therefore, uniform settlement of the refuse beneath the piggyback liner system is to be expected, but based on settlement calculations would not represent an identifiable threat to the structural integrity of the HDPE liner system. It should also be noted that the steep side slopes upon which the piggyback liner system were constructed would be great enough to allow for continued positive leachate drainage, mitigating the effects of uniform settlement as well as the potential for leachate ponding on the liner.

Conversely, differential settlement has the potential to represent a greater threat to the structural integrity of the HDPE liner system than uniform settlement. The potential does theoretically exist for a weakening of the subgrade beneath the piggyback liner due to differential settlement in the underlying refuse. In a worst case scenario, the 80 mil HDPE liner would no longer have sufficient subgrade support and the magnitude of the resultant strain could be great enough to elongate the liner beyond its yield point. This could result in a thinning and stretching of the HDPE material, conceivably allowing MSW piggyback leachate to percolate down to the underlying, upgraded 20 mil PVC cap system if a rupture in the HDPE developed. At this point the upgraded cap system would act to collect this leachate and transmit it to its perimeter leachate collection system. If this system were also breached then leachate could potentially infiltrate down into the underlying, existing refuse which is unlined. Because of this possibility for liner failure and potential leachate discharge to the environment, there was a need to identify an approach which could prevent such a scenario from occurring.

To assess the failure scenario, several subgrade reinforcement methods were reviewed. Numerous authors who have published on this topic have pointed out that tensile reinforcement fabrics (referred to as geogrids) have been one method successfully implemented to increase the stability of structures over weak foundation soils. Geogrids have been utilized since the 1950's for reinforcing railroad and road subgrades in areas of karst subsidence or wetlands, for example. Conceptually, polyethylene (PE) geogrids act to bridge areas of subsurface softening. There was no evidence in the literature indicating that geogrids had ever been specifically utilized in this type of landfill application, however. (References 3 - 7)

At Blydenburgh Landfill, time restraints, the presence of the gas collection system and the 20 mil PVC cap presented complications which appeared to rule out traditional subgrade improvement methods such as surcharging to accelerate consolidation and settlement, or grouting to stabilize the existing refuse prior to piggyback construction. The economics of surcharging or grouting also appeared to be prohibitive, in part due to the time requirements and the great thickness of the existing refuse. Conversely, geogrids have been reported in the literature to be easily installed and to have been utilized in similar type projects requiring subgrade reinforcement. The economics and required construction time appeared to be favorable and use of geogrids did not present any apparent complications with regard to the existing environmental systems.

Conservative design assumptions were developed so that the suitability of geogrids of varying tensile strengths could be evaluated. Figure 2 is a schematic representation of the design assumptions and the calculated magnitude of vertical settlement which could result if a void of 2.4 meters (8.0 feet) were to develop in the refuse beneath the geogrids. Under this occurrence the liner system would lose all means of subgrade support in that area similar to a punching or shear subgrade deformation. Malcolm Pirnie was aware of no evidence in the literature indicating that voids as great as 2.4 meters (8.0 feet) in diameter had developed in MSW landfill caps. But, as an apparent worst-case scenario for purposes of the model, such a void appeared to represent a rational maximum diameter void which could theoretically develop beneath the vertical piggyback liner system, considering the nature and composition of typical MSW. For example, a 2.4 meter (8.0 foot) diameter void would also be approximately equivalent to two adjacent house-hold appliances instantaneously crushing. The following design criteria were modelled:

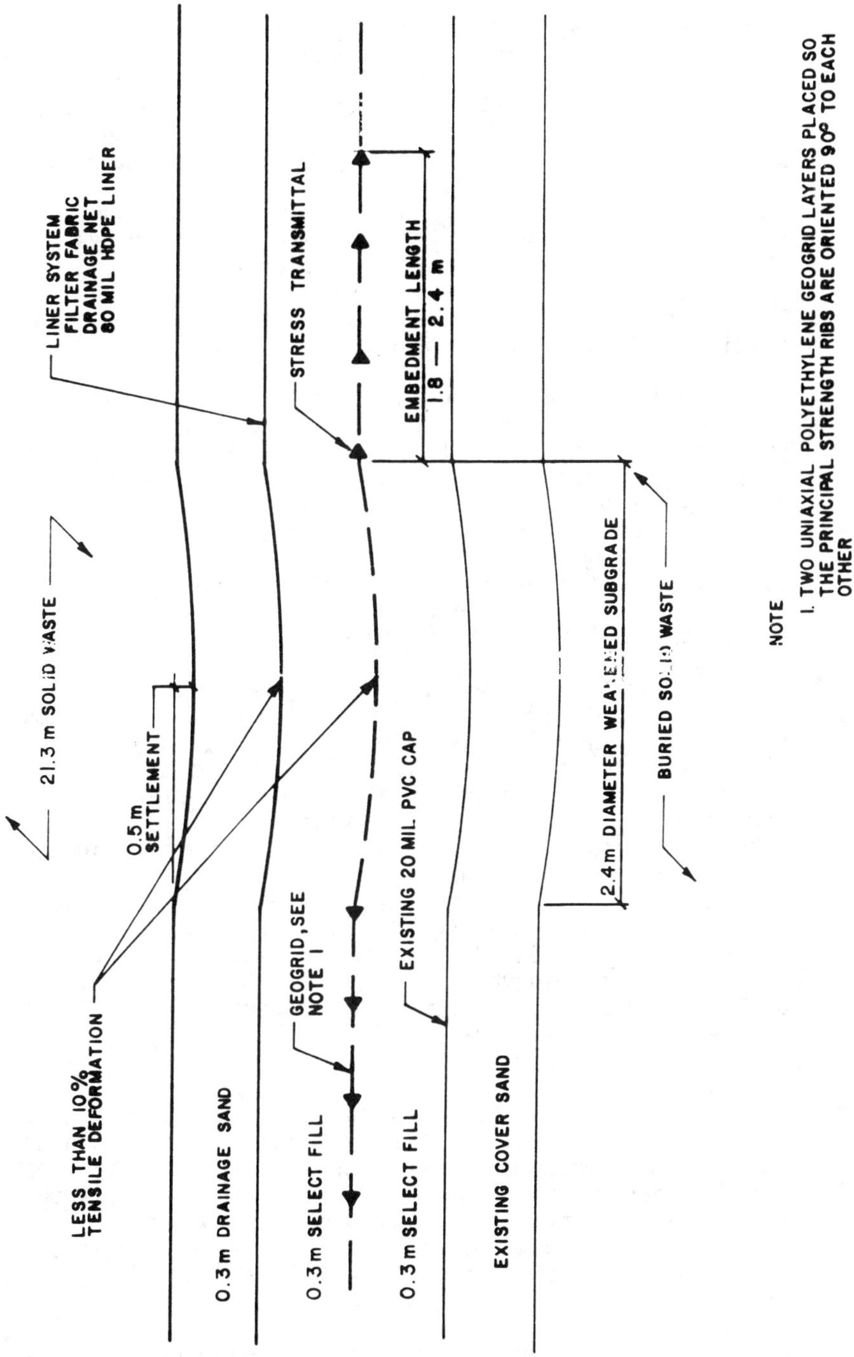

FIGURE 2. Geogrid Design Conditions.

- 10 percent or less reinforcement strain (elongation) over the 120 year design life for the geogrids;

- Geogrid reinforced soils, analogous to a soil beam, to be designed beneath the piggyback liner system acting as a non-rigid bridge spanning the lower, existing landfill;

- A circular 2.4 meter (8.0 foot) diameter void assumed to be of infinite depth to develop beneath the liner system. Although a circular void or depression was assumed, the actual depression might be nonsymmetric. Therefore, two layers of geogrid placed perpendicular to provide multidirectional support for nonsymmetric depressions;

- A void was assumed for analytical purposes, but based on the literature and personal field observations by the authors, the subgrade would be more expected to develop a localized decrease in strength, a softening, rather than an actual void;

- Inherent strength and resistance of the HDPE liner and associated geosynthetics of the liner system were assumed to be zero. Therefore the geogrids were assumed to provide all of the structural reinforcement;

- Uniform, maximum loading on the liner represented by 21.3 meters (70 feet) of MSW with a density of 993 kilograms per cubic meter (62 pounds per cubic foot), a long-term friction angle of 35 degrees, and a cohesion of 0 pounds per square foot. In actuality, loading would decrease towards the perimeter of the piggyback expansion where the refuse thickness would decrease to zero and some bridging of waste material could be expected to occur over a void. The density of refuse was also estimated very conservatively at 993 kilograms per cubic meter (62 pounds per cubic foot). Based on empirical data and operations at the vertical piggyback through December 1989, a value of 769 kilograms per cubic meter (48 pounds per cubic foot) may be more representative of the in-place densities being achieved; and

- Geogrid strength utilized was not the ultimate or yield strength, but instead was the creep strength as based on 10,000 hour maximum creep tests performed by the manufacturer; product durability data, and other environmental considerations. The creep strength was based on a 120 year design life because polyethylene, unlike steel, will gradually deform under loads less than the yield point.

In summary, two layers of uniaxial, polyethylene geogrid were placed perpendicular to each other in the soils between the vertical MSW piggyback liner system and the underlying 20 mil PVC cap. Calculations indicated that geogrids could span a 2.4 meter (8.0 foot) diameter void in the underlying MSW with a maximum downward vertical deflection to the piggyback liner system of 0.5 meters (1.8 feet). The resultant strain (elongation) to the liner system would be a maximum of 10 percent, or 30 to 50 percent less than the reported elongation at yield for the HDPE liner material used in the vertical piggyback expansion.

Conceptually, if a void greater than 2.4 meters (8.0 feet) were to develop in the subgrade beneath the piggyback liner there could be resultant displacement of the soil above and below the geogrid layers on the flanks of the void. This could result in an elongation of the HDPE liner beyond its 13 to 15 percent yield point, resulting in property changes to the HDPE liner. It should be noted though that the strain at rupture of a HDPE liner is a minimum of 500 percent (Reference 8). Therefore even under a larger diameter void scenario it is still anticipated that the liner system would continue to act as a barrier and inhibit the vertical flow of leachate. Even under this scenario, the steep side slopes would also act to continue the positive flow of leachate off the liner system towards the perimeter leachate collection system.

SIDE SLOPE STABILITY

The potential for sliding shear failure to develop in the soil and geosynthetic material between the existing landfill MSW and the new piggyback MSW was also assessed. The regions of primary concern on the site were the areas where the existing side slopes of the capped landfill ranged from 2.5:1 (23.6 degrees) to approximately 7:1 (8.2 degrees). An extensive review of the literature and discussions with various geotextile and geomembrane manufacturers suggested that the most critical interface friction angles would be those between the PE drainage net and smooth 80 mil HDPE liner. (References 9 - 18)

The computer modelling program STABL was utilized to assess potential side slope stability scenarios. Because the reported interface friction angles between smooth HDPE liner material and PE drainage net were even less than those reported for the textured HDPE liner material, the STABL analyses were based upon textured HDPE material being utilized on the lower, steeper side slopes. STABL was developed at Purdue University for the Indiana State Highway Commission in the mid-1970's and is a versatile stability program for this application. The soil strength parameters utilized in the analysis assume the long-term, drained condition and were based on empirical values, published technical literature and on-site soil borings. The values were as follows:

Soil types (cover and drainage layers	Coarse grained sands
Soil density	1,922 kilograms per cubic meter (120 pounds per cubic foot)
Effective soil cohesion	Zero

Effective angle of internal friction of soil	30 degrees
MSW density	769 kilograms per cubic meter (48 pounds per cubic foot)
Effective MSW cohesion	Zero
Effective angle of internal friction of MSW	30 degrees
Interface friction angle between sand and PP filter fabric	25 degrees (approximate)
Interface friction angle between PP filter fabric and PE drainage net	20 degrees (approximate)
Interface friction angle between smooth HDPE and PE drainage net	9 degrees (wet) and 11 degrees (dry)
Interface friction angle between textured HDPE and PE drain net	17 degrees (wet) and 19 degrees (dry)
Interface friction angle between HDPE and sand	20 degrees for smooth and greater than 20 degrees for textured (approximate)
Interface friction angle between sand and 20 mil PVC	25 degrees (approximate)

The initial stability condition simulated a sliding block type of failure analysis. Initially, the entire piggyback liner system with a portion of the overlying MSW material and the underlying soil/geogrid, and 20 mil PVC cap were analyzed. The initial runs verified that the most critical interface would occur in the piggyback liner system. The model was then optimized so that only potential shear failure surfaces in this weak zone were analyzed. Five landfill cross-sections, modeling the regions of steep side slopes, were then analyzed using the STABL program.

Interface friction angles between the drainage net and the smooth HDPE liner of 9.0 to 11.0 degrees were determined to lead to unacceptably low factors of safety. In a limit equilibrium analysis, values of 1.0 indicate that a state of imminent failure would potentially exist and that values of less than 1.0 would indicate that the structure was unstable and would fail. Values of 1.5 and greater are generally accepted by regulatory agencies as representing a long-term, stable condition. Because values of less than 1.0 were calculated for the smooth HDPE scenario, the initial iterations focused on textured HDPE liner material placed on the 2.5:1 to 7:1 side slopes.

Ultimately, as each cross-section was analyzed it became apparent that textured HDPE liner material need only be placed on the 2.5:1 to 6:1 side slopes. But factor of safety values of 1.5 and greater were only developed when the orientation of the primary drainage net rib was precisely controlled. Figure 3 is based on test data specific to the textured HDPE liner and PE drainage net material used for the construction of the Blydenburgh vertical MSW piggyback. The data summarized in Figure 3 is illustrative of the sensitivity which the interface friction angle has relative to the orientation of the primary drainage net rib. For example, point number 2, located at approximately 15.0 degrees represents a scenario where the drainage net would be rolled out lengthwise parallel to the slope; the normal procedure due to ease of installation. In this scenario, the primary rib of the drainage net would be approximately 28 degrees to the slope. Point number 3, located at approximately 23 degrees represents a scenario where the drainage net would be rolled out lengthwise perpendicular to the slope. Points 1 and 4, located at approximately 11 and 25 degrees respectively, represent a scenario where the drainage net would be placed lengthwise 45 degrees to the slope. As can be seen, varying the orientation of the drainage net during construction would apparently have the effect of varying the resultant interface friction angle from roughly 11 to 25 degrees.

Based upon these investigations, test data provided by various geosynthetic and geomembrane manufacturers, and the literature, the following observations regarding the combined use of HDPE geomembrane liner material and PE drainage nets for steep side slope applications were noted:

- Interface friction angles are reported to decrease by up to 2 degrees under wet conditions;

- Interface friction angles increase as overburden pressures increase; and

- Interface friction angles often vary between products due to differences in the manufacturers' resins and/or manufacturing processes. Therefore the specific manufacturers providing material for a project should demonstrate that their material will be compatible with the other synthetic and natural materials and achieve the desired minimum interface friction angles, as determined by detailed analysis.

Based on the STABL analyses, literature review, and discussions with the manufacturers, a conservative interface friction angle of 17 degrees between the textured HDPE liner material and the PE drainage net was estimated. This value was specific to the materials utilized in this project. Therefore, in the first reported significant use of textured HDPE liner material for a landfill application, this liner material was placed on the 6:1 and steeper side slopes. On these steeper slopes, the Contractor was also directed to place the drainage net on the textured HDPE liner so that the primary rib was perpendicular to the slope.

It was felt that the 17 degree value represented a safe design point, based on calculated factor of safety values of 1.5 or

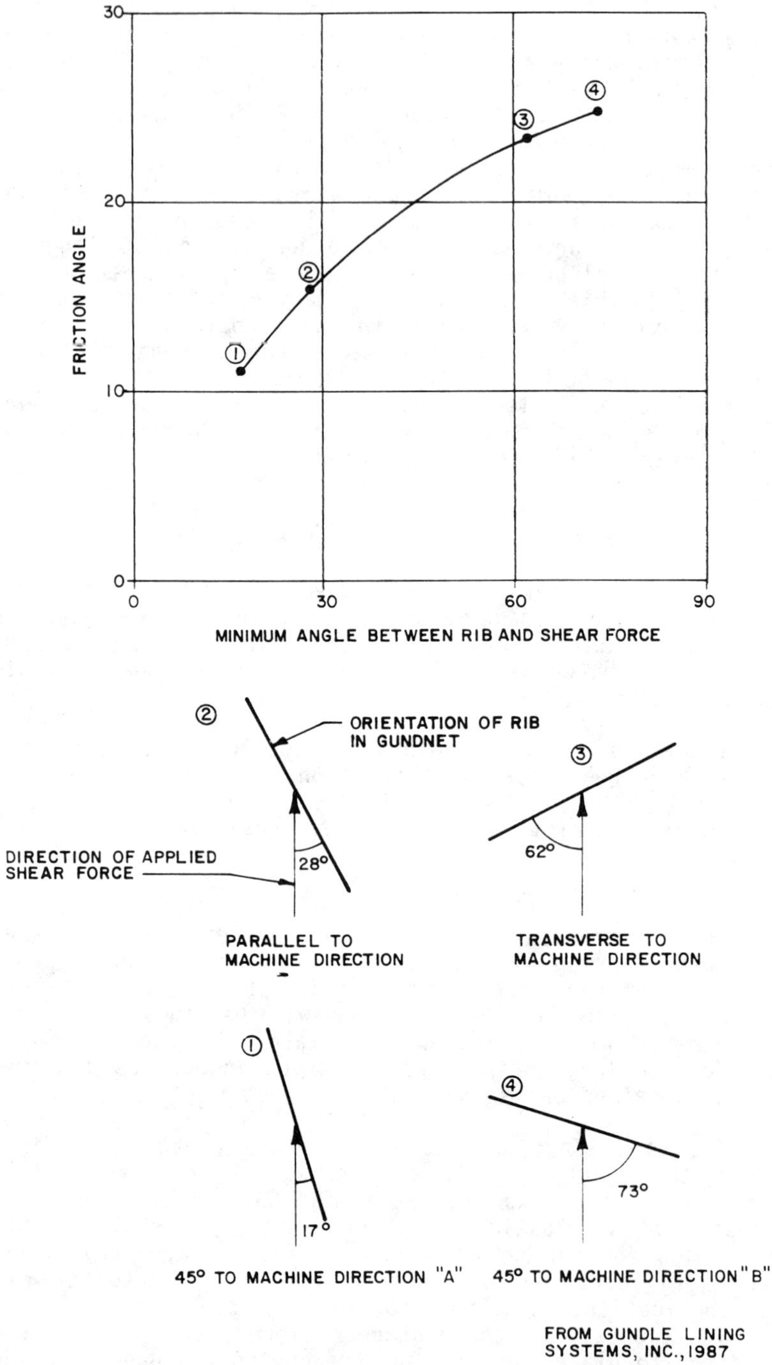

FIGURE 3. Drainage/Textured Liner Friction Factor.

greater, while allowing for variations in orientation during construction. For those slopes shallower than 6:1, smooth HDPE liner material was found to provide factor of safety values of 1.5 and greater. Therefore, the smooth material was placed on these slopes, realizing a cost savings over the textured HDPE material. Drainage net orientation relative to the smooth HDPE liner material on the smoother slopes was not a critical construction issue.

Imprinting of the PE drainage net into the surface of the HDPE liner, based upon literature review and discussions with the manufacturers, was not considered to represent a threat to the integrity of the 80 mil HDPE liner. This was partially considered to be a function of the rigid structural properties of the HDPE as compared to PVC material, as well as a reflection of the overall thickness of the liner material. Although there was no data specifically addressing increases in interface friction angle between the two materials due to imprinting, it is a possibility.

A second type of side slope analysis modelled by STABL was a deep, circular arc, rotational failure. Under this analysis, the five cross-sections were assessed to examine potential failure surfaces through the vertical piggyback, through the underlying landfill, and into the subsurface soils. The intent of the analysis was to determine whether the driving force generated by the piggyback loading was stable for this type of failure. Apparently reflecting the dense, compact nature of the sands the most critical factor of safety value calculated was 2.1.

It should be noted that New York regulations did not require an analysis of landfill stability relative to seismic hazards at this time. Therefore a seismic hazards analysis was not performed for the piggyback design. The STABL model does have the capability to perform a pseudo static seismic analysis for both the sliding wedge and circular failure scenarios.

LEACHATE COLLECTION

A goal of this project was to provide an upgraded cap for the existing unlined landfill, thereby decreasing the volume of rainfall infiltration into the old refuse and decreasing the rate of leachate generation. Such a positive effect could only be ultimately verified by long-term ground water monitoring at a site where a very detailed understanding of the hydrogeology is being developed. As of December 1989, additional ground water monitoring wells were still being installed at the site. In terms of the overall success of the vertical piggyback, it can only be inferred from such factors as apparent side slope stability, presence of voids in the surface, leachate in the lower secondary collection system, and total volume of leachate collected at the site.

As of December 1989, there had been no evidence of unusual voids developing in the surface of the refuse placed in the piggyback. Furthermore, there had been no observable shear failures in the identified weak zone resulting in observable movement of the piggyback down the slope. As of December 1989, the operator had reported that approximately 113,559 liters per day (30,000 gallons per day) of MSW and ash residue leachate were being collected by both leachate systems, conveyed to the on-site leachate storage tanks, and then transported off-site to a secondary wastewater treatment plant for disposal. Approximately 4,900 liters per day per lined acre (1,300 gallons per day per lined acre) was being collected. This is equivalent to the volume of leachate which would be expected to be collected from an uncapped landfill on Long Island with an efficient bottom liner system. Because there was no leachate collection system for the refuse underlying the piggyback, this 113,559 liters per day (30,000 gallons per day) value cannot as yet be compared to background data at the site. Both MSW leachate collection systems discharge by solid pipe into the same polyethylene header pipe; significant modifications at each individual junction manhole would be required to determine what if any leachate may be being collected by the upgraded 20 mil PVC cap perimeter collection system.

CONCLUSIONS

The vertical expansions at Blydenburgh Landfill, termed "piggybacking," have provided the Town of Islip, New York, with a cost-effective solution to their short-term solid waste disposal needs. Construction, including design changes, for both the MSW and ash residue piggybacks has represented a capital expense of less than $8,000,000. This expense must be compared to the estimated $174,000,000 which the Town would have had to expend to ship their waste off-Island during the period from May 1987 through December 18, 1990. All indications after 16 months of operation have been favorable and lead to the preliminary conclusion that the detailed analyses, a new application for geosynthetics and geogrids, and the first significant use of the recently developed textured HDPE liner material have apparently contributed to the environmental success of the vertical piggyback design. Finally, the use of geogrids for subgrade reinforcement, synthetics for flexibility, and the combination of textured HDPE liner material with a prescribed PE drainage net orientation in the piggyback's liner system have represented a viable method for addressing stability issues at this site. Therefore, permissible landfill capacity and economic benefit have been maximized.

This paper has illustrated some of the benefits of vertical piggybacking: maximization of existing landfill capacity, utilization of existing landfill environmental systems, and potential increased revenue from a site previously considered closed. But this paper has also illustrated that such an expansion can be much more complicated to design and construct due to various stability issues. At Islip, these included foundation subgrade reinforcement and side slope stability. In other regions, designing for seismic movement could be an issue. If these issues are accounted for early in the design process, then the potential for a successful piggyback expansion is good. This paper has illustrated some of the design issues and how they affected the design of the Blydenburgh Landfill piggyback expansion. However, each piggyback expansion will be unique with its own set of design and construction considerations.

ACKNOWLEDGEMENTS

We would like to acknowledge preparation of the figures by Tom Till and Tina Camporeale and typing of the manuscript by Lee Rudolph.

REFERENCES

(1) Rao, S.K., 1974. Prediction of Settlement in Landfills for Foundation Design Purposes, Ph.D. Dissertation, West Virginia University, Morgantown, WV, 281 p.
(2) Rao, S.K., L.K. Moulton and R.K. Seals, 1977. "Settlement of Refuse Landfills." in Proceedings of the Conference of Geotechnical Practice, Disposal of Solid Waste Materials, Ann Arbor, MI, June 13-15, 1977, p. 574-598.
(3) Berg, R.R., 1987. "TENSAR Geogrid Reinforcement of Membrane Liner, Islip, New York Landfill. Subgrade Stability Design with Calculations for Malcolm Pirnie," TENSAR, Corp., Morrow, GA, p. 23.
(4) Bonaparte, R. and R.R. Berg, 1987. "The Use of Geosynthetics to Support Roadways Over Sinkhole Prone Areas," in Proceedings of the Second Multidisciplinary Conference on Sinkholes and the Environmental Impacts of Karst," Orlando, FL, February 9-11, 1987, p. 437-445.
(5) Bonaparte, R. and R.R. Berg, 1987. "Long-term Allowable Tension for Geosynthetic Reinforcement," in Proceedings of the Geosynthetic '87 Conference, New Orleans, LA, p. 181-192.
(6) TENSAR, Corp., 1987. "Design Guideline for Subgrade Improvement Under Dynamic Loading with Tensar Geogrids," TENSAR Technical Note, Morrow, GA, p. 29.
(7) Whelton, W.S. and N.E. Wrigley, 1987. "Long-term Durability of Geosynthetics Soil Reinforcement." in Proceedings of Geosynthetic '87 Conference, New Orleans, LA, p. 442-455.
(8) National Sanitation Foundation, 1985. Standard Number 54, Flexible Membrane Liners, National Sanitation Foundation, Ann Arbor, MI, 69 pp.
(9) Brendel, B.F., 1986. "Liner Systems for Waste Disposal Sites, Geotechnical and Construction Considerations," presented at Pennsylvania Electric Association Structures and Hydraulics Committee Fall 1986 Meeting, Johnstown, PA, September, p. 24-28.
(10) CTC-Geotek, 1987. "Direct Shear Testing of Friction Liner Material. Prepared for GUNDLE Lining Systems, Inc.," Denver, CO, 67 pp.
(11) Duplancic, N., U. Dayal and J.C. Colella, 1987. "Hazardous Waste Landfill Cap System Stability," In Proceedings of the Geotechnical Practice for Waste Disposal '87, Geotechnical Special Publication No. 13, American Society of Civil Engineers, Ann Arbor, MI, June 15-17, 1987, p. 432-446.
(12) Koerner, R.M., 1986. Designing with Geosynthetics, Prentice-Hall, Englewood Cliffs, NJ, 424 p.
(13) Martin, J.P., R.M. Koerner and J.E. Whitty, 1984.

"Experimental Friction Evaluation of Slippage Between Geomembranes, Geotextiles, and Soils," in Proceedings of International Conference on Geomembranes, Denver, CO, p. 191-196.

(14) Richards, E.A. and J.D. Scott, 1985. "Soil Geotextile Frictional Properties," in Proceedings of Second Canadian Symposium on Geotextiles and Geomembranes, Canadian Geotechnical Society, Edmonton, Alberta, September 23-24, 1985, p. 13-24.

(15) Rowe, R.K., S.K. Ho and D.G. Fisher, 1985. "Determination of Soil-Geotextile Interface Strength Properties," in Proceedings of Second Canadian Symposium on Geotextiles and Geomembranes, Canadian Geotechnical Society, Edmonton, Alberta, September 23-24, 1985, p. 25- 34.

(16) Williams, N.D. and M.F. Houlihan, 1986. "Evaluation of Friction Coefficients between Geomembranes, Geotextiles and Related Products," in Proceedings of Third International Conference on Geotextiles, Vienna, Austria, p. 891-896.

(17) Williams, N.D. and M.F. Houlihan, 1987. "Evaluation of Interface Friction Properties Between Geosynthetics and Soils," in Proceedings of Geosynthetic '87 Conference, New Orleans, LA, p. 616-627.

(18) Soil and Material Engineers, Inc., 1987. "Geosynthetics Testing Services, Direct Shear Friction Testing, Gundle Friction Sheet," Prepared for Gundle Lining Systems, Inc., Cincinnati, OH, 5 pp.

Neno Duplancic

LANDFILL DEFORMATION MONITORING AND STABILITY ANALYSIS

REFERENCE: Duplancic, N., **"Landfill Deformation Monitoring and Stability Analysis,"** Geotechnics of Waste Fills - Theory and Practice, ASTM STP 1070, Arvid Landva, G. David Knowles, editors, American Society for Testing and Materials, Philadelphia, 1990.

ABSTRACT: An analysis of long term deformation monitoring data of a hazardous waste landfill is presented. A large number of field and laboratory tests provided an extensive data base for evaluation of the strength of the landfill materials. Slope stability analyses of the landfill's most critical section for static and seismic loading have shown that the landfill is stable.

KEYWORDS: landfill stability, landfill deformation, landfill monitoring, waste strength parameters.

In recent years geotechnical engineering has started to play an important role in designing and operating waste disposal facilities. One common problem that the geotechnical engineer faces in this new industry is the assessment of the stability of waste fills. The first question he must evaluate is whether standard geotechnical practice and techniques can be directly applied to estimate the strength properties of the waste materials. The problem becomes even more complicated when waste fills are underlined by a liner system usually consisting of multiple layers of natural and man-made materials designed to protect ground water.

Dr. Duplancic is the director of site operations and engineering at International Technology Corporation, 4585 Pacheco Boulevard, Martinez, California 94553.

This paper presents geotechnical evaluation of deformation monitoring data on a hazardous waste landfill. Included in discussion are the methods used to evaluate the strength parameters of landfill materials and the results of the landfill stability analyses. The stability of the landfill has been a subject of many investigations [1], [2], [3], [4], [5], and [6]. This paper presents the author's opinion on the issue and does not endorse or deny the conclusions drawn in any of the referenced reports. However, the comments and questions raised in these reports have helped to identify and clarify many important geotechnical issues associated with the landfill.

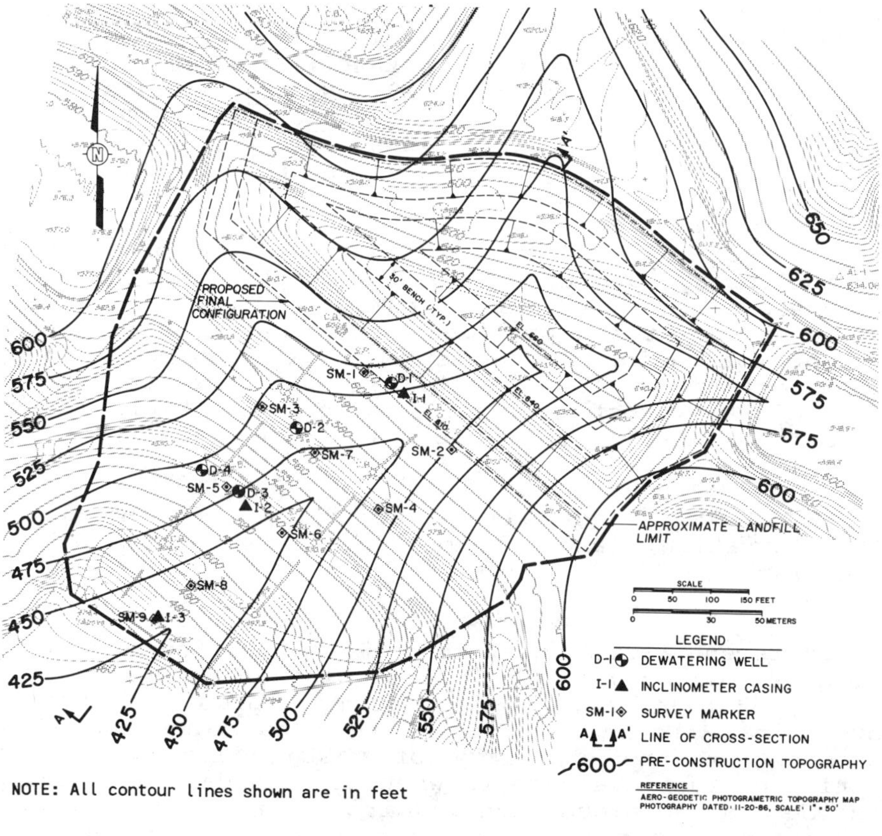

Figure 1: Plan of the landfill

BACKGROUND INFORMATION

The landfill is located in the northeast portion of IT's Corporation Panoche facility in Solano County, California. It encompasses an area of about ten acres and is roughly trapezoidal in shape (Figure 1). An earth fill toe embankment buttresses a narrow point in the "y" shaped canyon. From the toe embankment the landfill slopes upward at an approximate angle of 2.5:1 (horizontal to vertical) between benches with the overall slope of the landfill being less than 3:1. The landfill is unlined, having been designed in agreement with the regulations current at the time development started in 1979.

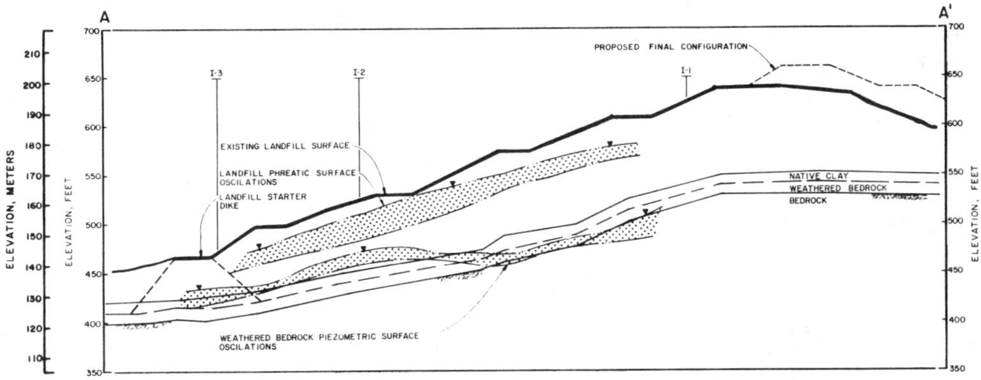

Figure 2: Landfill Cross-section

Hazardous waste placed in the landfill consisted primarily of contaminated soils, clean soils, and sludges. The origin of waste was from various industrial and municipal sources. The materials placed in the landfill have included soils from site remediations, inorganic contaminated soils and sludges, organic sludges, soils contaminated with gasoline, shredded currency, organic sludges, catalysts, and solids from the precipitation and treatment of hazardous waste. At the present state of development, the maximum thickness of the fill is about 40 meters (130 feet) beneath crest elevation.

GEOLOGICAL SETTING

The landfill is underlain by weathered and fractured shale of the Panoche Formation. The depth of intense weathering in the shale bedrock ranges from 1.5 to 6.7 meters (5 to 22 feet). Two to three meters (six to ten feet) of colluvium overlie the weathered bedrock. This colluvium zone (native clay) is composed of moist, stiff to hard clays and silty clays that contain varying amounts of bedrock and organic fragments [2].

An extensive network of observation wells and piezometers has been installed to study the ground water regime at the landfill. Recorded water levels indicate that the occurrence of ground water is limited to two separate zones: (1) an upper zone consisting of saturated zones within the landfill, and (2) a lower zone contained in the weathered and fractured bedrock (Figure 2).

The sources of water saturating the fill are the wastes placed in the landfill and infiltration of rainwater prior to landfill capping. Water entering the weathered bedrock beneath the landfill originates primarily from infiltration of rainwater at bedrock exposures at higher elevations to the north. Water flow in this zone is generally toward the south.

FIELD INVESTIGATION AND LABORATORY TESTING

A field investigation at the landfill was conducted to (1) obtain information on the subsurface stratigraphy and in-situ soil properties, (2) study the ground water regime, and (3) monitor the landfill performance. The investigation consisted of drilling and sampling 35 exploratory borings; installing 20 pneumatic piezometers, 25 piezometers, four dewatering wells, three inclinometer casings, and nine surface survey markers; and performing 10 Cone Penetration Tests (CPT) and four downhole seismic velocity surveys. The locations of the inclinometers and survey markers are shown in Figure 1.

A laboratory testing program was also conducted to determine the geotechnical properties of the fill materials, the native soils and the weathered bedrock underlying the landfill. The laboratory analyses of selected soil samples from the exploratory borings included index and strength tests. The index tests consisted of determination of natural water content, Atterberg limits, unit dry weight, specific gravity and grain size analyses. The strength tests included a total of 33 Isotropically Consolidated Undrained (ICU) Triaxial Compression Tests with pore pressure measurements. The results of these tests, along with eight ICU tests performed on soil samples from other borings in the vicinity of the landfill, were used to assess the strength characteristics of the landfill materials and the underlying native clays and weathered bedrock. Six sets of post-cyclic undrained triaxial tests were also performed on representative soil samples to evaluate the reduction in shear strength due to application of cyclic loading. The results of these tests were used for seismic stability analysis.

A critical point in evaluating slope stability is the determination of the strength properties of the landfill materials and subsoil. An interpretation of triaxial tests results was performed using a critical state soil mechanics theory. In this theory, the failure states of soil can be considered in terms of the octahedral effective stress parameters q' and p', which are appropriate stress invariant quantities for the study of the triaxial test results. Stress invariants are the magnitudes of stress parameters which are independent of the orientation of the reference axes. The stress parameters q' and p' are defined in Figure 3. Also shown in this figure are the triaxial test results for the landfill materials and weathered rock. The calculated angles of friction of landfill materials and weathered rock are 33 and 34 degrees, respectively.

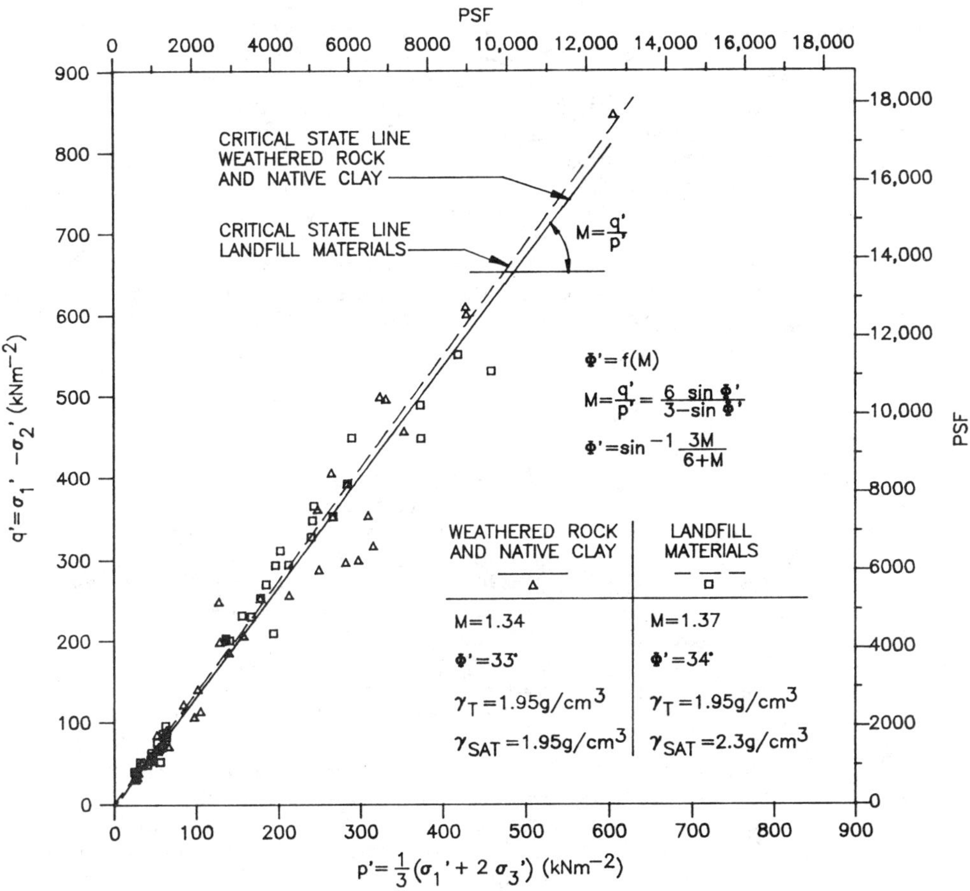

Figure 3: Triaxial Tests Results

GROUND WATER LEVELS AND DEFORMATION MONITORING

Failures of landfills, except for those caused by unanticipated events such as earthquakes, are almost always preceded by warning signals such as increased rate of deformation, strain discontinuities, cracking, or pore pressure buildup. These same warning signs may appear, yet be in no way associated with a potential failure. In order to detect significant changes in rate of deformation, and to evaluate the probable causes and consequences of such changes, continued monitoring of ground water levels, deflection, and settlement is often recommended.

Twenty-five piezometers and 20 pneumatic piezometers monitor ground water levels in the waste fill and weathered bedrock. The water levels in these piezometers have been measured monthly since 1985. The piezometers are screened in different zones within the landfill, weathered bedrock, and bedrock. The pneumatic piezometers are open to a selected zones within the landfill material and to the contact zone between the landfill and native clay. The extreme piezometric and phreatic surfaces recorded in the weathered bedrock zone and in the landfill zone, respectively, are shown in Figure 2.

Four dewatering wells have been maintaining ground water levels within the landfill since 1987. On average, these wells have produced less than 0.06 liters per second (1 gallon per minute). The locations of these wells are shown in Figure 1.

Three inclinometers installed at approximately the top, middle, and bottom elevations measure both the lateral and vertical deformation and provide data on potential creep, rotational movement and settlement. These inclinometers, which are numbered I-1, I-2, and I-3 in Figure 1, have depths of 48 (157), 34 (112), and 28 (91) meters (feet), respectively. Initial readings from the inclinometers were taken in late 1986.

For each inclinometer, measurements of lateral deformation are taken at intervals of two feet along two perpendicular axes. The orientation of one of the axes designates the bearing of the inclinometer. However, for analyzing the deformation of the landfill, the magnitudes of the resultant displacements and their directions are the most relevant parameters. Therefore, the displacements measured along both axes have been reduced to a resultant displacement with a corresponding bearing of true direction of movement. The Figure 4 shows the cumulative resultant deflections for the three inclinometers from January 1987 to October 1989 at four-month intervals. Four trends are evident from this figure: (1) the deflections at all depths

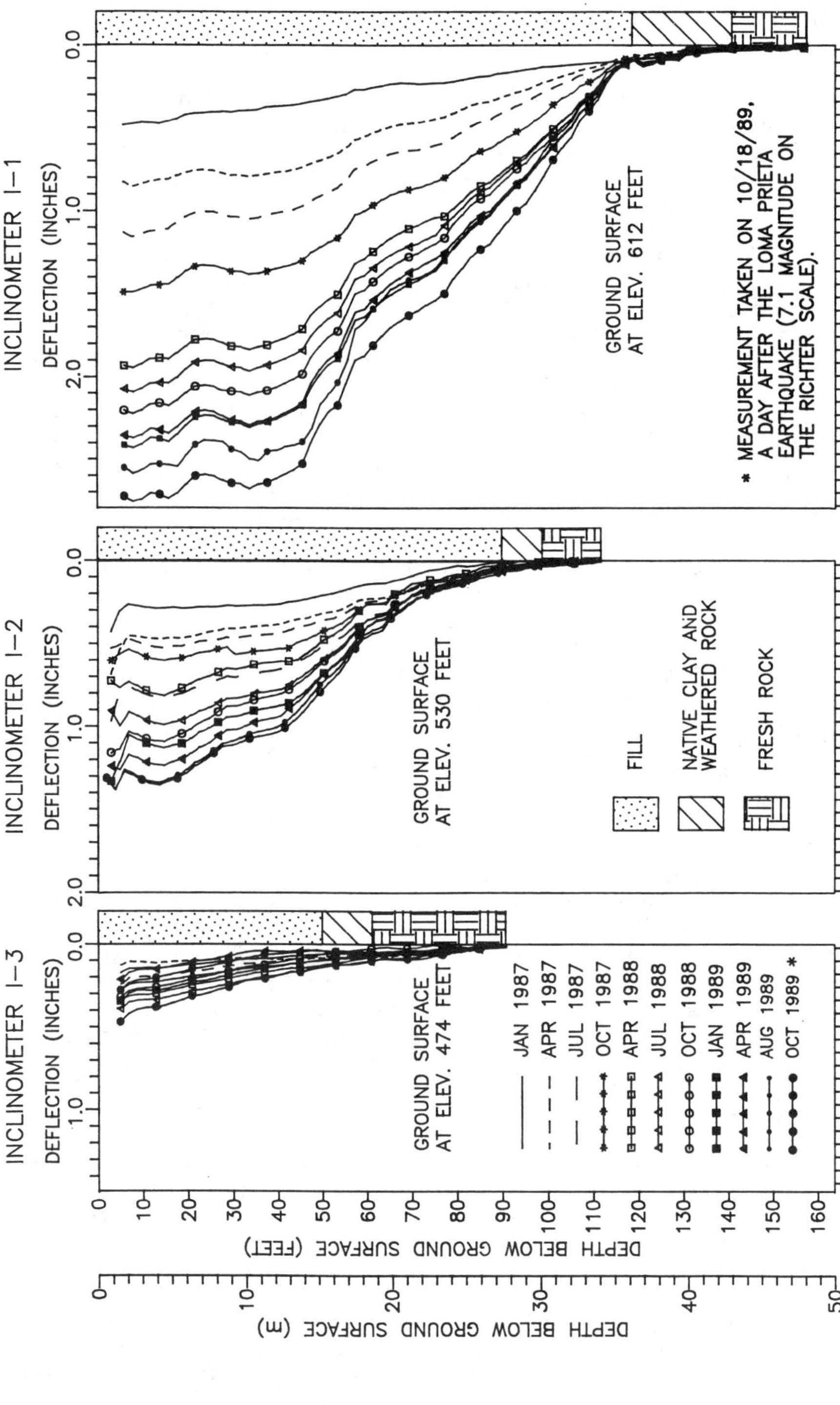

Figure 4: Deflections of Inclinometers

generally increase with time, (2) the deflections gradually increase from zero at the bottom to a maximum value near the top of the inclinometer casing, (3) the deflections within weathered bedrock and native clay are much smaller than the deflections in the fill, and (4) the rate of deflections decreases with time. There is no indication of a shear plane developing either in the fill or the colluvium below the fill.

The maximum recorded deflections of I-1, I-2, and I-3 are 6.3 (2.4), 3.3 (1.3), and 0.5 (0.2) centimeters (inches), respectively. The bearings of the resultant deflections of all three inclinometers lie close to S60°W, which is the direction of the landfill's downhill slope. The toe of the landfill is restrained by both the bedrock and an earth buttress and deflections increase with distance back from the toe. Inclinometer I-1 has the largest deflection because it is located in the deepest part of the landfill. The maximum measured deflection of 6.3 centimeters (2.4 inches) is within the range attributable to consolidation rather than sliding of the landfill. Fills on slopes commonly experience lateral deformation due to lateral force component imposed by the slope.

The measurements of the settlement of the landfill surface were obtained by surveying the elevation change and coordinates of each of nine survey markers shown in Figure 1. The survey markers SM-8, SM-9, and SM-5 are located over 18 (58), 14 (46), and 17 (55) meters (feet) of fill, respectively. These markers have settled 2.1 (0.84), 0.6 (0.24), and 5.5 (2.16) centimeters (inches), respectively, since November 1986, yielding a deformation of 0.04 to 0.33 percent. Survey marker SM-4 is located above approximately 30 meters (100 feet) of fill. Total recorded settlement at this location since November 1986 is about 5.8 centimeters (2.28 inches) yielding a deformation of only 0.19 percent. The largest observed settlement of 12.5 centimeters (4.92 inches) has occurred at survey marker SM-1 which is located above approximately 26 meters (85 feet) of fill. The total deformation associated with this settlement is 0.4 percent over 2.5 years period.

As expected, the settlements are larger at higher elevations and higher fill thicknesses and smaller near the starter dike. This is indicative of settlement due to consolidation which could have been accelerated due to the dewatering operations at the landfill. These operations started in late 1986 and are still in progress.

The deformation monitoring results cited above were compared with monitoring results reported for several earthfill dams [7]. Although dam deformations are not directly applicable to the deformations of the landfill (which is a canyon fill), the results can be used to demonstrate that the landfill lateral deformations may be those associated with vertical deformation caused by time-delayed consolidation. A comparison of landfill and dam deformations shows that (1) the observed settlement patterns are similar, (2) the observed horizontal deflection patterns are somewhat similar (except close to the ground surface), and (3) the ratio of horizontal versus vertical deflections is within acceptable limits of deviation, particularly considering the differences in geometry and material in the two cases. Thus, though a direct comparison of the observed deformations of the landfill and dams is not possible, enough evidence exists to suggest that the observed landfill deformations are similar to those expected to occur in an embankment undergoing deformation under its own weight.

In summary, the monitoring program has not revealed any unusual behavior of the landfill. All observed settlements are within acceptable limits and, along with inclinometer and piezometer data, indicate a stable condition.

STABILITY ANALYSES

Static Loading Condition

The stability of the landfill under static loading was evaluated by effective stress analyses using two-dimensional limiting equilibrium methods. The computer program STABL2 was used in the analyses [8]. Circular failure surfaces using the modified Bishop's method of analysis and non-circular failure surfaces using the simplified Janbu method of analysis were evaluated [2].

The slope stability model was developed for the most critical section along the centerline of the predevelopment canyon (Figures 1 and 2). This section represents only a small portion of the overall landfill. However, any other landfill section is less susceptible to sliding. The model assumes that native clay and weathered bedrock layers of constant thickness exist below the starter dike and the landfill. Strength properties of the landfill materials, structural fill of the starter dike, underlying native soils, and the weathered bedrock were assessed from triaxial test results performed on these materials and described earlier. A friction angle of 32 degrees was assigned to native clay and weathered bedrock layers to account for variability of strength parameters in these layers. A friction angle of 34 degrees was assigned to landfill materials based on the triaxial test results (Figure 3). Materials densities used in analyses are shown on Figure 3. Density increase of landfill materials with overburden pressure was not taken into account.

Two potentiometric surfaces were used in the analysis: (1) the phreatic surface within landfill saturated zone, and (2) the potentiometric surface within the weathered and locally fractured bedrock beneath the native clays. The piezometric surfaces used in the analyses corresponded to the highest anticipated water levels (Figure 2).

The stability analyses were performed for two cases: (1) the potential failure surface is restricted to the landfill and native clay and does not extend through the weathered bedrock, and (2) the failure surface passes through the landfill, the native clay and the weathered bedrock and is only restricted by the intact bedrock. The lowest factors of safety computed for the first case were 1.5 for a sliding block failure surface and 1.6 for a circular failure surface. For the second case, a theoretical factor of safety of 1.6 was computed for both the potential sliding block and circular failure surfaces [2].

Seismic Loading Condition

The performance of the landfill under maximum credible earthquake loading was evaluated by estimating earthquake-induced permanent deformations in the landfill. The deformation analysis approach for evaluating overall performance of an earth structure under seismic loading has been recommended by the International Commission on Large Dams [9].

The closest known active fault is located approximately 600 meters (2000 feet) from the landfill. A magnitude 6-3/4 earthquake at the closest approach to the fault would produce an estimated maximum peak horizontal bedrock acceleration of 0.68g at the landfill [2].

Permanent landfill deformations were evaluated using the approach based on a rigid body response procedure developed by Franklin and Chang [10]. The refinement of this procedure by Makdisi and Seed [11] and Sarma [12] was applied to account for the dynamic response of the earth structure. Both the rigid body and dynamic response methods of analysis require determination of yield acceleration. The yield acceleration of the landfill was computed as 0.16g using total stress strength parameters obtained from consolidated undrained tests and the most critical failure surface for block sliding. The effect of cyclic loading on the shear strength of materials along the potential sliding plane was evaluated through cyclic CU triaxial tests. The results of these tests indicated that there is no significant reduction in shear strength of the fill and native soils due to the anticipated levels of cyclic loading [2].

Based on the analysis performed, displacements of the landfill were calculated to be in the range of 10 to 50 centimeters (4 to 20 inches) for the postulated earthquake at the nearby fault [2]. The effects of these computed displacements on the landfill performance would not be detrimental. This is because the landfill is not equipped with a liner system or other displacement sensitive structures.

SUMMARY AND CONCLUSIONS

An analysis of deformation monitoring data of a hazardous waste landfill has been presented. The data indicate that the landfill is deforming similarly to earthfill dams. Deflections are larger in the fill zone, but almost negligible in weathered bedrock and native clay zones.

Static slope stability and seismic deformation analyses have been performed to assess the stability of the landfill. The results of field monitoring, exploratory borings, and laboratory tests on the various materials within and below the landfill form the basis of the analyses. Critical state soil mechanics theory has been used to evaluate effective strength parameters of the landfill and subsoil materials.

The minimum computed factor of safety under static loading condition is 1.5. The maximum induced landfill deformation resulting from estimated peak horizontal bedrock acceleration of 0.68g is about one foot. This displacement would not affect the integrity and performance of the landfill because there are no displacement sensitive structures within or beneath.

The analyses presented herein have shown that standard field and laboratory geotechnical techniques can be effectively used to monitor the performance of landfills, while standard geotechnical computational methods, when carefully applied, can be used for landfill stability analyses and deformation assessment.

REFERENCES

[1] Kleinfelder, J. H., and Associates, "Raising of Chemical Waste Fill," Benicia Facility, Benicia, California, Geotechnical Investigation Report, October 1981.

[2] IT Corporation, "Stability of Area 5 Landfill," IT Corporation Panoche Facility, Solano County, California, January 1987.

[3] Pyke, R., "Review of Slope Stability, Area 5 Landfill," Report to A. T. Kearny, Inc., September 28, 1986.

[4] Sitar, N., and R. Seed, "Analysis of International Technology Corporation's Report on Stability of Area 5 Landfill at Benicia Facility," Report issued to Shute, Mihaly and Weinberger, San Francisco, April 20, 1987.

[5] Yen, B. C., "Stability of Area 5 Landfill, Panoche Facility, CA. -- A Geotechnical Engineering Review ," Report by Bing C. Yen, Inc. to IT Corporation, Long Beach, California, May, 1988.

[6] West and Hansen Engineers, Inc., "Environmental Impact Report for proposed Area 5 Landfill Expansion," IT Corporation Panoche Facility, Solano County, California, June 1988.

[7] Wilson, D. S., "Deformation of Earth and Rockfill Dams," in Embankment-Dam Engineering, Casagrande Volume, J. Wiley & Sons, 1973, p. 382-416, edited by R. C. Hirschfeld and S. Poulos.

[8] Purdue University, Joint Highway Research Project, (JHRP-75-9), "Computer Analysis of General Slope Stability Problems," 1975.

[9] International Commission on Large dams, "A Review of Earthquake-Resistant Design of Dams," ICOLD, Bulletin 27, March 1975.

[10] Franklin, A. G. and F. K. Chang, "Earthquake Resistance of Earth and Rock-Fill Dams," Report 5 on Permanent Displacements of Earth Embankments by Newmark Sliding Block Analysis, Soils and Pavements Laboratory, U.S. Army Engineer Waterways Experiment Station, Vicksburg, Mississippi, November 1977.

[11] Makdisi, F. I., and H. B. Seed, "Simplified Procedures for Estimating Dam and Embankment Earthquake-Induced Deformations," Journal of the Geotechnical Engineering Division, Proc. of ASCE, Vol. 104, No. GT7, July, p. 849-867.

[12] Sarma, S. K., "Response and Stability of Earth Dams During Strong Earthquakes," Miscellaneous Paper GL-79-13, U.S. Army Corps of Engineers, Waterways Experiment Station, Vicksburg, Mississippi, 1979.

Case Histories

Francesco Belfiore, Mario Manassero and Claudio Viola

GEOTECHNICAL ANALYSIS OF SOME INDUSTRIAL SLUDGES

REFERENCE: Belfiore, F., Manassero, M., and Viola, C., "Geotechnical Analysis of some industrial sludges", Geotechnics of Waste Fills - Theory and Practice, ASTM STP 1070, Arvid Landva, G. David Knowles, Editors, American Society for Testing and Materials, Philadelphia, 1990.

ABSTRACT: The application of soil mechanics principles to the study of waste engineering behaviour involves an awareness of the similarities and differences between such material and soil. A suitable waste classification system is needed in order to standardize test procedures and assess the significance of the results. Since density is the most important waste property governing its overall behaviour, compaction procedures are important for the successful operation of a landfill from both a technical and an economical point of view. A traditional soil mechanics approach to waste investigation is presented in this paper, which emphasizes the necessity of adapting and integrating the usual geotechnical tools, also with the aid of a comprehensive performance monitoring program.

KEYWORDS: geotechnical engineering, waste, classification, laboratory tests, field tests, landfill, compaction, settlements, case history.

Mr. Belfiore is the Technical Development Manager of Browning-Ferris Industries Italia S.r.l., Via S. Orsola, 3, 20123 Milano, Italy, formerly at Studio Geotecnico Italiano. Mr. Viola is the Chief Geologist of Studio Geotecnico Italiano S.r.l., Via Ripamonti, 89, 20124 Milano. Dr. Manassero is the Managing Director of Ingegneria Geotecnica S.r.l., Corso Montevecchio 50, 10129 Torino, Italy.

In approaching the study of the engineering behaviour of some industrial sludges, traditional soil mechanics methods were applied which pointed out some differences and similarities between the two kinds of material.

The key objective of the study was to understand to what extent an improvement of landfilling operations can be achieved by adjusting the usual soil compaction procedures to take into account the specific characteristics of sludges.

This paper, which presents the analysis of some field and laboratory tests on sludges, is thought to be a first step in that direction, since actual data about the engineering properties of waste fill materials are still limited.

LANDFILLING OPTIMIZATION WITH RESPECT TO COMPACTION

It is becoming increasingly difficult to find suitable sites for waste disposal because of either environmental concerns or insufficient storage capacity. Therefore it is vital to exploit available storage volumes as much as possible by searching for more appropriate landfilling methods which can achieve higher waste densities.

In this respect, an actual case has been analyzed in order to give an estimate of the benefits that research in this direction can bring about. Considering a total landfill capacity $V_t = 350{,}000\ m^3$ and a percentage of cover and intermediate material equal to 15%, the net waste storage capacity is $V_n = 300{,}000\ m^3$. The total construction cost of the landfill is assumed to be $ 3 million. Under the further assumptions that the daily input of refuse is 4, 7 or 10 MN (meganewtons) and that the landfill is operated 300 days a year. Figure 1 shows the beneficial effect of increasing waste density, both in terms of landfill operational life and of unit cost of each MN disposed.

Besides, it is also important to compact waste to decrease permeability and improve surface water runoff, thus avoiding full saturation and stability problems due to pore pressure build up.

According to the usual soil engineering practice, the factors controlling compaction effectiveness are [1, 2]:

- water content;
- surface slope;
- thickness of compacted layers;
- number of roller passes.

With respect to waste compaction additional controlling factors may be the presence of liquids other than water (e.g. oils) and the optimum surface slope for the stability of the waste front during disposal operations. Moreover, the choice of compaction equipment and procedure should take into account such factors as type and amount of waste, density requirements, landfill type and development, weather conditions and economical aspects [3].

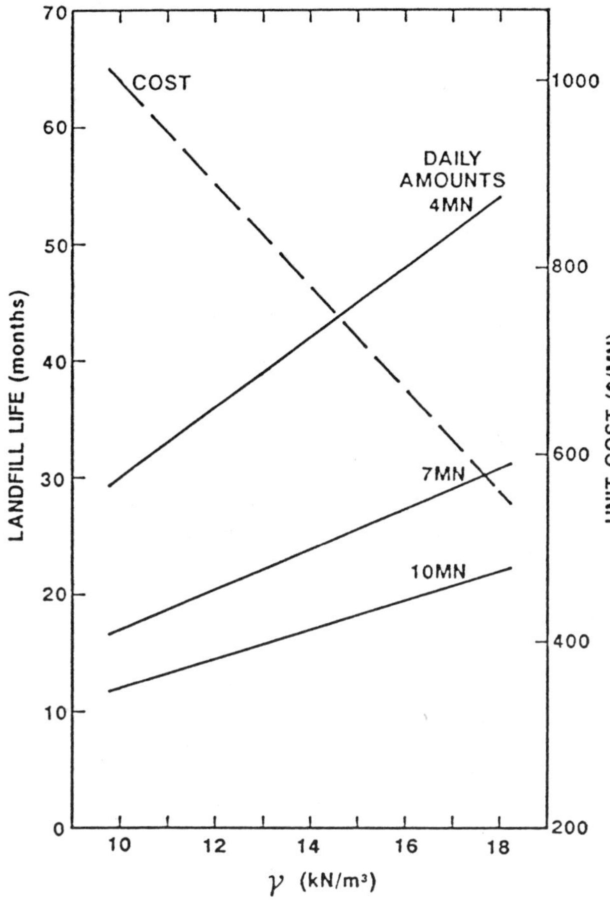

FIG. 1: Effect of increasing waste density on landfill operational life and unit cost of volume stored.

At present, the authors are gaining experience in the treatment and disposal of sludges from industrial manufacturing, described in detail in the following, and attention has been paid to the optimization of compaction methods.

Several Standard Proctor tests have been performed to determine the optimum compaction parameters. The corresponding results, though quite scattered, show that compaction allows an increase in dry density by as much as 100 percent, from about 5 kN/m^3 to a maximum of 9.6 kN/m^3. The corresponding average total unit weight is about 14.6 kN/m^3. Using a Modified Proctor test, these figures are slightly higher (see Table 1). It can be noted that in the wide range of water content values dry density does not change significantly, indicating that other factors can control the material response to compaction.

TABLE 1 -- Laboratory tests on compacted samples of Site 2 sludges (<1.00 mm fraction)

Sample Identif.[a]	γ_t (kn/m^3)	w (%)	γ_d (kN/m^3)	S (%)	k (cm/s at σ'=150 kPa)	CR[b]	c_u (kPa)	c′ (kPa)	ϕ' (°)
SP 1	12.7	34	9.5	49.2	–	–	620	40	39
SP 2	13.9	43	9.7	64.8	–	–	520	–	–
SP 3	15.1	56	9.7	84.8	–	–	380	0	35
SP 4	14.4	49	9.7	73.3	–	–	470	–	–
SP 5	14.1	49	9.5	70.7	–	–	494	–	–
SP 6	13.8	47	9.4	67.9	–	–	526	–	–
SP 7	14.4	51	9.5	73.8	–	–	434	–	–
SP 8	15.4	56	9.8	86.3	$7.5\text{x}10^{-7}$	0.245	–	0	38
SP 9	14.7	53	9.6	78.0	–	–	–	–	–
MP 1	16.9	42	11.9	89.1	–	–	830	–	–

[a] SP = Standard Proctor; MP = Modified Proctor

[b] CR = $\Delta\epsilon_v/\Delta \ell og\sigma'_v$

Observations of bulk in-situ density of a landfill have also been made, with the aid of high precision topographical surveys. In this case, before landfilling, the industrial sludges had an initial unit weight of 10.5 kN/m^3, measured by weighing containers of known volume filled with sludges. The volume of the empty cell was also precisely measured. During disposal operations no special compaction was performed, but a continuous action was developed by earthmoving equipment. At a certain stage of landfilling, when 14,000 m^3 were stored and the incoming quantities had been measured, an average in-situ unit weight of 15.8 kN/m^3 was calculated, which increased to 16.8 kN/m^3 after two months of operation (about 20,000 m^3 stored). These figures are significantly higher than those measured by Proctor tests, possibly due to the more effective compaction produced by tyred equipment, the self-weight consolidation of waste and the occasional supply of material with a higher specific gravity.

It was believed that an even higher waste density could be obtained, due to the high specific gravity of the metals present in the sludges. A research program was then initiated, with the aim of determining the most suitable compaction method in order to achieve a better use of the available volume. The program consists of a trial waste embankment, compacted with different equipment and procedures, which currently is still under planning.

SETTLEMENT OF WASTE FILLS

From a geotechnical point of view, settlement of waste is a major issue which affects several design and management aspects, such as the integrity and functionality of the cover cap and the possibility of land re-use. In analyzing this problem, it is usually assumed that the traditional soil mechanics theories also apply to waste material. In fact the settlement mechanisms are quite different and are both of physical and of bio-chemical nature.

Even though experimental records show some similarity between the waste settlement-time relationship and the primary and secondary stages of soil consolidation, the application of consolidation theory to waste materials is inevitably approximate because some of the assumptions of the classical theory are not satisfied. In particular, the presence of gas causes incomplete saturation, both the solids and the fluids are not incompressible, the validity of Darcy's law is not verified and chemical processes may prevail on the consolidation itself [4].

From the data available in literature, it seems that the mechanical compression of the waste occurs in the first few months after the disposal of waste, while the rate of settlements due to chemical activity is more dependant on the particular conditions at a site [5].

Differential settlements can arise from the heterogeneity of waste composition and density, changes in disposal procedures and potentially unstable areas within the landfill. As noted previously, differential settlements impair the proper operation of a landfill after its completion, both with respect to the integrity of the protective cover cap and the possibility of site reclamation and re-use.

Effective landfilling procedures can reduce settlements significantly but can also slow down biological stabilization processes due to the lower available porosity, thus decreasing settlement rate. This aspect should be thoroughly examined at the planning stage involving the final intended use of the site [4].

CASE HISTORIES

The authors are presently involved in the design and management of two landfills (class II-B and II-C respectively, according to the Italian classification system; see Appendix A) for the disposal of industrial waste.

The first site, which in the following will be referred to as Site 1, is located in central Italy, between Florence and Pisa, and accommodates sludges from nearby leather manufacturing plants. Prior to landfilling, sludges are partially dewatered by means of filtering-pressing machines, which reduce the moisture content from about 95% to 40%-75%, depending on the intensity of the treatment procedure, with an average of 63% at the time of disposal.

The content of organic matter ranges from 25 to 62%, with an average of 40%, while the pH is always ≥ 8.

The prevailing metal in the sludges is lead, having a concentration which varies from about 200 to 600 ppm, with significant traces of copper, cadmium and some hexavalent chrome.

The geotechnical investigation consisted of both laboratory and in-situ tests, performed on sludges of the same type, dumped in a nearby disposal facility for periods of time up to 3 years.

In the laboratory, at Studio Geotecnico Italiano, Milan, and Technical University, Turin, total and dry unit weight were determined, as well as water content and Atterberg limits of indisturbed samples of waste, taken with an Osterberg sampler during the drilling of boreholes. The samples were quite non-homogeneous, showing clods and fibrous materials. The most representative parts were selected for testing and the specimens were obtained very slowly (up to 3 hours per specimen) in short tubes, 76 mm high, in order to minimize their disturbance. Unconsolidated-undrained triaxial and shear box tests were also performed, and consolidation parameters were obtained from oedometer tests. The results of these tests are summarized in Table 2, in which it can be seen that the average value of unit weight is 12.5 kN/m^3, while water content, liquid limit and plasticity index are exceedingly high. Some difficulties arose in determining the grain size distribution by hydrometer and specific gravity of solid particles, due to flocculation phenomena, therefore only sieve analysis was completed. However, it was observed that 31 to 35% of the waste is composed of particles greater than 0.06 mm, some of which are jointed finer particles.

Consolidation tests show that compression ratio is quite constant, $CR = \Delta\epsilon_v/\Delta \log \sigma'_v = 0.32$, and that the preconsolidation pressure, σ'_p, seems to be higher than the in-situ overburden pressure, probably due

TABLE 2 -- Laboratory tests on indisturbed samples of Site 1 sludges

Sample identif.	depth (m)	γ_t (KN/m^3)	w_n (%)	γ_d (KN/m^3)	LL (%)	PI (%)	k (cm/s at σ'=150 KPa)	CR	c_u (KPa)	c' (KPa)	ϕ' (°)
S1-1	1.10	12.7	117	5.8	163	105	1×10^{-6}	0.32	28	-	-
S1-2	2.50	12.2	145	5.0	152	107	-	-	23	-	-
S1-3	3.50	12.4	132	5.3	139	97	8×10^{-7}	0.32	39	0	37
S2-1	2.00	13.0	113	6.1	167	108	-	-	34	-	-
S2-2	3.50	12.2	166	4.6	179	111	-	-	39	-	-
S3-1	2.50	11.5	153	4.6	-	-	2×10^{-7}	0.22	-	-	-

to the effect of compaction and to some kind of chemical bonding (Fig. 2). Shear strength parameters of $c' = 0$ and $\phi = 37°$ were measured, in the stress range of 0 to 300 kPa, where c' is the effective cohesion and ϕ' is the drained angle of friction, respectively. As an index of the material response to quick loading, an average undrained shear strenght C_u of 33 kPa was measured.

In-situ data include piezocone penetration tests, CPTU, standard penetration tests and falling head field permeability tests (Lefranc type). The CPTU results are shown in Figure 3, in which q_{ct} is the cone resistance corrected to take into account the pore pressure behind cone tip. The penetration resistance is generally remarkably low and the hydraulic response is quite variable. Dissipation tests, performed during piezocone penetration, seem to confirm that the waste is partially saturated.

Due to the similarity between the waste index properties and consolidation behaviour and those typical of a clay, it was decided to try to go further in applying soil mechanics methods to this material. By using the method proposed by Lunne et al. [6], it was possible to derive the C_u vs. depth profile from the piezocone penetration data. Figure 4 shows the interpreted results as well as the laboratory data. The comparison of the two sets of data indicates a good agreement, even though field values are generally lower than laboratory ones, probably due to problems in pore pressure measurements in a partly saturated material.

Hydraulic conductivity k is also shown to be quite variable, both in-situ and in the laboratory, ranging from 1×10^{-8} to 1×10^{-5} - cm/sec, and measurements made on landfills of different age seem to show a tendency of k to decrease with increasing age (Fig. 5).

The second case study (Site 2) concerns an abandoned borrow pit located in the alluvial plain surrounding Turin in northern Italy which is being landfilled with a mixture of sludges coming from steel mills, chrome-plating and painting plants. Recently, other kinds of solid wastes are also being disposed in this landfill. Relevant design and management details regarding the disposal facility are reported by Bortolami et al. [7] and Bonvicini et al. [8].

Geotechnical data, collected by testing undisturbed samples of the same waste already being disposed in other landfills, are shown in Table 3. Some waste samples were also compacted in the laboratory, according to the Standard and Modified Proctor procedures, and results of the relevant tests were presented and discussed in the previous section on waste compaction.

As far as the undisturbed samples are concerned, it can be seen that the unit weight is generally slightly lower than in the case of Site 1, and the water content w_n is of the same order of magnitude for samples from Site 2 - location A) and significantly higher for Site 2 - location B). Atterberg limits are considerably lower in samples from location A) than for Site 1, while they were not measured for location B). Two grain size distributions indicate that 8% and 37% of the particles, respectively, are greater than 0.06 mm. In the latter case, the coarser fraction consisted of cemented lumps of finer particles. Almost all the samples were not fully saturated.

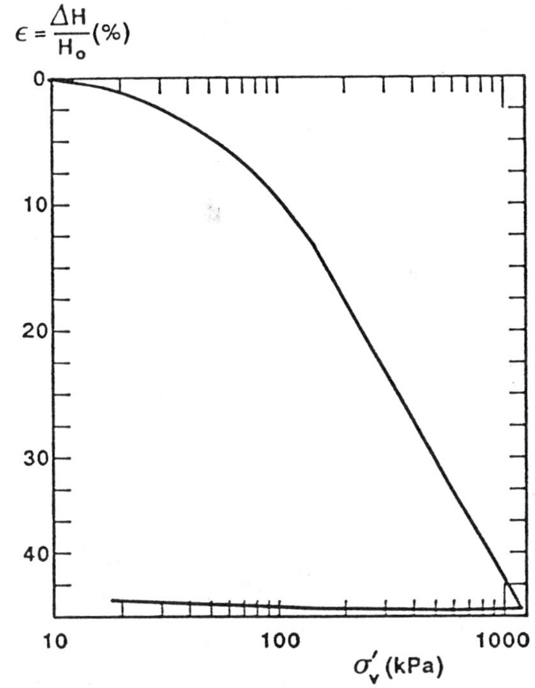

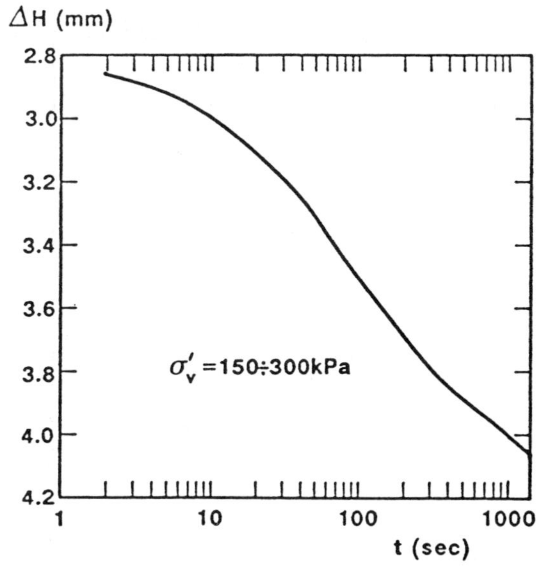

FIG. 2: Typical stress - strain and time - settlement curves from oedometer tests on waste samples.

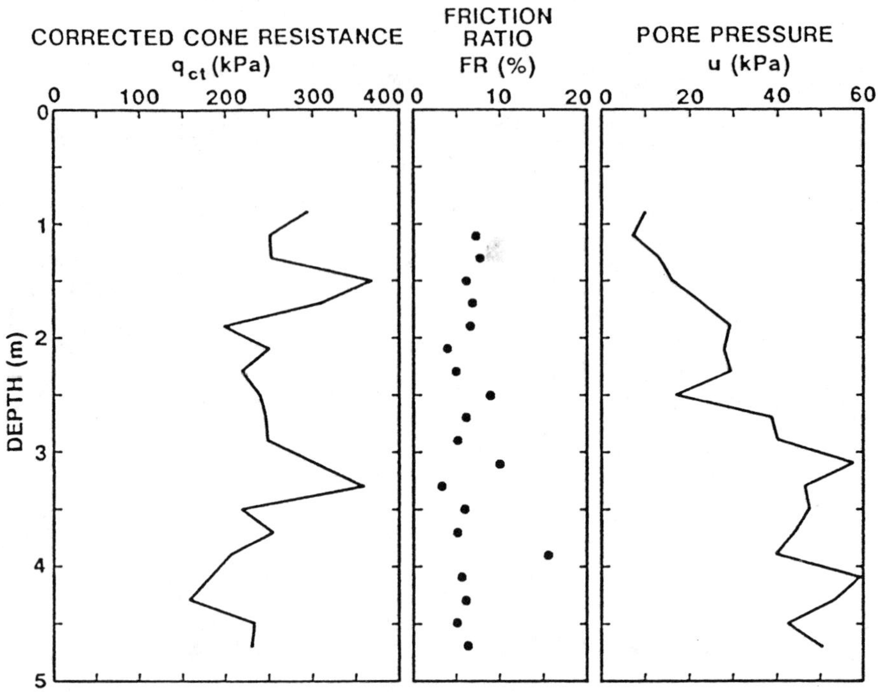

FIG. 3: Piezocone penetration test data.

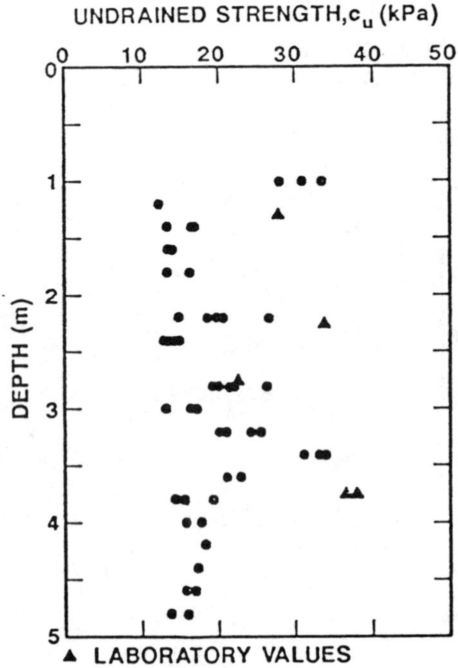

FIG. 4: Undrained shear strength profile.

TABLE 3 -- Laboratory tests on indisturbed samples of Site 2 sludges

Sample identif.	depth (m)	γ_t (KN/m^3)	w_n (%)	γ_d (KN/m^3)	γ_s (KN/m^3)	e_o	S (%)	LL (%)	PI (%)	k (cm/s at σ'=150 KPa)	CR	c_u (KPa)
Loc. A												
S1	0.50	12.7	167	4.8	-	4.7	97	105	41	-	-	-
S2	0.50	11.4	105	5.5	27.4	4.0	73	63	18	-	-	-
S1÷S2	0.50	12.0	147	4.9	-	4.6	88	-	-	-	-	-
Loc. B												
S3	3.00	11.3	286	2.9	26.2	8.0	98	-	-	1.5×10^{-7}	0.32	65
S4	3.00	11.3	285	2.9	27.9	8.6	91	-	-	5.5×10^{-8}	0.47	46
S5	5.00	11.0	235	3.3	25.4	6.7	99	-	-	7.0×10^{-7}	0.37	82

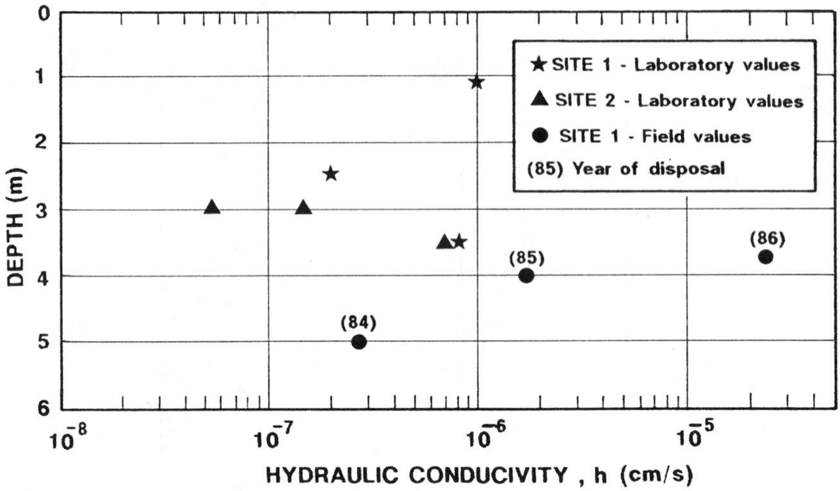

FIG. 5: Summary of hydraulic conductivity results.

The average value of compression ratio, CR, is 0.38 for undisturbed samples, and decreases significantly to 0.20 for compacted samples.

The undrained strength of undisturbed samples varies from 46 to 82 kPa, while compacted samples show increased C_u values of 380 to 620 KPa and drained shear box tests give $c' = 0 \div 40$ KPa and $\phi' = 35° \div 38°$ (Table 1).

It is interesting to note that the measured drained shear strength parameters are typical of soils with a coarser grain size distribution than the waste under examination.

CONCLUSIONS

A comprehensive program of geotechnical field and laboratory tests on industrial sludges has been carried out to assess the applicability of soil mechanics principles and methods to the study of the engineering behaviour of waste. Particular attention has been devoted to compactability characteristics in order to optimize landfilling procedures.

The comparison between laboratory test results and field behaviour of the sludges already landfilled suggests the following remarks:

- the high drained strengths measured in the laboratory are confirmed by the long term behaviour of the sludges, landfilled with front slopes of up to 35° to the horizontal without the occurence of any instability, either general or local;

- as far as deformability is concerned, laboratory tests show some similarity between sludges and natural soft plastic silts and clays but a direct comparison with field behaviour is not yet possible since a comprehensive settlement monitoring program has not been undertaken so far;
- both laboratory tests and field measurements agree in showing the beneficial effect of waste compaction, which can give rise to significant volume reduction and improvement of strength and deformation properties.

Further studies and experiences are needed in order to collect more relevant and reliable data on field behaviour of sludges and to obtain additional information on issues such as waste origin and characteristics, chemical properties, microstructural features (e.g. by means of electron microscope photographs) and mechanical behaviour, including static and dynamic loading response. A specific classification system is also needed to act as a general framework for standard testing and analysis.

REFERENCES

[1] Kelley, M., "Equipment Selection for Landfills", Waste Age, January 1986.

[2] Forssblad,L., "Vibratory Soil and Rock Compaction", Robert Olsson Tryckeri AB, Stockolm, 1981.

[3] Sanna, M., and Floccia, M., "La discarica di rifiuti", Edizioni delle Autonomie, Rome, 1986.

[4] Cancelli, A., and Cossu, R., "Problemi di stabilità negli scarichi controllati", Ingegneria Ambientale, Vol. 13, 1984, pp. 632-642.

[5] Sowers, G.F., "Settlement of Waste Disposal Fills", Proceedings of the Eighth International Conference on Soil Mechanics and Foundation Engineering, Vol. 2, Part 2, Moscow, U.S.S.R., 1973, pp. 207-210.

[6] Lunne, T., Christoffersen, H.P., and Tjelta, T.I., "Engineering Use of Piezocone Results in North Sea Clays", Proceedings of the Eleventh International Conference on Soil Mechanics and Foundation Engineering, Vol. 2, San Francisco, 1985, pp. 907-912.

[7] Bortolami, G.C., Buonomo, L., Di Molfetta, A., Manassero, M., and Viola, C., "Design Aspects of a Toxic Waste Landfill in an Industrial Area", Proceedings of the Twelfth International Conference on Soil Mechanics and Foundation Engineering, Rio de Janeiro, 1989.

[8] Bonvicini, A.C., Manassero, M., Street, A., and Viola, C., "Lining, Leachate Collection and Monitoring of a Large Toxic Waste Landfill", Proceedings of the Second International Landfill Symposium, Alghero, Italy, 1989.

APPENDIX A

ITALIAN CLASSIFICATION SYSTEM FOR WASTES AND LANDFILLS

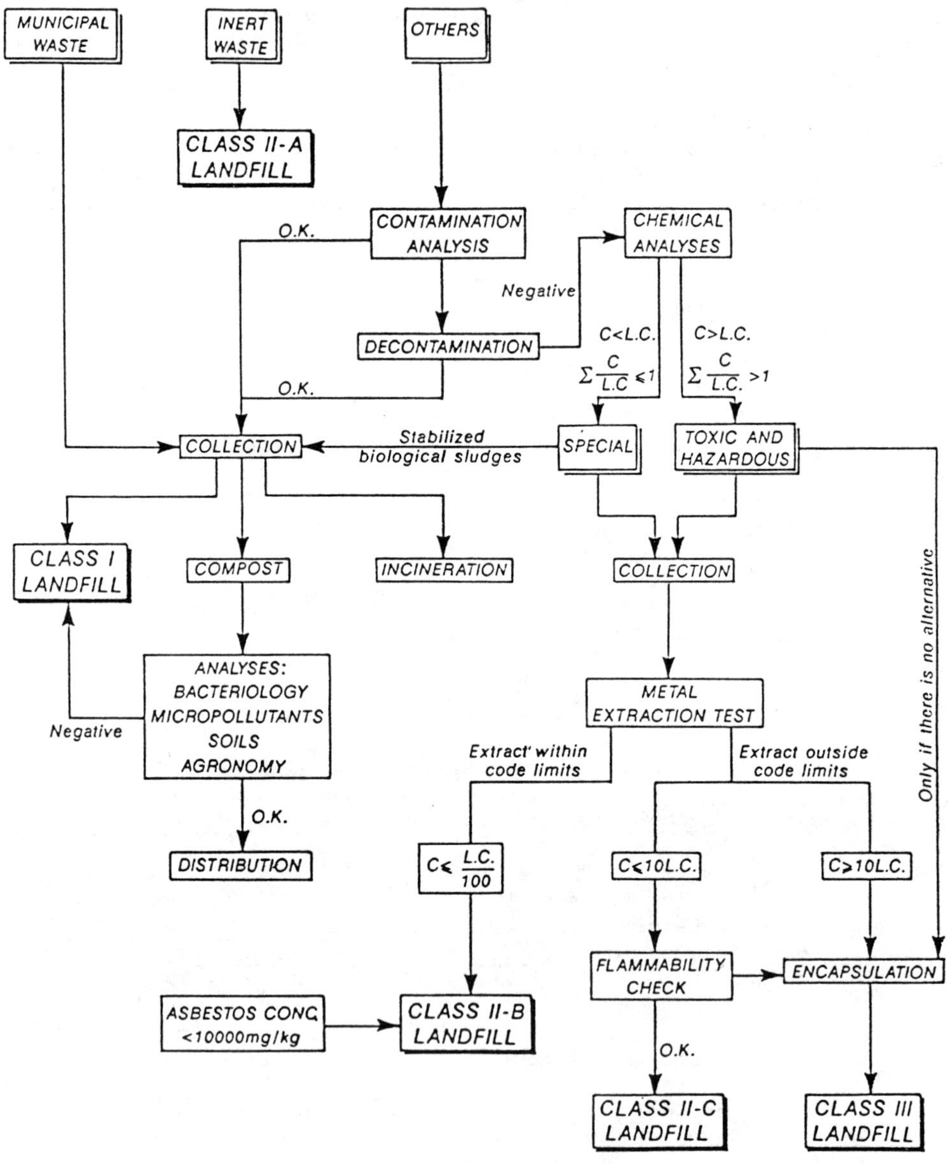

$$\Sigma \frac{C}{L.C} \leq 1 \qquad \Sigma \frac{C}{L.C.} > 1 \qquad C \leq \frac{L.C.}{100}$$

C=CONCENTRATION OF TYPICAL SUBSTANCES

L.C.=LIMIT CONCENTRATION (ACCORDING TO THE RELEVANT CODE)

R. D. Hinkle[1]

LANDFILL SITE RECLAIMED FOR COMMERCIAL USE AS CONTAINER STORAGE FACILITY

REFERENCE: Hinkle, R. D. " Landfill Site Reclaimed for Commercial Use as Container Storage Facility," Geotechnics of Waste Fills-Theory and Practice, ASTM STP 1070, Arvid Landva, G. David Knowles, editors, American Society for Testing and Materials, Philadelphia, 1990.

ABSTRACT: An abandoned 38 acre sanitary landfill approximately 100 feet (31 m) deep was closed, covered, sealed, and converted to a marine container storage and repair facility. An impermeable cover material was developed using crushed miscellaneous base, silt and asphalt emulsion. A large number of field and laboratory permeability tests were performed to verify the material properties. The landfill surface is used to support heavy container movers and two permanent structures. The project shows that landfill property can be reclaimed and put into profitable high demand use while meeting all environmental regulations.

KEYWORDS: landfill, cover, permeability, waste strength parameters, settlement, structures on landfill

INTRODUCTION

The landfill site is located in the Wilmington area in the City of Los Angeles. The location, near the growing Los Angeles Harbor, Figure 1, causes the land to be valuable enough to economically reclaim the abandoned landfill. The site was originally a low lying, poorly

[1]President, Dale Hinkle P. E. Inc., 15510 B Rockfield Boulevard, Irvine, CA 92718.

drained area of the City of Los Angeles. During the 1940's and 1950's the area served as a borrow pit for soil used to construct the Los Angeles harbor. After the maximum depth of borrow was reached, the site was turned into a landfill and filled with rubble, used tires and residential waste.

The total depth of landfill is about 95 feet (30 meters). The lower 45 feet (14 meters) is rubble and tires and the upper 50 feet (15 meters) is rubbish. The fill was randomly placed and covered each day with a thin soil cover.

After filling was complete, a 1.5 foot (0.5 meters) soil cover was placed over the rubbish and the site used for storage of coke piles until approximately 1985. A gas extraction system was installed prior to 1985. The closure was designed to meet all of the requirements of the EPA and State of California.

FIGURE 1 - SITE VICINITY

Purpose

The State of California is requiring the closure of all inactive landfills. The purpose of the landfill project was to permanently close the site and make it available for a new use. This consisted of sealing the site to contain the gas as well as divert any runoff water and provide drainage. Because of the location of the landfill, the owner decided to reclaim the site for use as an industrial facility.

A container servicing company (lessee) has agreed to lease the site after closure for their marine container storage and maintenance activities. The use of the landfill as a container storage facility created some very difficult constraints for the cover design. The container mover is a forklift type vehicle with a 68,000 pound (302 kN) single axle load on the front drive wheels. The rear steering axle load is 38,000 pounds (169 kN). These loads are 3 to 4 times that of trucks designed for highway use. The loads are very similar to those required for aircraft tire loads for a L-1011 or DC-10.

Scope

It was necessary to develop a surface which was flexible enough to withstand the deformation of a settling landfill, impermeable to methane gas and water, and stable enough to withstand repeated wheel loads up to 4 times heavier than imposed on a freeway.

It was decided to use an asphaltic surface made with a special cold-mix, low-permeability, asphaltic concrete covered with an asphaltic concrete wearing surface. Special permeability test apparatus and procedures were developed to verify permeability of on-site mixed asphalt. Figure 2 shows some of the permeability test data for various materials tested, these data are included to show some of the materials evaluated during feasibility studies.

The project required special design to provide flexible landfill surface which would be compatible with improvements such as:

a) Paving of surface
b) Drainage structures and outlet structures
c) A 50 X 150 foot (45.8 X 115.2 m), five story warehouse structure
d) Utilities for electricity and water conduits crossing the surface

Measured settlement of the surface ranges from 3 to 12 inches (8 to 31 cm) per year which must be tolerated by all of these improvements.

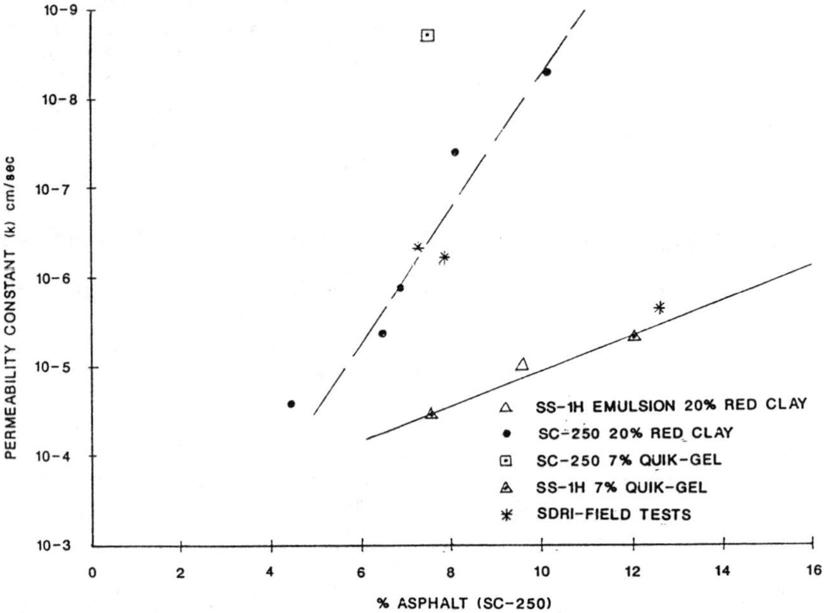

FIGURE 2 - PERMEABILITY TEST DATA - EARLY DEVELOPMENT

Lines are best fit interpretation of measured data.

Site Conditions

As previously discussed, the site was a borrow pit and was subsequently filled with residential and other rubbish. The general site, prior to excavation and landfilling, contained silty fine sand and alluvial deposits. The site was mined for borrow soil until the ground-water level was reached, then abandoned for several years. The landfilling operation began in 1964 and continued until 1981. Approximately 3 million cubic yards (2.6 million M^3) of waste was placed. Between 1981 and 1986 the surface was used to store coke piles up to 40 feet (12 m) high which partially preloaded the area. Between the period of 1965 and 1981 the water table in the area rose approximately 20 feet (6 m) as a result of a sea water barrier injection system south of the site, so the lower portion of the landfill is saturated. The present surface of the landfill is approximately elevation plus 60 feet (18 m) msl and the bottom is approximately elevation minus 40 feet (12 m) msl.

The site was completely regraded to accomplish final closure and construction of this project. The final project is shown on Figure 3. The filled surface had 1 to 2 feet (0.3 to 0.6 m) of soil cover. Approximately 230,000 cubic yards (198,000 m^3) of silty fine sand fill was imported to the site to regrade the surface. The entire surface was recompacted and fill added to provide a minimum 2 feet (0.6 m) of soil cover over the rubbish. The actual fill over rubbish thickness ranged from 3 to 10 feet (1 to 4 m) over the site with the average being approximately 4 feet (1.2 m). The soil fill was all compacted to 90 percent of maximum density as obtained by ASTM D1557 a method of compaction. The soil was compacted using a sheepsfoot compactor and a heavy rubber tired loader.

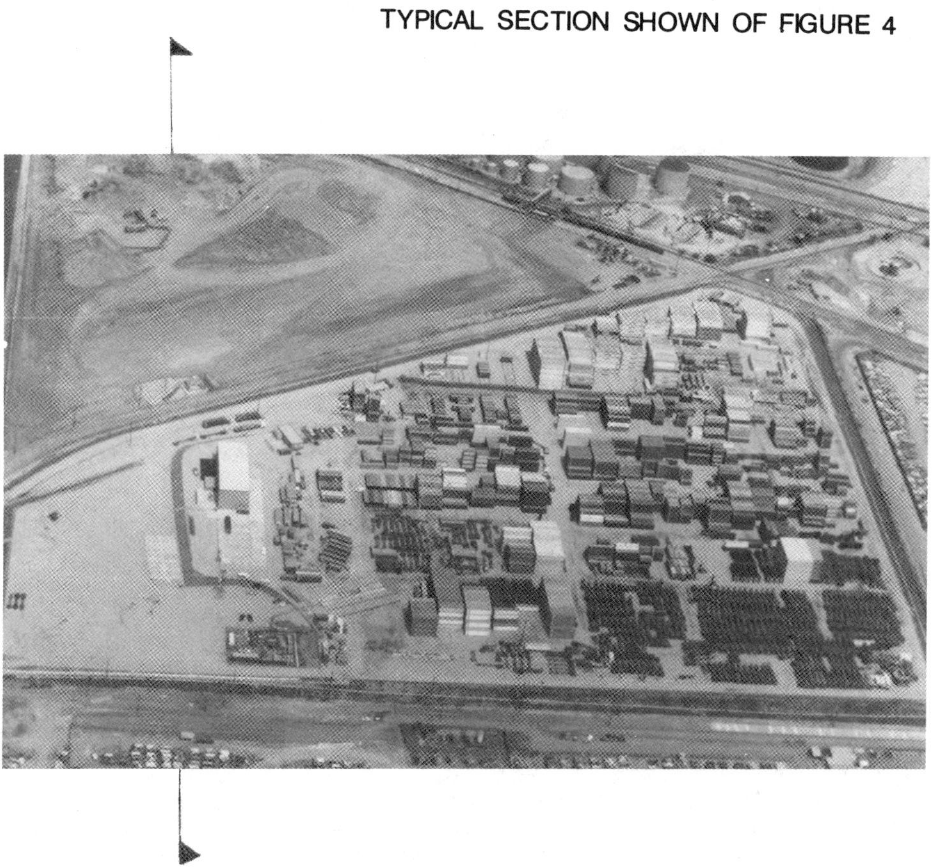

FIGURE 3 - COMPLETED PROJECT

The filled and compacted surface was covered with a low permeability asphalt cover 12 inches (0.3 m) thick with 2 to 3 inches (5 to 8 cm) of asphalt macadam wearing surface. The surface is designed to slope at a minimum one percent to provide drainage. A typical section of the landfill is shown on Figure 4. This section is typical of the warehouse and repair building areas and shows an 80 mil (2 mm) HDPE membrane and concrete slab floors at finished grade.

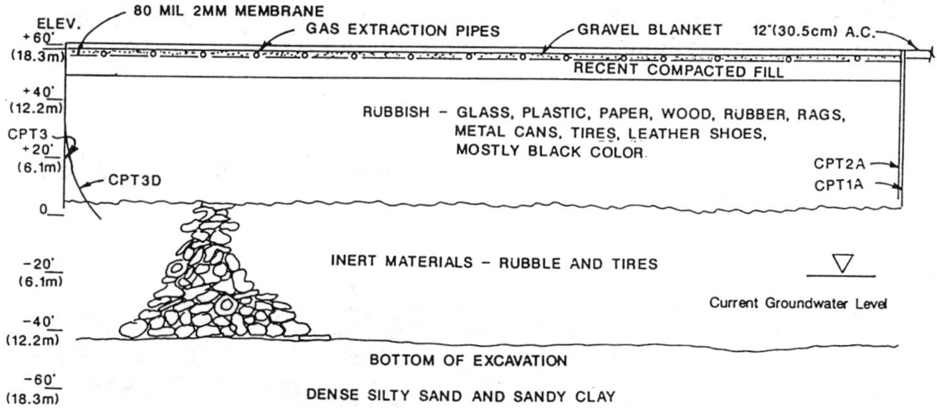

FIGURE 4 - TYPICAL SITE CROSS SECTION

The exterior earth and rubbish slopes were left at approximately 4:1 (horizontal to vertical) approximately 20 feet (6 m) high. The slopes were completed by placing compacted sand fill to provide 2 feet (0.6 m) of soil cover, 12 inches (0.3 m) of compacted clay (k) $<10^{-6}$ cm/sec then placing fill to the desired finished grade. The clay which was available approximately 5 miles away, was actually a red-brown silty clay with a PI of approximately 20. The final slopes were at a gradient of 2:1 and landscaped to prevent erosion. A compacted clay cutoff approximately 3 feet (0.9 m) deep by 1 1/2 feet (0.5 m) wide was placed at the interface between the clay cover, natural ground, and rubbish. Approximately 30,000 cubic yards (26,000 m^3 of clay was imported to the site for the slope cover and cutoffs.

Design Conditions

The design conditions for this project were developed by a series of compromises between the developer-owner, the prospective user, and the State of California-Los Angeles Regional Water Quality Control Board. The design guidelines and applicable laws are stipulated in California Administrative Code, Title 23, Chapter 3, Subchapter 15.

The basic design requires a permanent cover material which is impermeable to the penetration of runoff water from the exterior and methane landfill gas from the interior. The cover must be flexible enough to absorb the deformations caused by the settling landfill and still retain its impermeable nature.

The user-imposed constraint was that the surface must be strong enough to withstand wheel loads from a large forklift type loader. A second condition was that the slope was to be kept as flat as possible to avoid overturning the loaded forklifts. The State closure laws requires a 3 percent slope but the user needed a 1 percent slope. A special asphalt double seal system was designed so that a 1 percent slope could be used and still be accepted by the State.

The asphalt cover was designed to have a permeability coefficient (k) less than 10^{-6} cm/sec, yet have a pavement rigidity slightly less than road grade pavement. The material specification is as follows:

Asphalt Content SC-800	6.0 - 6.5 %
Residual Asphalt in Aggregate	1.0 - 1.5 %

Sieve Size	Percent Passing
1/2	100
#4	75 - 100
#16	40 - 70
#50	20 - 40
#200	10 - 20
HVEEM Stability	Over 30

The material was physically created by crushing waste concrete and asphalt in a cone crusher, then blending with waste created from cleaning railroad ballast for added fines. Approximately 100,000 tons (91,000 kg) of aggregate was produced for the project. The oil was added in a pugmill type mixer, the material placed and compacted in 6 inch (15 cm) layers and allowed to cure 30 to 60 days. The material was compacted to 95 percent modified Proctor compaction (ASTM D1557 method) [4] to provide the required permeability and density. Tests showed that at less than 5 percent air voids, the permeability was less than 10^{-6} cm/sec. The field permeability verification data is presented in Figure 5.

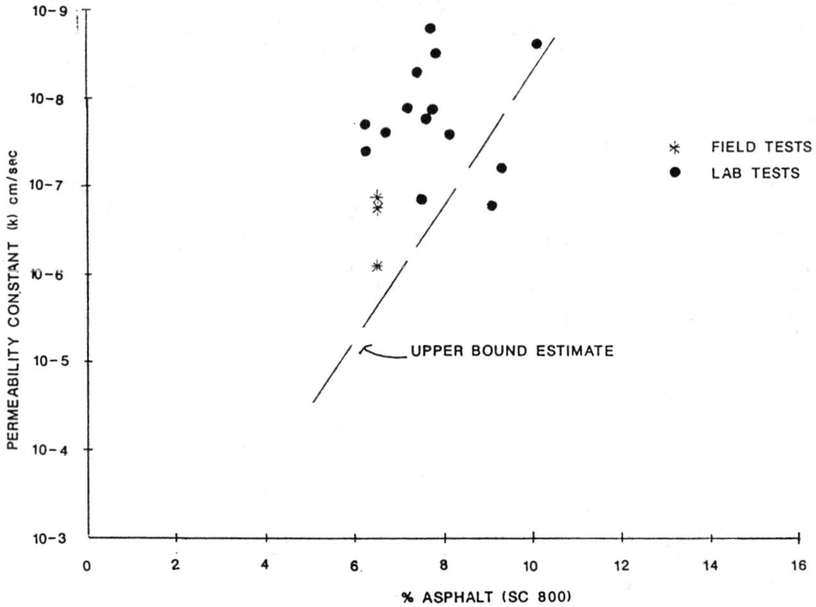

FIGURE 5 - FINAL PERMEABILITY TEST DATA - FIELD MIXED AND PLACED ASPHALT

Building Design

Two structures were required for the project: a small one-story office approximately 1,500 square feet (139 m^2) and a large warehouse approximately 40 feet (12 m) high and 7,500 square feet (698 m^2). The warehouse structure is a steel framed structure with steel siding and is open on one side with a 150 X 50 foot (46 X 15 m) clear opening for container mover access. It was impossible to penetrate the rubbish due to concrete rubble in the lower 50 feet (15 m) so deep foundations were impossible.

It was decided to place the structures on reinforced concrete mat foundations. The foundations consist of 3 sections of mat approximately 50 X 50 feet (15 X 15 m) by 18 inches (0.5 m) thick. The mat sections are connected by post-tension cables to allow movement between segments. The mats are also fitted with permanent pipe sleeves to re-level the mat with cement grouting. The structure is designed with leveling pads at the column connections and joints to allow movement. The structure is designed to allow up to 6 inches (15 cm) differential settlement.

The area beneath the structures has a gas collection system and an 80 mil (2 mm) HDPE methane barrier. Beneath the barrier is 12 inches (0.3 m) of 3/8 inch (1.0 cm) pea gravel and approximately 5 feet (1.5 m) of compacted fill over 95 feet (29 m) of rubbish. Settlement monitoring is underway but incomplete.

Settlement Measurements

In 1981, a compressor station resting on rubbish was installed approximately 300 feet (92 m) west of the new structure. The compressor was founded on a reinforced concrete mat foundation approximately 12 inches (0.3 m) thick. The differential settlement of the compressor station was monitored from 1983 through 1988, because several stages of releveling were required. The compressor area was approximately 50 X 100 feet (15 X 31 m) and differential settlement rates at the corners of the 3 slabs ranged from approximately 1 to 1.6 inches per year (2.5 to 4.0 cm/yr).

Total settlement rates were not measured but later measurements of the non-structural finished surface revealed total settlements of 3 to 12 inches (8 to 31 cm) in 1989.

A surcharge settlement test was performed by Woodward-Clyde Consultants [1] on the adjacent 24 acre landfill of similar composition but with only 60 feet (18 m) of rubbish. These tests showed that using a 20 foot (6 m) fill surcharge, settlements of 12 to 39 inches (0.3 to 1.0 m) could be induced immediately with no predictable amount of settlement or location. Approximately 40 to 80 percent of the settlement occurs in the first 10 days after loading with a rather linear settlement rate continuing in a non-ending pattern.

Settlement rates appear to vary in a random pattern. Long term behavior shows some heave or rebound behavior at some locations. It appears that the deep fills tend to squeeze and spring upward in random locations.

Settlement data in similar landfills was collected by the Los Angeles County Sanitation Districts at their SPADRA Landfill [2] showing total settlements of approximately 7 to 12 inches/year (0.2 to 0.3 m) in a 100 foot (31 m) rubbish fill. Ultimate settlement can be up to 30 percent of the height of rubbish.

Unique Testing Procedure

Several applications of existing technology were adopted for this project.

Cone Penetrometer

Because of the susceptibility to settlement, it was decided to test the consistency of the rubbish. Drilling and sampling is relatively dangerous and unreliable, so cone penetrometer test (CPT) probes were attempted at the corners of the structures. Typical CPT data are shown on Figure 6.

Approximately 40 percent of the CPT attempts penetrated the rubbish to the desired depth without encountering an obstruction. All probes stopped at the rubble interface.

The data showed the rubbish to be relatively consistent with a tip resistance of approximately 50 tons/square foot (5 MPa) and a skin friction of approximately 0.5 to 1 tons/square foot (50-100 kPa). The data were not used to predict ultimate settlement but verified that conditions were as recorded and that the rubbish has considerable shear strength.

Permeability Testing

In order to install the cover, it was necessary to conduct numerous permeability tests to verify the material properties. The tests were a combination of field and remolded laboratory permeability tests.

It was necessary to have a permeability coefficient (k) less than 10^{-6} cm/sec. During placement, random samples were obtained from the batch plant and returned to our laboratory for testing. Since no ASTM standard for falling head permeability tests exists, we adopted a soil permeability procedure for asphalt samples. Samples were compacted into a 4 inch (10 cm) diameter mold approximately 1 inch (2.5 cm) thick, allowed to cure 14 days then confined and approximately 8 feet (2.4 m) head of water applied. Results were measured daily and plotted until a steady-state rate achieved. Usually, approximately 30 days were required to achieve consistent results. The test apparatus and procedure are described in Bowles [3].

Field permeability tests were required so sealed double ring infiltrometer apparatus as shown in Figure 7, were fabricated and installed after surface curing. These tests required approximately 30 days to complete.

During the project 16 laboratory permeability tests and 3 field permeability tests were performed on the material. The average permeability coefficient (k) of the laboratory tests, saturated and near 95 percent relative compaction, was 9.1×10^{-8} cm/sec and the field tests results averaged 4.2×10^{-7} cm/sec. It can be seen from Figure 7 that the field test data is slightly higher than

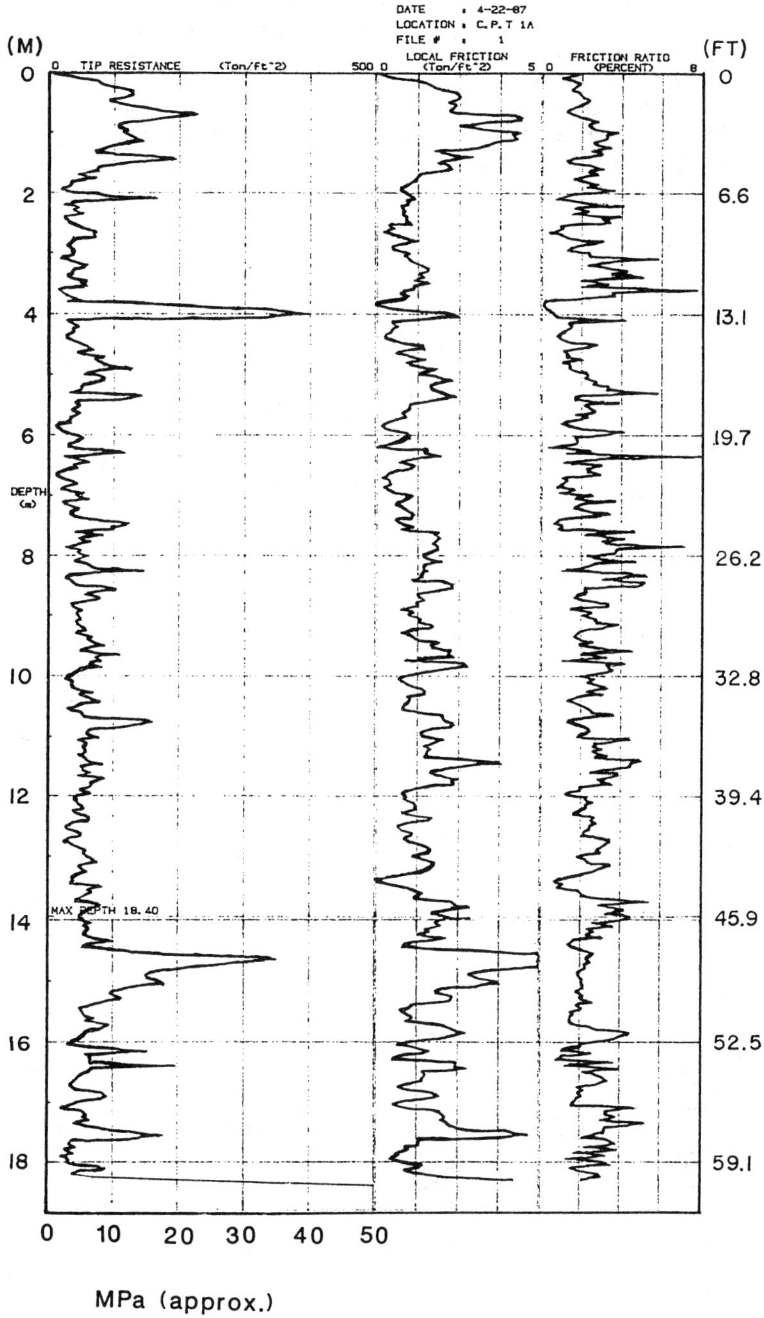

FIGURE 6 - TYPICAL CONE PENETROMETER DATA IN RUBBISH

the laboratory data. This can be explained by the fact that the laboratory tests were performed on higher density samples (96 to 98 percent compaction) versus field tests at 95 percent compaction. Permeability is dependent upon air voids. Air voids can be reduced by more compaction, more asphalt or more fines and each reduction in air voids reduces permeability.

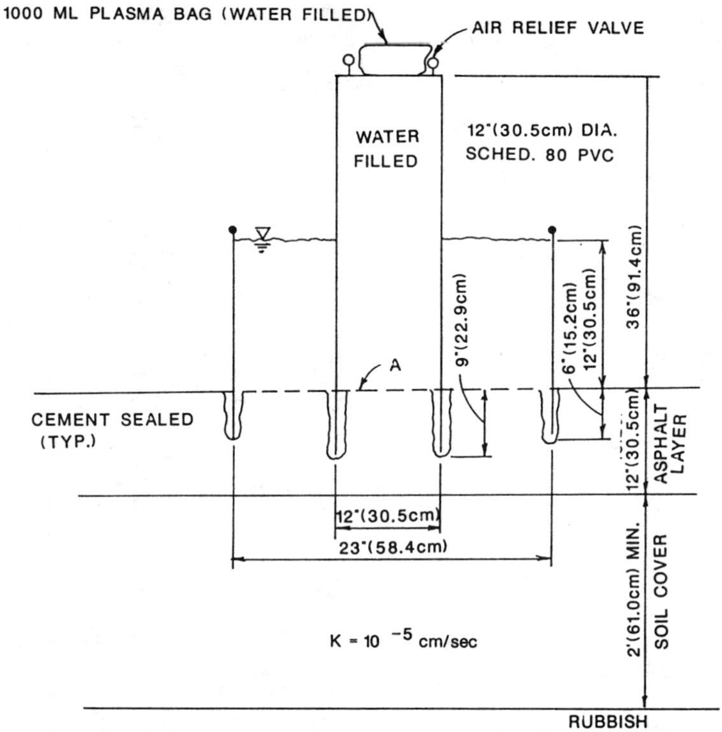

LEGEND

Q=kiAt
k=Q/At
I=HEAD/COVER THICKNESS
=42"(106.7cm)/12"(30.5cm)
=3.5
k=PERM. COEFFICIENT (CM\SEC)
A=AREA OF BOTTOM OF TUBE=700CM²

t =TIME BETWEEN FILLING AND READING BAG (SEC.)
Q=VOLUME LOST BETWEEN READINGS IN GRAMS (CM³)

FIGURE 7 - FIELD PERMEABILITY TEST DEVICE

Conclusions and Recommendations

This project demonstrates that landfill property can be reclaimed and put to profitable use. The cover and structures have now been in service for approximately one and one-half years. Several lessons have been learned during this process.

The seal, as placed, has been highly effective. Prior to cover completion, the methane gas extracted from the landfill was up to 50 percent oxygen from the atmosphere. After closure the methane was less than 1 percent oxygen.

The system, as installed, is probably only effective in a very dry environment such as the southwestern United States. The asphalt mix during placement is very moisture sensitive and cannot be placed or cured during rainy weather.

The asphalt cover requires extreme quality control measures. Precise measurements of asphalt and moisture content must be made 4 to 6 times daily to ensure a consistent, stable impermeable mix.

A batch type mixer is required for asphalt mixing. A continuous pugmill type mixer was used on the this project which created several weak areas caused by too much cutback asphalt in the mix. The range of 6.0 to 6.5 percent could not be exceeded without a significant decrease in surface stability and resistance to traffic loads.

The landfill slope in excess of 1 percent can be tolerated for safe operation of loaders. The critical item is sudden bumps. A slope of 2 percent will greatly improve surface drainage and reduce maintenance costs to repair low areas caused by differential settlement.

The maintenance cost of the surface over a landfill is approximately twice the maintenance cost of normal parking lots. The maintenance cost of this project is approximately 2 to 3 cents/square foot/year.

The closure cost for the project was approximately 10 to 20 percent higher than the closure costs for normal landfills without a planned recycled use. It is hoped that the information developed in this project will assist in recycling other landfill properties.

REFERENCES

[1] Woodward-Clyde Consultants, "Surcharge Test Program, 24 Acre Site, ASCON Landfill, Wilmington, California, May 1987.

[2] Los Angeles County Sanitation Districts, "Private Files of Test Data, 1982.

[3] Bowles, J. E., Engineering Properties of Soils and Their Measurement, McGraw Hill, New York, 1978.

[4] ASTM, 1987 Annual Book of ASTM Standards Volume 4, Soil and Rock; Building Stones; Geotextiles.

Richard E. Oakley, III

CASE HISTORY: USE OF THE CONE PENETROMETER TO CALCULATE THE SETTLEMENT OF A CHEMICALLY STABILIZED LANDFILL

REFERENCE: Oakley, III, Richard E., "CASE HISTORY: Use of the Cone Penetrometer to Calculate the Settlement of a Chemically Stabilized Landfill," Geotechnics of Waste Fills-Theory and Practice, ASTM STP 1070, Arvid Landva, G. David Knowles, Eds., American Society for Testing and Materials, Philadelphia, 1990

ABSTRACT: A waste fill containing chemically stabilized waste materials has been subjected to the overburden loading of a final cap for a period of over two (2) years. Prior to placement of the multicomponent cap, Cone Penetration Test (CPT) soundings were performed in the waste fill to determine the in situ condition of the materials. Settlement calculations developed from CPT and other data are compared to actual settlement survey data and discussions of the accuracy of the predictive method are presented.

KEYWORDS: chemical stabilization, Cone Penetrometer Test (CPT), elastic settlement, consolidation deformation, time rate of settlement

INTRODUCTION

A facility in the mid-western United States received municipal and industrial wastes for treatment and disposal from pre-1972 until about September 1983. Two (2) waste management units at this facility were used for the disposal of chemically-stabilized industrial wastes. In the latter stages of the operating life of this facility and through the closure era, these two (2) particular units were commonly referred to as the "Old Basin" and the "New Basin."

The Old Basin is a unit of rectangular shape, approximately 137 m (450 ft.) by 76 m (250 ft.). The lined floor of this basin slopes to the east to a maximum depth of about 7.5 m (25 ft.). The liner in this basin consists of glacial till and clayey loess deposits of nominal 1 m (3 ft.) thickness, placed and compacted to an average 95 percent of

Mr. Oakley is the Divisional Vice President in charge of Environmental Remediation for Browning-Ferris Industries, P.O. Box 3151, Houston, Texas 77253

maximum standard Proctor density (ASTM D698-78). The Old Basin received freshly mixed stabilized wastes for full hydration and solidification. This unit operated from about August 1979 to November 1980, when the unit was closed and a sloped, compacted clay cap was placed over the area.

The New Basin is a square-shaped unit approximately 125 m (400 ft.) per side. The lined floor of this unit slopes to the east-northeast to a maximum depth of about 10 m (30 ft.). The liner in this basin consists of glacial till and clayey loess deposits of nominal 1 m (3 ft.) thickness, placed and compacted to an average 95 percent of maximum standard Proctor density (ASTM D698-78). This basin also received freshly-mixed stabilized wastes for full hydration and solidification. This unit operated from about December 1980 until September 1983. An east-west aligned cross-section of both the Old and New Basins is presented on Figure 1.

CHEMICAL FIXATION PROCESS

Both the Old and New Basins were used for deposition of freshly-mixed chemically stabilized wastes. Placement in these basins allowed the freshly mixed product an opportunity to fully hydrate and solidify. These basins (in conjunction with a final cap) ultimately became the final depository for these materials.

Typical waste streams (raw state) received by these units for chemical fixation were American Petroleum Institute (API) separator sludges, washwater inks, wastewater treatment sludges, lead tank bottoms, copper wire drawing solutions, biotreated petroleum sludges, flexographic ink wastes, and common industrial floor sweepings. These wastes were initially deposited in on-site receiving basins for transfer to the patented solidification unit. In the solidification unit the wastes were thoroughly blended with a dry reagent (Type I Portland cement) and a liquid reagent (sodium silicate solution) in proprietary portions. Type I Portland cement has been widely used for waste solidification due to its availability and low cost [1]. Soluble silicates, such as sodium silicate, are additives which will generally prolong suspension and accelerate the "set" of Portland cement to produce a solid by-product. Research with soluble silicates indicates these materials are beneficial in reducing the interference from metal ions in the waste solution [2, 3].

Once mixed with the solidification reagents, the waste was converted to a thixotropic suspension. This suspension was then discharged down a gravity sluiceway to one of the two (2) basins described previously. Over a period of about two (2) to three (3) days, the semi-solid suspension hydrated to a damp solid. Over a period of about thirty (30) days, the material typically set-up to a more rigid state. Some of the stabilized wastes were similar in appearance to cemented sand while the majority took on the consistency of a stiff to very stiff clay.

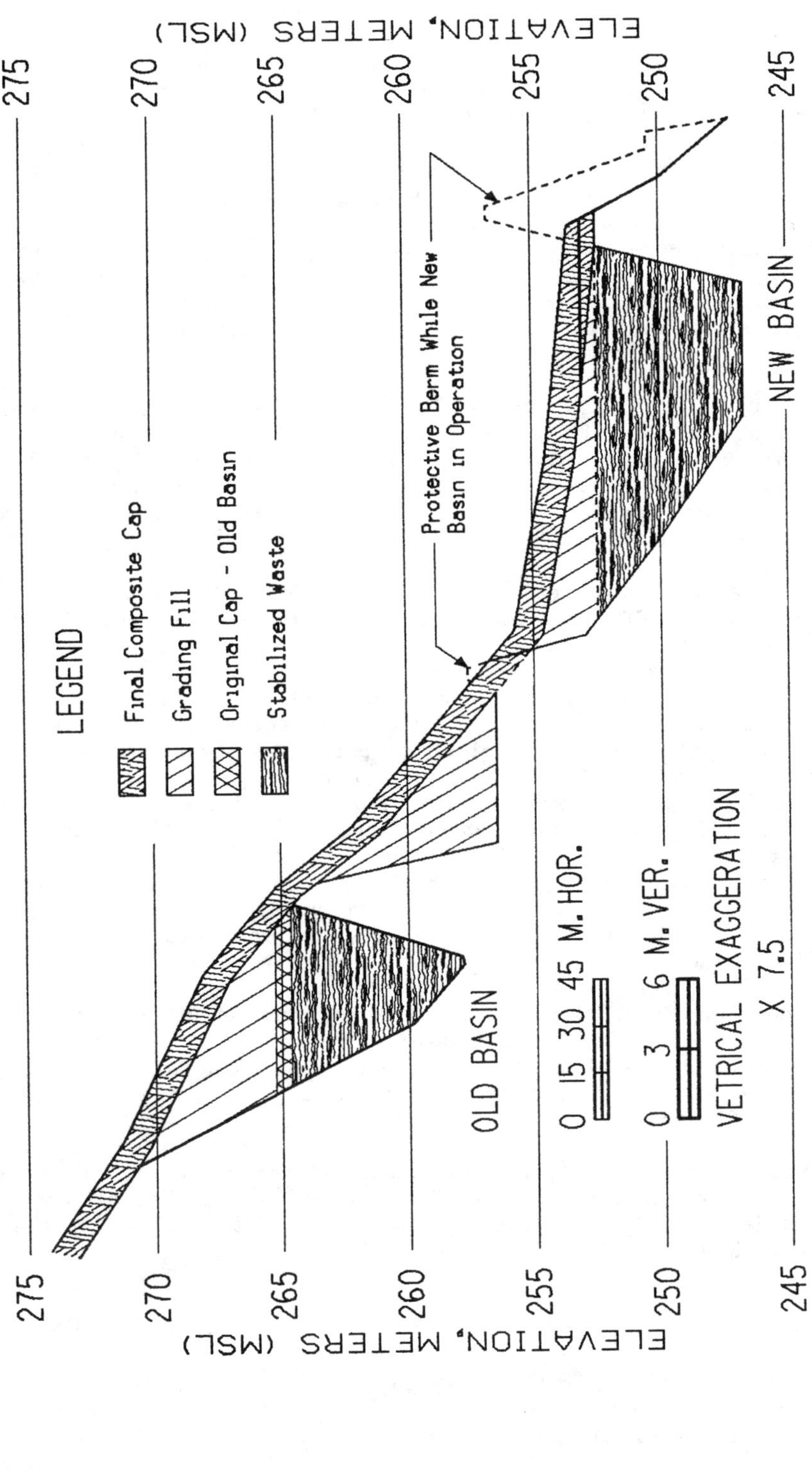
LEGEND
Final Composite Cap
Grading Fill
Original Cap - Old Basin
Stabilized Waste
Protective Berm While New Basin in Operation
OLD BASIN
NEW BASIN
0 15 30 45 M. HOR.
0 3 6 M. VER.
VETRICAL EXAGGERATION
X 7.5
ELEVATION, METERS (MSL)
275
270
265
260
255
250
245

FIGURE- 1

FIELD INVESTIGATION

The field investigation for this study was conducted on May 14 and May 15, 1985, prior to commencement of final closure activities. A total of eleven (11) cone penetration tests (CPT's) were performed at selected locations to evaluate the depth and mechanical characteristics of the solidified waste materials. Some CPT locations were selected to provide correlation with existing borings previously performed in a 1984 field investigation. All CPT's were performed with a truck-mounted cone penetrometer in general accordance with ASTM D 3441-86. Nine (9) tests were performed with a 15 cm^2 three-channel electric piezocone, and two (2) tests (CPT's 1 and 5A) were performed with a 10 cm^2 two-channel electric friction cone. The CPT locations and adjacent boring are shown on Figure 2. A typical CPT sounding log is shown on Figure 3.

CPT's 1, 2, 3, 5, 5A, 6 and 7 were performed in or near the Old Basin. CPT 5 was duplicated and labeled 5A because of an electronic data recorder failure. Only two (2) CPT's (CPT 8 and CPT 9) were accomplished in the New Basin because of poor accessibility around the edges of this open basin. CPT's 4 and 10 were performed for reference outside each of the basins.

Measurements of tip resistance (q_c) and sleeve friction (f_s) were obtained at all 11 CPT locations. In addition, measurements of dynamic pore pressure (u) were obtained at the 9 test locations where the three-channel piezocone was used. During piezocone testing at locations 4, 8, 9 and 10, the porous element in the piezocone became blocked due to the "smearing" characteristics of the materials encountered. With the filter blocked, pore pressure measurements were unreliable and were therefore, not reported. In each of the CPT locations where pore pressure values were obtained, there were zones of zero (0 kg/cm^2) pore pressure value (indicating a lack of saturation) and other zones where positive pore pressure measurements were registered (indicating total or near-total saturation). Since saturation was discontinuous the CPT pore pressure data was only used to qualitatively assess waste conditions and was not used in quantification technics.

In addition to CPT work, three (3) undisturbed waste samples were obtained during the 1985 field studies. These samples were taken at a depth of about 30 cm (1 ft.) in the New Basin using a 7.6 cm (3 in.) diameter Shelby tube. The locations of these samples are shown as S-1, 2, and 3 on Figure 2.

Previous field investigations conducted for final closure design activities in early 1984, resulted in three (3) samples of interest. These samples are S-48 in the New Basin and S-54 and 55 in the Old Basin, all shown on Figure 2.

LABORATORY INVESTIGATION

A very limited laboratory testing program was performed on the three (3) undisturbed samples recovered from the New Basin in 1985. Total unit weight (total density) of these waste materials ranged from

7.8 to 10.8 kN/m^3 (50 to 68.5 $lb./ft.^3$). Previous field investigations conducted in 1984 (samples S-48, S-54, and S-55) demonstrated total unit weights of 9.3 to 11.4 kN/m^3 (59 to 73 $lb./ft.^3$), with moisture contents ranging from 151 to 259 wt.%.

In the 1984 laboratory program, one-dimensional consolidation tests were also performed on the three (3) recovered waste samples. Sample numbers S-54 and S-55 were recovered from the Old Basin at depths of 1.4 and 0.4 m (4.5 and 1.3 ft.), respectively. Sample S-48 was recovered from a depth of 1 m (3.2 ft.) from the New Basin. Pertinent data from the consolidation tests are presented in Table 1.

TABLE 1 -- Consolidation Test Results

Parameter	S-48	S-54	S-55
Initial Water Content, wt.%	259	151	170
Initial Total Density, gm/cm^3	1.08	1.18	0.97
Initial Dry Density, gm/cm^3	0.30	0.47	0.36
Degree of Sat. (S_r), vol.%	97	99	79
Final Water Content, wt.%	135	147	165
Specific Gravity (G_s)	1.50	1.64	1.49
Compression Index (C_c)[a]	1.76	0.07	2.14
m_v, kPa^{-1}	$2.2x10^{-3}$	$5.8x10^{-3}$	$4.9x10^{-5}$

[a] C_c was determined from the steepest part of the consolidation plot

[b] m_v = coefficient of volumetric compression over the stress range of 0.5 to 1.0 kg/cm2

FINAL CLOSURE

The Old Basin ceased operation in November 1980. Shortly thereafter, a sloped clay cap of minimum 60 cm (24 in.) thickness was placed and compacted. The New Basin ceased operation in September 1983. This basin remained open, exposed to the elements and impounding rainwater and runoff for a period of about three (3) years, from September 1983 to about August 1986. This impounded water was periodically transported off-site to a permitted facility for treatment. Even so, the frequent presence of ponded water over the stabilized waste materials tended to saturate and soften the materials over the prolonged period of time.

A final closure plan for the facility was submitted to the governing regulatory agency in April 1984. By the time final regulatory approval of the closure plan was obtained and a contract let, closure activities did not commence until June 1985. The New Basin was one of the final waste management units to be closed in October 1986.

The final closure of this facility was a multifaceted activity. The principal activity was the placement of a continuous, multicomponent final cap over about 28 hectares (70 acres) covering all of the waste management units present at the facility. The final cap consisted of a passive gas venting system composed of geosynthetics; 45 cm (18 in.) of low permeability compacted clay; 0.75 mm (30 mil) polyvinyl chloride (PVC) geomembrane; high density polyethylene (HDPE) drainage net; polyethylene geotextile; 60 cm (24 in.) of compacted cover soil; and well established vegetative cover. Exclusive of any fill emplaced to establish drainage grades, this final cap exerted an overburden pressure of about 0.22 kg/cm^2 (455 psf.) on the underlying materials. The maximum overburden pressure exerted on the solidified waste contents of either basin, including grading fill, was about 1.95 kg/cm^2 (3995 psf).

CALCULATED SETTLEMENT

In selection of an evaluation technique, both elastic theory and consolidation approaches were considered. Visual inspection of recovered waste samples and construction excavations into some areas of the solidified waste, revealed a moist but solid-looking material, appearing partially cemented. This classification suggests the material might be analyzed using elastic theory, such as for granular soils. Soil classification using the electric friction cone [4] produced a largely predominant classification of "clay" or "clayey silt and silty clay" indicating the materials could deform consistent with consolidation theory.

A number of published techniques [5, 6, and 7] were considered in choosing a method to calculate settlement. Elastic deformation techniques as developed by Schmertmann [5 and 6] and Sangrelat [7] were not used in this evaluation. The majority of elastic deformation settlement usually occurs shortly after application of the overburden load. Since the settlement markers were not set and surveyed until about 140 days following completion of the final capping, this type of settlement calculation would not be comparable to the measured values.

Deformation through pore fluid consolidation was chosen as the method to calculate settlement, via the classic consolidation equation presented below.

$$\Delta H = \sum_{1}^{n} [H_1 \cdot \frac{C_c}{1+ei} \cdot \log_{10} \frac{P_1 + \Delta P}{P_1}] \qquad (1)$$

Where:
- n = number of consolidating layers
- C_c = compression index
- e_i = initial void ratio
- P_1 = initial vertical effective stress
- ΔP = expected increase in stress
- H_1 = initial thickness of consolidating layer
- ΔH = total consolidation deformation

One of the principal advantages of cone penetrometer testing is the recovery of near-continuous in situ data. Therefore, values for C_c in Equation 1 were determined from empirical relationships with q_c values

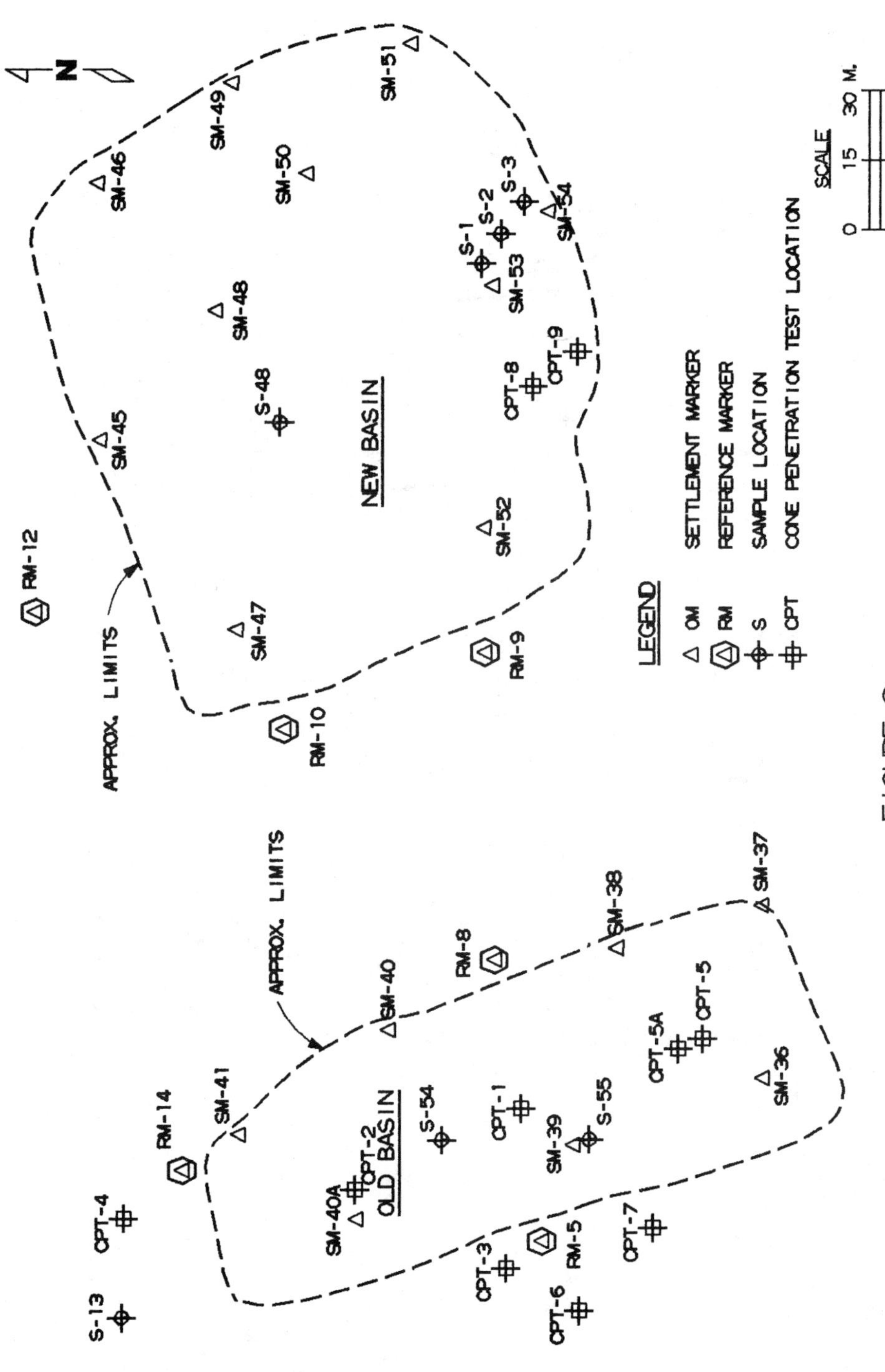
N
NEW BASIN
OLD BASIN
APPROX. LIMITS
APPROX. LIMITS
RM-12
RM-10
RM-9
RM-14
RM-8
RM-5
SM-45
SM-46
SM-47
SM-48
SM-49
SM-50
SM-51
SM-52
SM-53
SM-54
SM-36
SM-37
SM-38
SM-39
SM-40
SM-40A
SM-41
S-1
S-2
S-3
S-48
S-54
S-55
S-13
CPT-1
CPT-2
CPT-3
CPT-4
CPT-5
CPT-5A
CPT-6
CPT-7
CPT-8
CPT-9
LEGEND
SM SETTLEMENT MARKER
RM REFERENCE MARKER
S SAMPLE LOCATION
CPT CONE PENETRATION TEST LOCATION
SCALE
0 15 30 M.

FIGURE 2

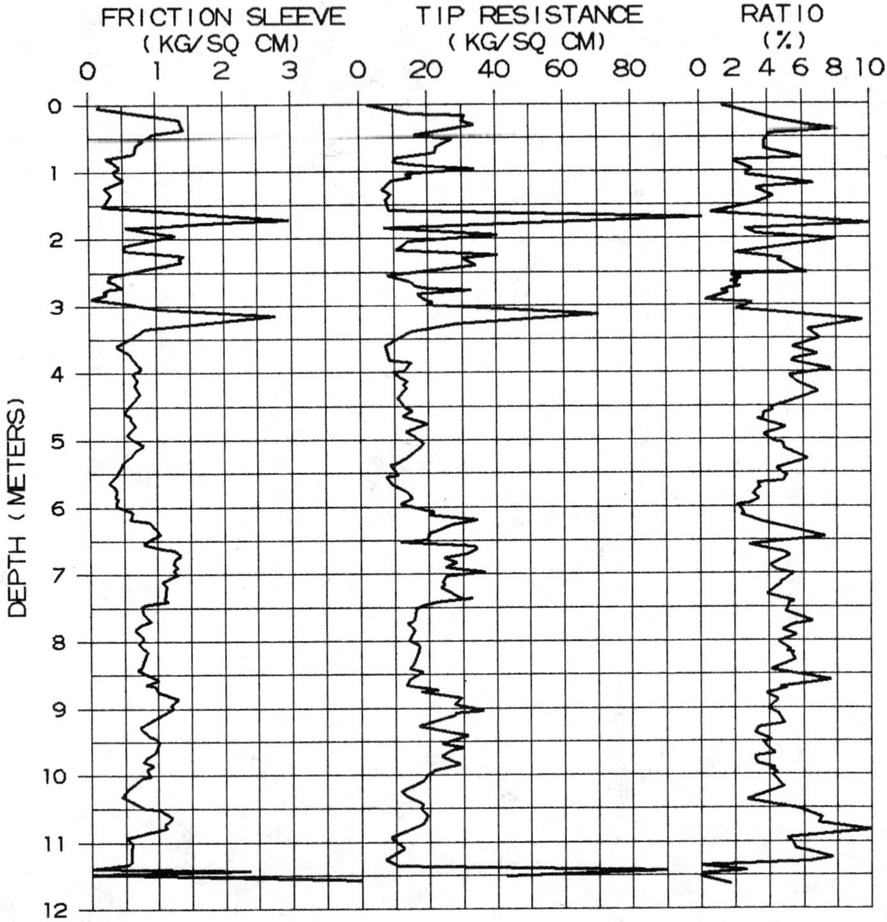

CPT-1
FIGURE 3

obtained from the CPT data [7]. This approach allowed the use of individual values of C_c for discrete waste sublayers, as opposed to generalized values for C_c (determined from the laboratory testing program), applied to the entire thickness of waste. Values for e_i in Equation 1 were determined from the laboratory consolidation tests.[1]

MEASURED SETTLEMENT

Placement of the final engineered cap over the waste management units of the facility was begun in June 1985 and completed by October 1986. A system of settlement and reference markers were established across the 28 hectare (70 acre) waste management area. In the area of the New and Old Basins, there are six (6) adjacent reference markers. Settlement of the Old Basin area is monitored by seven (7) settlement markers. The New Basin is monitored by ten (10) settlement markers. Settlement markers are typically marked corners of concrete foundation mats for passive gas vents or fluid collection sumps. These points are directly over the waste materials and can be used to measure cover subsidence. Reference markers are tradition 10 cm by 10 cm (4 in. by 4 in.) concrete survey monuments imbedded in the upper 60 cm (24 in) of the final cap. These markers are located outside the bounds of the waste management units and are typically stable, except for settlement of natural foundation soils under the loading of the new cap and any grading fill.

All of these markers were initially surveyed on February 19, 1987, with follow up surveys on August 25, 1987, February 19, 1988, and August 25, 1988. The survey schedule is now on an annual basis to post closure year five, when the survey interval expands to once every two (2) years. Actual settlement values determined from survey data are presented in Tables 2 and 3.

TABLE 2 - Old Basin Settlement[a]

Settlement Marker No.	Cumulative Settlement, cm.[b] 08/25/87	02/19/88	08/25/88
SM 36	0	0	2.0
SM 37	0	0.5	0.5
SM 38	0	1.0	1.5
SM 39	2.5	2.5	4.5
SM 40	0	1.5	2.0
SM 40A	1.5	1.5	2.5
SM 41	0	0	1.5

[a] Baseline survey established February 19, 1987

[b] All measurements rounded-off to the nearest 0.5 cm.

TABLE 3 - New Basin Settlement[a]

Settlement Marker No.	Cumulative Settlement, cm.[b]		
	08/27/87	02/19/88	08/25/88
SM 45	0	1.0	1.0
SM 46	0.5	2.5	4.0
SM 47	2.0	5.5	7.0
SM 48	1.0	5.0	5.5
SM 49	2.0	3.5	4.0
SM 50	5.0	8.0	12.0
SM 51	0	2.0	2.0
SM 52	0.5	1.0	2.5
SM 53	2.0	3.5	5.5
SM 54	0	2.0	2.0

[a] Baseline survey established February 19, 1987

[b] All measurements rounded-off to the nearest 0.5 cm.

The average settlement of the final cap overlying the Old Basin is about 2 cm. while the average cap settlement over the New Basin is about 4.5 cm. This is probably attributable to the preconsolidation of the waste materials in the Old Basin due to overburden loading exerted by the interim cap, which has been in existence since November 1980. In addition, the interim cap sealed the Old Basin's contents from precipitation and inundation facilitating an environment for continued solidification and hardening. The waste materials contained in the New Basin were exposed to the elements and subjected to periodic inundation for a period of about three (3) years. This situation probably moistened the waste materials keeping them in a softened, pliable condition, thus resulting in greater settlements compared to the Old Basin.

COMPARISON OF CALCULATED AND ACTUAL VALUES

The original design calculations (circa 1985) for settlement in these two (2) basins were intentionally broad and conservative, to account for the numerous unknown or unqualifiable conditions. Total settlement for the Old Basin was predicted to be between 15 cm (0.5 ft.) and 91 cm (3 ft.), with 90 percent of the consolidation settlement occurring at eight (8) years after final closure. Total settlement for the New Basin was predicted to be between 15 cm (0.5 ft.) and 122 cm (4 ft.), with 90 percent of the consolidation settlement occurring at three (3) years after final closure.

For this evaluation, comparisons of calculated and actual settlement values were performed for three (3) points in the Old Basin and four (4) points in the New Basin. Settlement markers SM-36, SM-39, and SM-40A were chosen for the Old Basin because: (a) they are in close proximity to CPT-5A, 1, and 2, respectively; and (b) they are in the interior of the basin where the waste thickness is the greatest and, therefore, settlements are expected to be the largest. Settlement markers SM-48, SM-50, SM-52, and SM-53 were chosen for the New Basin because they are located in the interior of the containment unit where waste thicknesses are the greatest. Due to site accessibility (e.g. ponded water and untrafficable conditions), there were no CPT tests performed in the immediate vicinity of SM-48, SM-50, SM-52, or SM-53. CPT-8 and 9 were used for settlement calculations for markers located in the New Basin.

Those settlement markers located around the perimeters of the basins were not used in the comparisons because the magnitude of their actual settlements was generally small, probably due to the limited thickness of waste materials present at the edges of the bowl-shaped containment units. Additionally, there is little CPT data available in the vicinity of these perpherial settlement markers.

Comparisons of actual and calculated settlement values are presented in Table 4.

TABLE 4 - Actual and Calculated Settlement[a]

Location/ Marker	Actual Settlement, cm.[b]	Calculated Values			
		Total, cm.	% of Actual	Time Rate, cm.[d]	% of Actual
Old Basin					
-SM 36	2.0	6.0	200%	1.0	-50%
-SM 39	4.5	14.0	211	7.0	+56
-SM 40A	2.5	13.0	420	2.0	-20
New Basin					
-SM 48	5.5	7.5	36	2.0	-64
-SM 50	12.0	15.0	25	8.0	-33
-SM 52	2.5	9.0	260	2.0	-20
-SM 53	5.5	14.0	155	6.5	+18

[a] All values rounded off to the nearest 0.5 cm. or the nearest whole %.

[b] Measured settlement as of the August 25, 1988 survey.

[c] % of Actual is the percent difference between the actual and calculated settlement values.

[d] Coefficient of Consolidation (c_v) of 2×10^{-3} cm^2/sec. was used.

The time rate value for consolidation settlement presented in Table 4 was calculated using an average value for the coefficient of consolidation (c_v) of 2×10^{-3} cm^2/sec., determined from the laboratory consolidation tests. The time rate settlement calculations presented above represent only that percentage of the total calculated consolidation settlement that takes place between the baseline survey obtained in February 1987 and the latest survey of August 25, 1988 (553 elapsed days).

FINDINGS AND OBSERVATIONS

The author offers the following findings and observations for the readers consideration.

1. Average settlement of the cap over the Old Basin was about half the average settlement of the cap over the New Basin, probably attributable to the preconsolidated condition of the waste materials present in the Old Basin.

2. Markers located around the periphery of either the Old or New Basins generally experienced less settlement than the interior settlement markers; with the exception of SM 47 for which there is no apparent explanation.

3. Calculated values of total consolidation settlement developed using CPT data appear reasonable compared to the original design calculations and the actual settlement data obtained to date.

4. Calculated values of time-dependent consolidation settlement (i.e. the percent of the total consolidation settlement occurring at a point in time) were generally within about ± 50 percent of the actual field measured settlement.

ACKNOWLEDGEMENTS

The author acknowledges with appreciation the assistance of Messrs. Thierry R. Sanglerat and Wayne D. Smith for their professional performance of field, laboratory, and engineering services on the original project. The author further acknowledges with appreciation the help of Ms. Kay Gilbreath and Messrs. John Magness, Randy Morgan, Roman Stockton, and Robert Blake in the preparation of this paper.

REFERENCES

[1] Cullinane, J., Jones, L.W., and Malone, P.G., "Handbook for Stabilization/Solidification of Hazardous Wastes," Report EPA/540/2-86/001, Environmental Laboratory, U.S.A.E. Waterways Experiment Station, Vicksburg, Mississippi, June 1986, pp. 2-12 - 2-15

[2] Columbo, P., and Neilson, R.M., Properties of Wastes and Waste Containers, Progress Report No. 7, BNL-NUREG 50837, Brookhaven National Laboratory, Upton, New York, 1978

[3] Falcone, J.S., Jr., Spencer, R.W., and Reifsnyder, R.H., Chemical Interactions of Soluble Silicates in the Management of Hazardous Wastes, Draft Report, The PQ Corp., Lafayette Hill, Pennsylvania, 1983

[4] Douglas, B.J. and Olsen, R.S., "Soil Classification using Electric Cone Penetrometer," Cone Penetration Testing and Experience, ed. Norris, G.M. and Holtz, R.D., Proceedings from American Society of Civil Engineers (ASCE), National Convention, St. Louis, Missouri, October 26-30, 1981

[5] Schmertmann, J.H., "Static Cone to Computic Static Settlement over Sand," ASCE Journal of Soil Mechanics & Foundation Engineering, Vol. 96, SM3, May 1970, pp. 1011-1043

[6] Schmertman, J.H., "Guidelines for Cone Penetration Test; Performance and Design," Report No. FHWA-TS-78-209, U.S. Department of Transportation, Federal Highway Administration, Washington, D.C., July 1978

[7] Sanglerat, G. The Penetrometer and Soil Exploration, Development in Geotechnical Engineering, Elsevier Scientific Publishing Company, Amsterdam, Holland, 1979

Donald P. Coduto and Raymond Huitric

MONITORING LANDFILL MOVEMENTS USING PRECISE INSTRUMENTS

REFERENCE: Coduto, D. P. and Huitric, R., "Monitoring Landfill Movements Using Precise Instruments," Geotechnics of Waste Fills - Theory and Practice, ASTM STP 1070, Arvid Landva, G. David Knowles, editors, American Society for Testing and Materials, Philadelphia, 1990.

ABSTRACT: Investigations of landfill movements usually consist of measurements of settlement at the ground surface. This project consisted of installing instruments inside vertical borings drilled through a landfill in order to monitor both vertical and horizontal movements at various depths. Following two years of monitoring the data suggests that vertical strain rates are independent of depth while horizontal movements on slopes are greatest near the surface and diminish with depth. No permanent displacements occurred during a magnitude 6.1 earthquake.

KEYWORDS: sanitary landfills, settlement, deformation, instrumentation, earthquakes

INTRODUCTION

Sanitary landfills gradually settle for several years after closure. The rate and magnitude of this settlement varies with many factors, including the time after placement and the thickness and composition of the refuse.

Engineers must be aware of future settlements when designing structures, surface facilities and landforms. Permanent structures may need to be equipped with costly foundations; piping, drainage channels and other surface

Donald Coduto is a professor of civil engineering at California State Polytechnic University, Pomona, CA 91768; Raymond Huitric is the supervisor of research and monitoring in the solid waste management department of the County Sanitation Districts of Los Angeles County, P.O. Box 4998, Whittier, CA 90607

facilities will probably require articulated and telescoping connections; and the landform needs to be constructed in anticipation of the changing topography, especially in light of maintaining good surface drainage patterns.

Accurate settlement estimates are needed to best design these systems, but our ability to predict landfill settlements is lacking. This paper briefly reviews some of the past efforts by the Los Angeles County Sanitation Districts to study settlement behavior and our recent efforts to monitor movements using precise instruments embedded within the refuse. The ultimate objective of these efforts is to develop more precise methods of predicting settlements.

INITIAL EFFORTS: MONITORING SURFACE SETTLEMENTS

The earliest efforts characterized settlement through measurements of benchmarks using conventional surveying techniques. One such project is the 25-year effort carried out at the Mission Canyon Landfill by the Los Angeles County Sanitation Districts.

The Mission Canyon Landfill in Los Angeles County, California is composed of three adjoining fills known as Canyons 1, 2 and 3. These are situated in the rugged terrain of the Santa Monica Mountains. Each of these canyons was filled with about 1.5 million metric tons of commercial and residential refuse between 1960 and 1964. A total of 29 settlement monuments consisting of brass tags set in concrete were placed in 1964 and 1965. The results of this project are summarized below. Huitric [1] provides a more detailed analysis.

The observed settlements from 1964 to 1981 ranged from 5 to 27 percent of the initial fill depth and averaged 15 percent. Monuments with the greatest settlement showed a rate of settlement that decreased with time. Figure 1 presents typical time-settlement data from two of the monuments.

The observed settlement rates generally decreased with time, and continue to be quite large. It appears that the ultimate settlement will be in the range of 30 to 35 percent of the refuse thickness. More recent monitoring continues to show results consistent with these findings.

The Mission Canyon work also addressed the mechanisms controlling the settlement. Three mechanisms were considered: Consolidation, compaction, and shrinkage.

Consolidation is the squeezing of air and water from saturated material under an applied load and is the primary mechanism controlling the settlement of soils. However, since little of the Mission Canyon refuse was saturated, this mechanism was not considered to be a major factor. This conclusion was confirmed by laboratory tests on partially saturated refuse.

Compaction is the decrease in volume under an applied load resulting from mechanisms other than the squeezing of water (such as strain in the solids). Laboratory tests on refuse samples suggest that compaction may be responsible for settlements of up to 25 percent of the refuse thickness.

Shrinkage is settlement due to the loss of solids due to decomposition. Empirical shrinkage models which have been used to estimate landfill gas generation can also be used to estimate settlements. Based on the composition of wastes typically received at Los Angeles County landfills, the ultimate settlement due to shrinkage is probably between 18 and 24 percent of the refuse thickness.

Laboratory research suggests that the combined ultimate settlement resulting from compaction and shrinkage will probably be less than 50 percent of the refuse thickness, which generally agrees with the Mission Canyon Data.

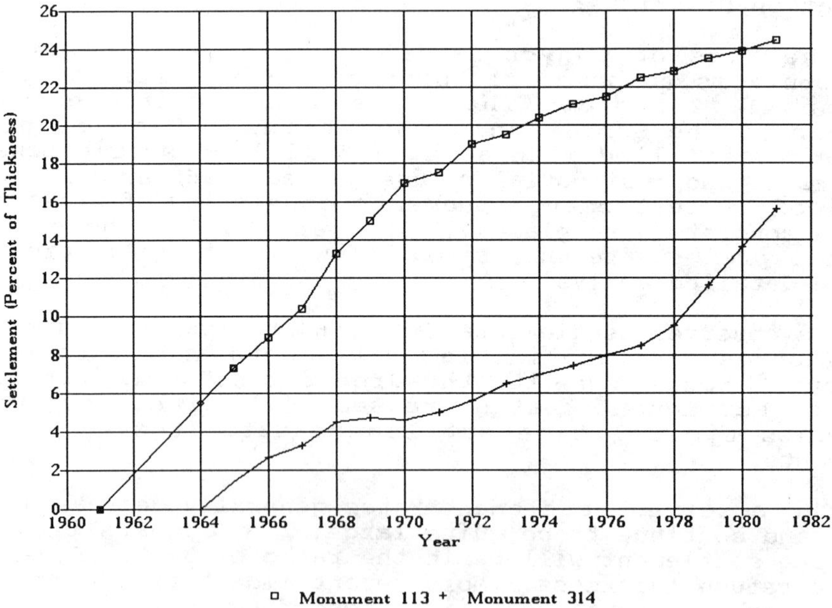

Note: Monument 113 was placed in 1965 and monument 314 was placed in 1966. Time-settlement data before these dates are estimates.

FIG. 1 -- Surface Settlements at Mission Canyon

FOLLOW-UP WORK: MONITORING MOVEMENTS AT VARIOUS DEPTHS

Measuring the settlement of surface monuments, such as those at Mission Canyon, is a useful way of accurately monitoring landfill movements over a large area. Similar studies can also be made using topographic maps generated from aerial photographs. However, neither of these methods gives us any insight into the distribution of movements throughout the depth of the landfill.

How do strain rates vary as a function of depth? The deeper portions of a landfill are subjected to higher stresses and are also older than the shallower portions. If we could measure movements at various depths, we might be able to gain more insight into these relationships and understand more about the processes which control landfill settlements. When combined with surface settlement data, this type of information should also help us predict settlements more accurately.

Use of Sondex Device

We have embarked on a project to investigate settlement as a function of time at various depths in a landfill. We have done this using the Sondex device manufactured by the Slope Indicator Company of Seattle, WA. The principle of operation of this device is shown in Figure 2.

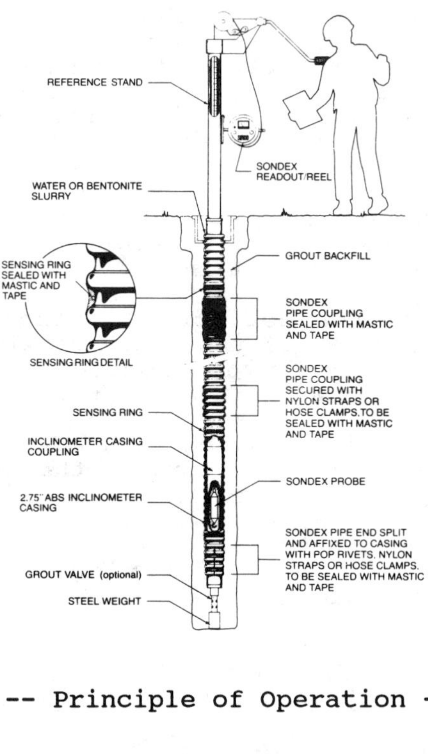

FIG. 2 -- Principle of Operation - Sondex

The Sondex was developed in order to monitor settlements in soft soils, but is ideally suited for our purposes as well. The permanently installed part of this system consists of a 76 mm (3 in) diameter corrugated plastic casing installed in a 350 mm (14 in) diameter vertical boring. The annular space around the casing is backfilled with pea gravel. As the landfill settles, the corrugated casing will compress and the pea gravel will move radially outward into the refuse.

Small steel wire rings are embedded into the casing at approximately 300 mm (12 in) intervals. The locations of each of these rings can be determined by lowering an inductive probe into the casing and moving it up and down until the induction signal peaks. We can then measure the depth from the top of the casing to each ring. Subtracting this depth from the depth to the lowermost ring gives the elevation above the bottom of the casing. Periodic monitoring gives us the elevation (settlement) of each ring as a function of time.

The manufacturer claims a precision of ~ 1.3 mm, but our experience suggests that it is closer to ~ 3 mm. However, considering the very large magnitudes of settlement we are measuring, this precision is quite sufficient.

<u>Use of Inclinometer</u>

An inclinometer is a geotechnical instrument used to measure horizontal movements in the ground, such as in landslides. The permanently installed part of this system consists of a 70 mm (2.75 in) diameter plastic casing with longitudinal grooves on the inside. This device is installed in a vertical boring and it moves horizontally as the ground around it moves.

The alignment of this casing can be measured by inserting a probe as shown in Figure 3. This probe is mounted on wheels which travel in the casing grooves. These wheels are 610 mm (24 in) apart. Sensors within the probe are able to measure the horizontal offset (both "north-south" and "east-west") of the upper wheels as compared to the lower wheels. These two offsets are displayed on a digital indicator as shown in the figure.

A set of readings is obtained by lowering the probe to the bottom of the casing and recording the two offsets. The operator then raises the probe in 610 mm (2 ft) increments, recording the offsets at each location. The accumulated data can then be reduced to give the horizontal position of the casing as a function of elevation. Periodic readings will disclose horizontal movements as a function of both elevation and time.

Although we are primarily interested in the vertical movements (settlement) in a landfill, we would also like

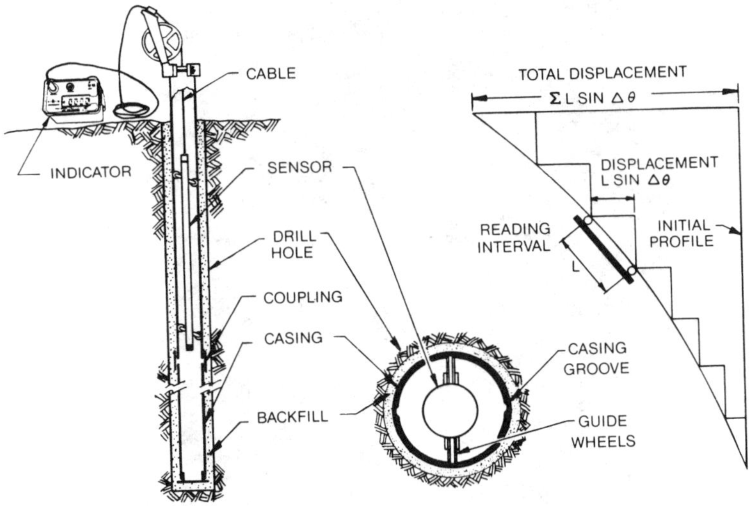

FIG. 3 -- Principle of Operation - Inclinometer

to understand more about horizontal movements. Since the inclinometer and Sondex are made to work together (one casing fits inside the other), it is a simple matter to install both instruments together.

DETAILS OF THE TWO PILOT INSTALLATIONS

We have tested this technology by constructing two pilot installations in a slope at the Spadra Landfill in Pomona, California. The primary purpose of these two installations was to demonstrate the workability of this technique. Many more installations will be needed in order to develop a more complete understanding of landfill movements.

The refuse in this portion of the landfill was placed between 1976 and 1978. It consists of general household and commercial wastes. A methane gas recovery system was being installed at the same time as the two instruments.

A cross section of a typical installation is shown in Figure 4.

FIG. 4 -- Typical Installation at Spadra Landfill

These two installations are known as instrument number 1 and instrument number 3. A third installation (number 2) had to be abandoned. The refuse at instrument no. 1 is about 16 m thick. At instrument no. 3 it is about 22 m thick.

These installations were completed and we began monitoring them in May 1987. The results of this monitoring obtained through May 1989 are described below.

VERTICAL MOVEMENTS (SETTLEMENT)

The results of our measurements of vertical movements are shown on figures 5 through 8.

Figure 5 shows settlement vs. time at six points along the casing in instrument no. 1. Note that the settlement at the ground surface has been about 370 mm over a period of two years. Likewise, figure 6 shows comparable data for eight points along instrument no. 3. The settlement here was greater because the refuse is correspondingly thicker. These results agree very well with surface settlements measured by others.

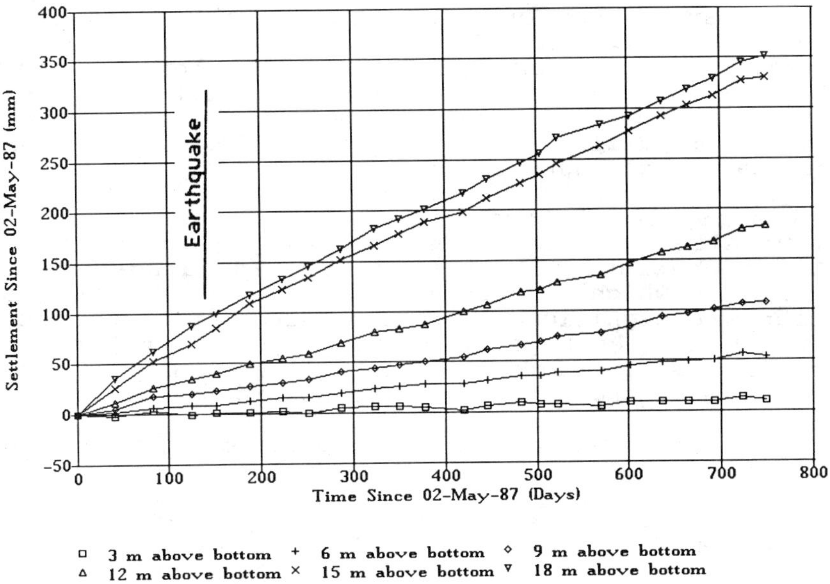

FIG. 5 -- Settlement vs. Time - Instrument No. 1

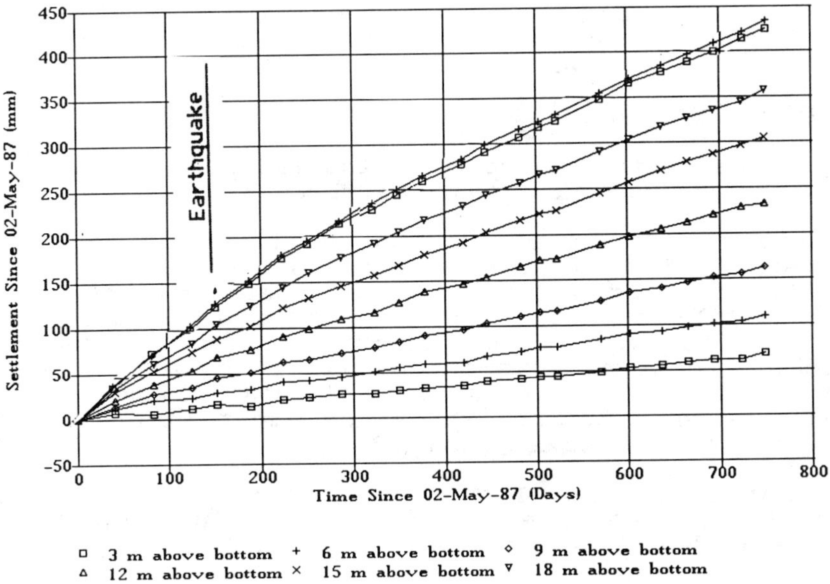

FIG. 6 -- Settlement vs. Time - Instrument No. 3

Figures 7 and 8 show the vertical strain rates as a function of elevation for instruments 1 and 3, respectively. Due to the non-uniformities in the makeup of the refuse we would expect this plot to be fairly erratic, and it is.

The greatest anomaly occurs between elevation 12.5 and 15.0 in instrument no. 1. It is interesting to note that the strain rates in this zone were much lower during the second year, which suggests that the first year's data in this zone may not be representative of the true movements within the landfill and may instead reflect local spreading of the pea gravel backfill.

At a couple of locations the plots indicate a small negative strain rate. This would suggest that at some locations the landfill is actually expanding rather than consolidating which is extremely unlikely. However, at no location are significant negative values indicated for more than one year. Perhaps small portions of the casing are moving in a slip-stick fashion and these negative values represent local short-term adjustments in the casing

We can draw the following conclusions from these four plots:

1. The average vertical strain rate is about 1 percent/year for 11 year-old refuse.

2. The vertical strain rate within the refuse appears to be independent of depth.

3. The rate of settlement appears to be slowly decreasing with time.

4. The bottom of the casing is compressing, suggesting that we may not have bottomed it in firm natural ground.

HORIZONTAL MOVEMENTS

The horizontal movements in the landfill during the past two years are shown in figures 9 through 11.

Figures 9 and 10 show the horizontal movement as a function of elevation for instruments 1 and 3, respectively. In both of these plots the terms "parallel to slope" and "perpendicular to slope" refer to movements in a horizontal plane as they would appear from above. Both of these plots indicate almost no movement parallel to the slope and as much as 110 mm of movement perpendicular to the slope. The negative sign on the horizontal movement figures indicate that the slope is moving outward.

At instrument no. 1 this outward movement is confined primarily to the upper 12 m. However, in instrument no. 3 it extends down to about 18 m.

Figure 11 shows the horizontal movement perpendicular to the slope at instrument no. 1 as a function of time. This

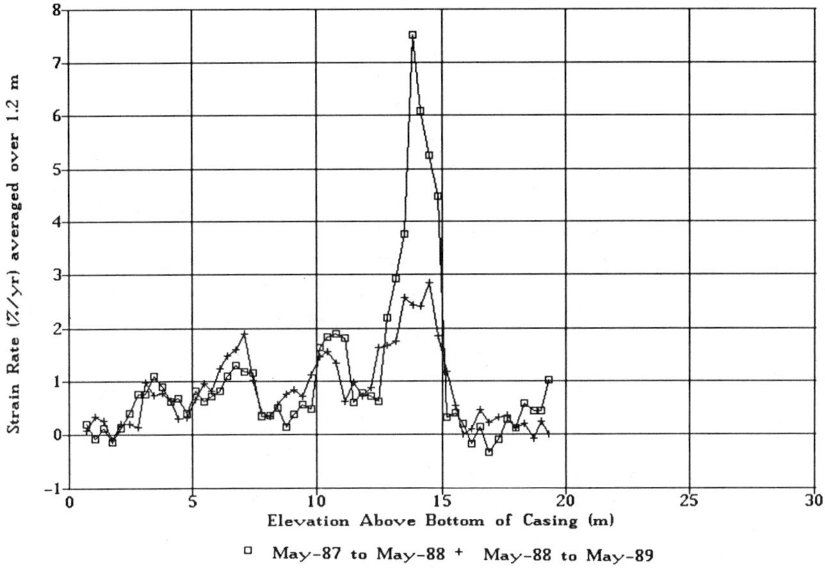

FIG. 7 -- Vertical Strain Rate vs. Elevation Instrument No. 1

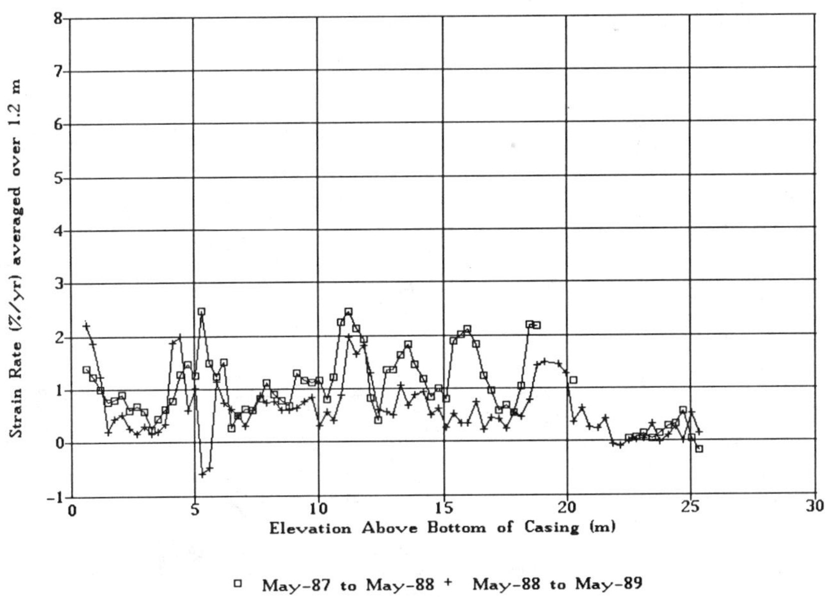

FIG. 8 -- Vertical Strain Rate vs. Elevation Instrument No. 3

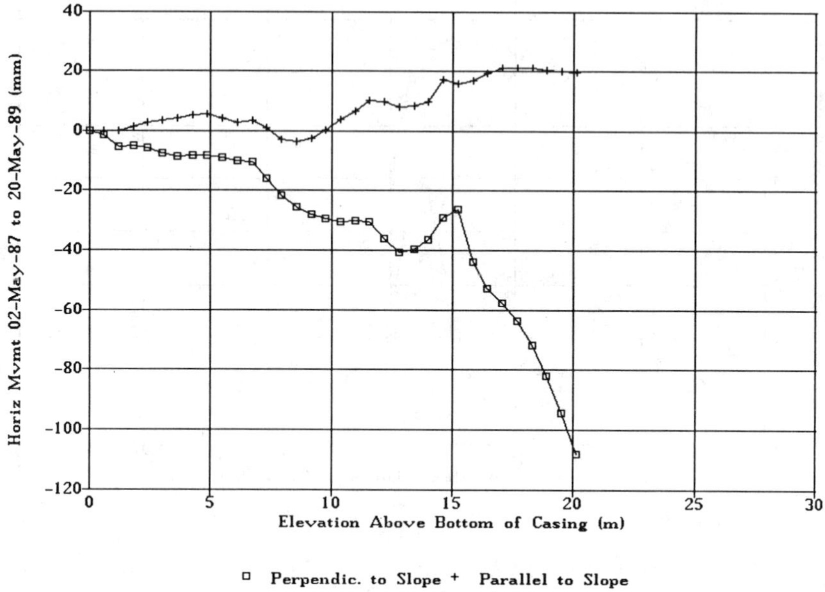

FIG. 9 -- Horizontal Movement vs. Elevation Instrument No. 1

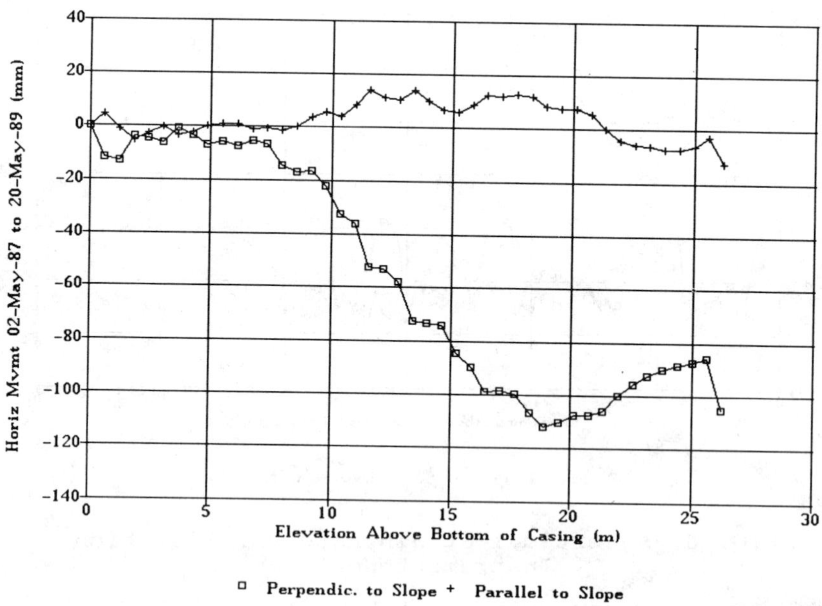

FIG. 10 -- Horizontal Movement vs. Elevation Instrument No. 3

movement seems to be occurring at a fairly constant, or perhaps slowly decreasing rate.

It is interesting to note that the curves on figure 11 are not perfectly smooth. We are not certain if this is because the movements are occurring in a "stick-slip" fashion, or if it just represents random error in the measurements. These irregularities do not appear to correlate with the seasons of the year (and therefore rainfall), which suggests that they may be random measurement errors.

We can draw the following conclusions from these three plots:

1. The ground surface is moving horizontally away from the slope at a rate of about 52 mm/yr. This motion gradually diminishes with depth and becomes nearly zero at depths of 12 to 18 m.

2. The rate of horizontal movement does not appear to correlate with rainfall.

3. No clear slip surface is present which suggests that the horizontal movement is a creep phenomena rather than a classical slide.

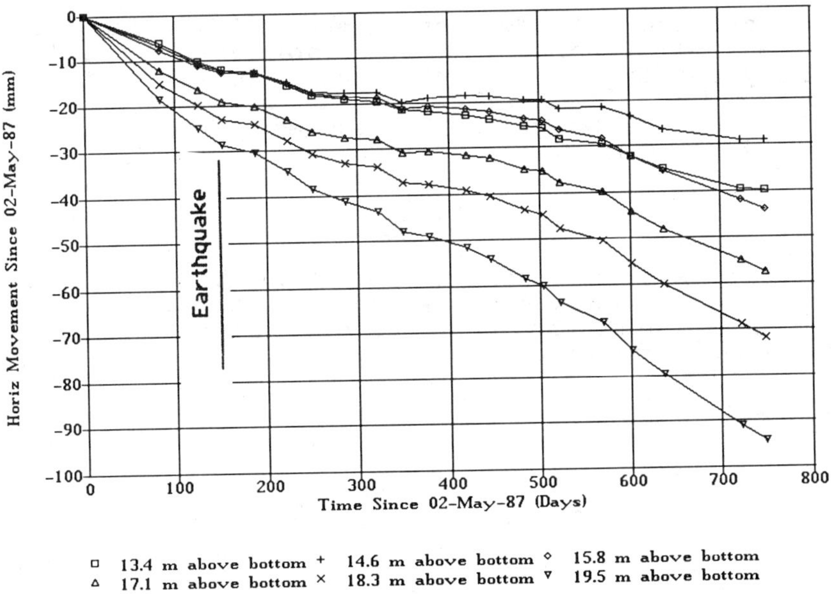

FIG. 11 -- Horizontal Movement Perpendicular to Slope Instrument No. 1

RESPONSE TO THE OCTOBER 1, 1987 EARTHQUAKE

The October 1, 1987 Whittier Earthquake occurred after these instruments were in place. We had obtained five months of seemingly reliable data before the earthquake, so this event provided a rare opportunity to accurately measure the effect of the tremor on the stability of the landfill. The earthquake had a Richter magnitude of 6.1 and was centered 22 km (14 mi) west of the site.

The plots on figures 5, 6 and 11 clearly indicate no perceptible permanent movements in the landfill during the earthquake. Although this does not necessarily imply that a larger earthquake would not cause movements in the refuse, it is a helpful point of reference.

PLANS FOR THE FUTURE

These trial installations have demonstrated the workability of the Sondex and inclinometer instruments for measuring movements in landfills. The County Sanitation Districts plan to install additional instruments at Spadra and other landfills in order to monitor movements of various ages and compositions.

ACKNOWLEDGMENTS

This project was made possible through funding from the Spadra Landfill LandLab project sponsored by Cal Poly University and from the County Sanitation Districts of Los Angeles County. Mr. Walter Christensen of Cal Poly collected the field data and the success of this project is due in no small part to his diligent work.

REFERENCES

[1] Huitric, R., "Sanitary Landfill Settlement Rates", Verlängerung Der Nutzungsdauer Von Deponien (Conference on the Prolongation of the Capacity of Sanitary Landfills), Technische Universität Berlin, 1981.

Author Index

Subject Index